# Technical Writing

# Technical Writing

THIRD EDITION

*Gordon H. Mills*
*John A. Walter*
*The University of Texas*

HOLT, RINEHART AND WINSTON, INC.
*New York Chicago San Francisco Atlanta*
*Dallas Montreal Toronto London Sydney*

Library of Congress Catalog Card Number: 71-107877
ISBN 0-03-078680-0
Printed in the United States of America
56 038 9876

# Preface to the Third Edition

Our objectives in the third edition of this book have been to take account of changing concepts about writing, to introduce fresh illustrative materials, and to improve the content and clarity of the text wherever possible.

The preparation of this edition has been a special pleasure because of the particularly interesting changes recent years have seen in attitudes toward language. Developments that have been occurring over a long period of time in linguistics, in the philosophy of science, and in the profession of technical writing are now making themselves evident in the classroom and in the general practice of technical writing. Although linguistics has not displaced traditional methods of instruction in writing, it has, through its phenomenological approach, significantly changed the relationship of the writer to his language. At the same time, scientists and philosophers have been cultivating a new and, as it seems to us, a more realistic and genial concept of science. The scientist's role as a person is heightened, and some of the intuitiveness and enthusiasm of art are accepted as a natural part of his science.

Examination of current scientific reports and articles confirms the expectation that in this climate of opinion evidences of a new flexibility of style and outlook are becoming apparent in technical writing. Shibboleths like "data are" are giving way to a point of view that is both more relaxed concerning what is socially acceptable and more insistently curious about the resources available to a writer who is trying to capture in language the elusive complexities of scientific theories and mechanisms. Oddly enough, the most revealing single indication of this changing point of view is probably found in an increasingly permissive attitude toward the use of the first-person pronoun. The significance of "I," and sometimes of "we," arises, of course, from fundamental issues encountered in decisions about their appropriateness. These issues include the philosophical problem of the relationship of the experimenter to his experiment, the social problem of agreeing upon what will be acceptable in a given situation, and the linguistic problem of determining the actual dialect of a given group. For ideas about the philosophical and cultural implications of the use of the first-person pronoun we found three sources of special interest: Percy Bridgman's *The Way Things Are,* where he defends the use of "I" in scientific writing; Chapter Ten in Michael Polanyi's *Personal Knowledge*, where he distinguishes between the concepts of the personal and the subjective as these concepts are related to scientific research; and the preface in Lionel Trilling's *Beyond Culture*, where he defends the use of "we" in humanistic writing. For information about the dialect of the group, if we may continue to put it that way, we have studied a great many examples of current technical writing, and of current opinion about technical writing.

The maturing of the profession of technical writing is evident in the many instances of superb work now appearing in scientific and industrial publications. It is also evident, if our impression is correct, in the widening influence of its principles and standards on the writing done by scientists and engineers themselves, and on the instruction available in college classrooms. The fact that in this edition, as in the first two, we continue to make pointed criticisms of materials we have quoted does not indicate any pessimism about the future of this maturing process. As we have remarked before, we are very much aware that many of the materials we quote were written under great pressure. We also remember about other writers, as we beg everybody to remember about ourselves, that nobody writes well all the time.

We have sought fresh illustrative materials, but novelty for its own sake has not been one of our criteria. Certain materials that appeared in earlier editions and that have been retained in this one are, to the best of our knowledge, unique, and we believe they continue to serve a purpose. Examples are the excerpts from the General Electric Company's report, "What They Think of Their Higher Education," and from the U.S. Navy's "A Study of the Use of Manuals."

Now we would like to say something about how the sections and chapters in this book are organized. Over the years since it was first published, we have received a great many helpful and perceptive suggestions about both its organization and its content. Indeed, we have begun to feel, and with good reason, that in some ways our book is virtually a community project, and we have a deep sense of appreciation for the many kindnesses we have been offered. With respect to organization, however, we are, frankly, in the clutches of the infamous double bind. The suggestions tend to cancel each other out. We cannot turn one way without excluding another way that is equally good. When preparing this edition, we discussed at length several different approaches to organizing the text. In the end we decided that the opinion we had expressed in the preface to the first edition was still sound in principle, and we determined to leave the organization fundamentally unchanged. Our opinion was then, as it is now, that the sequence in which an instructor presents the material in this text is best decided by him rather than by us. We are confident that the sequence of chapters as they stand in the text is practical and workable because we have tried it. But we know it is not the only good sequence because we have tried others too, with satisfying results. And, as we said, we have also become aware of many new possibilities through well-considered suggestions we have received.

Many of our obligations for assistance are indicated in the text. In addition, we want to express our appreciation here to a few people whose kindness requires a special acknowledgment: Clarence A. Andrews, of the University of Iowa; Norman S. Cameron, of Vancouver City College; Herman Estrin, of the Newark College of Engineering; John Sterling Harris, of Brigham Young University; James F. Jones, of Management Research International, Incorporated; W. Paul Jones, of Iowa State University; Mrs. H. E. McKinstry; Thomas E. Pearsall, of the University of Minnesota at St. Paul; F. Bruce Sanford, of the Bureau of Commercial Fisheries, United States Department of the Interior; Mary E. Weimer, of the Indiana Institute of Technology; and George H. Sullivan, of Holt, Rinehart and Winston, for whose sense of humor and editorial skill we are most grateful.

*Austin, Texas* G.M.
*February 1970* J.W.

# Preface to the First Edition

This book had its inception in our need for a logical bridge between the professional writing of scientists and engineers and the content of a course for students of technical writing. Certain widespread practices had developed in such courses, as we knew both from personal experience and from such published studies as A. M. Fountain's *A Study of Courses in Technical Writing* (1938), the American Society for Engineering Education's report on *Instruction in English in Engineering Colleges* (1940), and M. L. Rider's *Journal of Engineering Education* article, "Some Practices in Teaching Advanced Composition for Engineers" (1950). We felt that many of these practices were unquestionably proving their value, but about others we weren't sure, and there seemed to be no clearly established basis upon which to decide about them. The difficulty was partly that the limits of the subject were uncertain; apparently nobody had ever seriously explored the concept of technical writing with the purpose of trying to say precisely what technical writing is. There were, of course, numerous systems of classification of articles and reports; but, unfortunately, these systems were dissimilar at many points and were often more puzzling than helpful in relation to our question of what materials and instruction were most needed by our students.

In an effort to find practical solutions to the problems just noted, as well as to others not mentioned, we undertook three investigations. We began by seeking examples of reports and articles, and expressions of opinion about important problems; altogether we incurred an indebtedness to over three hundred industrial and research organizations in making our survey. We also worked out, in writing, a theory of what technical writing is (later published as Circular No. 22 of The University of Texas, Bureau of Engineering Research, under the title, *The Theory of Technical Writing*). Thirdly, we studied the content and organization of college courses in the subject. The content of this book rests primarily upon these investigations, together with numerous other studies of a more limited scope. Perhaps it is proper to say here that these investigations did not constitute our introduction to the subject, since we had both had considerable experience in the field, in the capacity of teachers and editors. On the other hand, we did try hard to avoid letting the particularities of our personal experience affect the conclusions we drew from these systematic studies. We realize, of course, that the nature of our own experience, both academic and nonacademic, has no doubt been reflected in our text; and if in spite of the good counsel and abundant materials furnished to us we have fallen into error, the fault is entirely our own. We do believe, however, that our methods have been sound; we hope that our book is sound too.

Perhaps we should add, about ourselves, that our collaboration has extended to all parts of the text. Almost every page of it represents a joint effort.

A few comments on the text itself need to be made here. As we said, the organization of the book was determined by a study of the needs and practices of courses in technical writing, as well as by the internal logic of the subject matter. One problem, however, resisted solution: we could not find any clear grounds on which to decide when to introduce certain elements of our subject that would not themselves usually be the basis of writing assignments. Section Three (Transitions, Introductions, and Conclusions) and Section Five (Report Layout) are chiefly involved, although the same difficulty exists with Chapter 3 (Style in Technical Writing). We have no pat answers as to how these elements should be introduced into a course. On the contrary, we believe that a suitable decision can be made only by the instructor.

We should also like to remark that we are aware we have sometimes been blunt in criticizing quoted materials. We hope all readers of the book will understand that these materials were not prepared especially for our use. They are, instead, routine products, and many of them were doubtless written under great pressure. We have been critical in order to help students learn, not because of any fancied superiority to the writers whose work we criticize.

We regret that a complete list of those organizations and persons who have helped us is too long to present here. We are deeply grateful to all of them, and we have acknowledged our specific indebtedness to many in the text. A few have requested anonymity. Our greatest single debt is to John Galt, Manager, Phenolic Products Plant, Chemical Materials Department, General Electric Company, Pittsfield, Massachusetts. Mr. Galt permitted us to quote the extremely interesting manuscripts in Appendix B. We should like to mention also The Civil Aeronautics Administration, Technical Development and Evaluation Center; and the Research Laboratories Division, General Motors Corporation. Dr. W. E. Kuhn, Manager of the Technical and Research Division, The Texas Company, deserves special thanks for repeated favors.

*Austin, Texas*
*January 1954*

G.M.
J.W.

# Contents

## *Section Two* SPECIAL TECHNIQUES OF TECHNICAL WRITING



## *Section Three* TRANSITIONS, INTRODUCTIONS, AND CONCLUSIONS

## *Section Four* TYPES OF REPORTS



## *Section Five* REPORT LAYOUT

# Technical Writing

# SECTION ONE

# Preliminary Problems

*Everybody knows what a problem is, but the idea of "preliminary problems," as we have entitled this first section of our book, calls for explanation.*

*Some problems are more obviously preliminary to a study of technical writing than others. Most obvious of all is no doubt the problem of determining what the term "technical writing" means. In Chapter 1, therefore, technical writing is defined, its major subdivisions are designated, and some evidence is examined as to its role in science and engineering.*

*Closely related to the problem of defining and subdividing the subject of technical writing is that of identifying what might be called its basic concepts. That is, can the practice of technical writing be reduced to a few general propositions? Probably not; on the other hand, five concepts about technical writing are so fundamental that recognition of them does help give perspective and meaning to later study of the many different aspects of the subject. These five fundamental concepts are presented in Chapter 2—which is limited to one short page of text.*

*The third chapter in this section is concerned with style. Of course the style of any piece of writing is an integral part of that writing and should not be thought of as preliminary to it. Nevertheless, style also reflects a writer's responses to a broad array of problems that present themselves the instant he picks up his pen, before he has written a word. One of these problems, for example, is whether a writer should seek what he may think of as self-expression in his style, as opposed to giving his attention to the needs of his reader. The importance in technical writing of a number of such problems draws the subject of style into the area of preliminary consideration. Style is also given repeated attention in subsequent sections of the book, as occasion requires.*

*The fourth and last chapter in Section One is a review of the logic of organization, in the form of outlining and abstracting. This problem of the elements of organization is preliminary to a study of technical writing only in the practical sense that it is common to almost any study of writing and that in some degree it will already be familiar to any reader of this book.*

# 1
# Introduction

The purpose of this book is to discuss the principles and practice of the kind of writing required of engineers and scientists as part of their professional work. The reader to whom the book is directed is primarily the technical student who has had enough training in the fundamentals of composition to profit from consideration of some of the problems of technical writing.

In this chapter we shall first explain what technical writing is, and then go on to discuss the importance of writing as compared with other elements of technical work, what kinds of writing engineers and scientists are expected to do, and what aspects of writing they particularly need to study. At the end of the chapter will be found a series of statements by experienced professional men on the part that writing plays in their work.

## What Technical Writing Is

Although one of the obvious characteristics of technical writing is its technical subject matter, it would be very difficult to say precisely what a technical subject is. For our purposes, however, it will be sufficient to say merely that a technical subject is one that falls within the general field of science and engineering.

Technical writing has other characteristics besides its subject matter, of course. One of these characteristics is its "formal" aspect—a term hard to define but easy to illustrate. There are, for example, certain forms of reports, like progress reports, that are used in technical writing. There are also certain forms of style and diction used and certain forms of graphic aids (e.g., sketches, graphs, flowsheets).

Another characteristic of technical writing is its scientific point of view. Ideally, technical writing is impartial and objective, clear and accurate in the presentation of facts, concise and unemotional. In practice, naturally, some of these qualities are often lacking, particularly clarity and conciseness. An additional fact about point of view is that technical writing is usually designed for a specific reader or group of readers, perhaps the staff of a certain research group, rather than for a great mass of readers, as is newspaper writing.

The last major characteristic can be called the special techniques of technical writing. What this means can easily be explained by an analogy. When a person decides he would like to write short stories, he soon finds himself studying, among other things, how to write dialogue. No one questions the logic of this. But the short-story writer is not the only one who uses dialogue; probably we all occasionally write down some conversation, in a letter or elsewhere. Nevertheless, the short-story writer uses dialogue more than most people, and it is very important in his writing. Therefore he must know all he can about it. Similarly, there are certain techniques that the technical writer uses particularly often. They appear in other kinds of writing, but not so frequently, and not so often as such important parts of the whole. Consequently, the technical writer should learn all he can about these techniques. The most important are description of mechanisms, description of processes, definition, classification, and interpretation. Each one of these writing problems is complex enough to need careful attention, and each one of them appears frequently in technical writing.

It should be clearly understood that these special techniques are not types of technical reports. Several of them may appear in a single report; but for an entire report to be nothing more than, say, the description of a mechanism would be unusual. Again, it is like dialogue in a short story, which may take an important part, but is seldom the whole story.

In summary, then, technical writing can be defined as follows:

A. Technical writing is writing about scientific subjects and about various technical subjects associated with the sciences.

B. Technical writing is characterized by certain formal elements, such as its scientific and technical vocabulary, its use of graphic aids, and its use of conventional report forms.

C. Technical writing is ideally characterized by the maintenance of an attitude of impartiality and objectivity, by extreme care to convey information accurately and concisely, and by the absence of any attempt to arouse emotion.

D. Technical writing is writing in which there is a relatively high concentration of certain complex and important writing techniques, in particular description of mechanisms, description of a process, definition, classification, and interpretation.

## What the Technical Man Is Required to Write

Technical men are called upon for a considerable variety of writing: reports of many kinds, memoranda, technical notebooks, proposals, professional papers and magazine articles, patent disclosures, letters, promotional brochures, specifications, technical bulletins, instruction manuals, handbooks, and sometimes even books. As you would expect, the young college graduate is not often called upon for more than what might be called routine writing, not greatly different from the simple reports he wrote in school. He is frequently used primarily as an instrument for the collection of data or as an overseer of simple operations.

He might, for instance, be assigned to check some aspect of the quality control in the production of a certain type of carburetor. If his data show that the quality control is not satisfactory, then what? Perhaps his boss will examine the data and then write a report to his own superior recommending corrective measures. The top man, the executive, may be scarcely aware of who it was that collected the basic information, and if he does know he may feel that any one of a dozen of his other young men could have done as well. On the other hand, the young man might be asked to include in his own report or memorandum suggestions for correcting the situation. If his suggestions are sound, it becomes a matter of considerable interest to him whether his report goes up to the top executive with his own name on it, or is rewritten and goes up with somebody else's name in addition to, or in place of, his own. He naturally would like to have his good work recognized, and not find himself ignored just because he hasn't yet learned to write effectively. On the other hand, it is also only natural that nobody likes to spend hours struggling to figure out the meaning of an incoherent report.

With the passing of time, this kind of problem is likely to become more acute for the technical man, because in the normal course of events he often finds himself concerned with decisions, sometimes decisions about action to be taken, sometimes decisions about the soundness of conclusions having to do with theoretical problems.

Customers of his company write him letters asking for advice about their technical problems, and he writes letters in reply. For other men in his own organization he must frequently write letters and reports, and both informally and formally carry on oral discussions of joint problems. His superiors call for progress reports at regular intervals on the work he is doing, and for long reports at important stages. He is asked to address chapters of professional organizations. If he is ambitious to establish a reputation, he submits articles to professional journals. Of course circumstances differ, but among all these kinds of writing the demand for decisions about action may well prove most significant, as is suggested in the statement from the Ethyl Corporation on pages 11 – 12 below.

It was pointed out that the technical man will probably have to write reports and letters, and he may write articles and books. But the bulk of his writing is usually in the form of reports and other routine documents such as memoranda. What is a report? There is little point in attempting an exact definition. Perhaps as good a definition as any is that a report is a piece of technical writing designed to meet a specific need. In the introduction to Section Four you will find a list of 30-odd "types" of reports. Many of these types differ from each other only in minor details, however, and in some cases probably in none at all. What happens is that a group of technical men decide that they need to have information about certain types of projects written up in a certain form, and perhaps at certain stages of progress. They make up some rules and give this "type" of report a name—perhaps "preliminary," or "partial," or "shop," or "test." That is exactly what they should do. If the form of report they devise serves their purposes, no one can ask for more.

There are, nevertheless, a few types of reports that are pretty well standardized. Three that deserve mention are the progress report, the recommendation report, and the form report. They will be discussed in detail in Section Four.

## Basic Aspects of Technical Writing

In the most elementary terms, technical writing can be broken down into two parts, or aspects: (1) the "end products" (like reports and letters), the concrete "package" that you deliver; (2) the skills that enter into the preparation of the end product. This distinction is useful in pointing out specific aspects of writing that are of particular importance to the technical man and which we shall accordingly be concerned with in this book.

The important "end products" of technical writing are these:

1. Business letters
2. Various kinds of reports
3. Articles for technical journals—and possibly books
4. Abstracts
5. Oral reports
6. Graphic aids
7. Instruction manuals
8. Handbooks
9. Brochures
10. Proposals
11. Memoranda
12. Specifications

It is quite possible that you may never be interested in writing for technical journals, but the other items in the list above are all routine work. Oral reports seem less tangible than the others, perhaps more like a skill than a "product," but the spoken word has as real an existence as the written. And we should add that "oral reports" refers not only to formal speechmaking, but also to informal discussions of technical problems. The heading "graphic aids," by which we mean graphs, drawings, and other nontextual supplements, also looks a little odd in this list. Are graphic aids a skill, an end product, or neither? Whatever they are, it is good to know about them. It doesn't particularly matter what we call them.

The skills which deserve particular attention are the following:

1. Special techniques of technical writing
2. Style
3. Introductions, transitions, and conclusions
4. Outlines (or organization)
5. The layout, or format, of reports

The special techniques of technical writing have already been commented on. The other items in the list need a brief explanation.

The word "style" usually suggests an aesthetic quality of prose, a quality determined by the relative smoothness or awkwardness with which sentences are put together. Many eminent scientists and engineers have developed a splendid prose style. Naturally we should like to encourage you to develop a good style; but above all else we will emphasize clarity. Since technical writing is by definition a method of communicating facts, it is absolutely imperative that it be clear. At the same time, the nature and complexity of the subject matter of technical writing often involve the writer in particularly difficult stylistic problems. These problems will be discussed in the chapter on style. One other important aspect of style in technical writing is point of view. In brief, the point of view should be scientific: objective, impartial, and unemotional.

The third item is introductions, transitions, and conclusions. The problem here is to learn to tell your reader what you're going to tell him, then to tell him, then to tell him what you've told him. This skill is one of the most important a technical writer, or any writer, can possess.

The fourth element is outlines. A more accurate phrase might be "the theory of organizing your writing," because that is what we are really interested in. Outlines, like dentists, are popularly associated with pain, but both serve an admirable purpose.

And the fifth and last element, the layout of reports, has to do with such matters as margins, spacing, subheads, the title page, and the like.

Doubtless you have noticed that grammar and punctuation were not included in either of the two lists above. The reason for the omission of these fundamentals is that they are not properly part of the formal subject of technical writing. Constant attention to these fundamentals is nevertheless a necessity. We suggest that you get a good handbook of English, if you do not already have one, and—if you have not already done so—develop the habit of using it. Professional writers do no less.

Altogether, the topics which have been listed are those that are most important to the beginner in technical writing. It should be understood, however, that they are not the only aspects of technical writing that deserve attention. Others, for instance the handling of footnotes and bibliography, and the use of the library, will be discussed in the appropriate place.

Our purpose has been to state in the simplest terms what the study of technical writing involves. Setting up a practical course of study naturally requires rearrangement and regrouping of the topics listed. Reference to the table of contents will indicate how that has been done in this text.

## The Place of Writing in Technical Work

Finally, you have probably been wondering what place technical writing might take in the special area you plan to work in.

The particular circumstances of your own job will determine how much writing you will have to do, of course, as well as how much importance writing will have in your career, but there are some facts available about what you are likely to encounter. For example, one large corporation has prepared the following account of how its young technical men spend their time:

| *Type of Work* | *Percentage of Time* |
|---|---|
| 1. Collection and Correlation of Data | 26 |
| 2. Calculations | 34 |
| 3. Writing Reports and Letters | 20 |
| 4. Selling Results of Their Work | 12 |
| 5. Other (literature reviews, attendance at meetings, consulting with others, etc.) | 8 |

According to these figures, the college graduate entering this particular corporation can, on the average, expect to spend a fifth of his working time in writing, or the equivalent of at least one whole day each week. As a matter of fact, if we include the spoken word as well as the written word, he will evidently spend a good deal more than a fifth of his time communicating ideas in words, since items 4 and 5 clearly require talking.

These figures represent only the averages reported by one corporation. Let us approach the problem in another way. Suppose we ask how much importance writing will have in determining success in your career.

A significant answer to this question can be found in the results of a questionnaire that General Electric Company gave to 7000 engineers in its employ. When asked to select from their college courses the ones that had made the most important contributions to their success with the Company, these men ranked English second only to their technical subjects. A nearly equal number of non-engineers in the Company ranked English first. Further details can be found on pages 219–241.

When the Esso Research and Engineering Company conducted a somewhat similar survey, their young technical employees selected English as the most important one of a group of 15 nontechnical subjects. They placed public speaking and report writing second and third, respectively.

We hope these few figures will encourage you to think about the part that writing is likely to play in your own future work. The best thing you can do will be to take every opportunity to ask questions on this subject when talking with men who are actively engaged in your professional field. The illustrative material with which this chapter is concluded will provide a further insight.

## ILLUSTRATIVE MATERIAL

Many examples of technical writing are presented in this book, both within the text of the discussion and in special sections at the end of chapters. These special sections are labeled, as this one is, "Illustrative Material." Some of these examples were written by students; more of them were written by scientists and engineers as part of their regular work. Examination of the latter will help you acquire an understanding of what you might expect in your own career.

For the pages that close this introductory chapter, we have chosen a few items which give direct expression to attitudes toward writing held by men in science and industry. As you read these materials, we suggest you ponder a sentence from a recent pamphlet entitled *Stromberg-Carlson "Accucode" Supervisory Control Systems.* The sentence reads:

> Stromberg-Carlson is staffed to undertake the "end-to-end" (engineer, furnish and install) system responsibilities to solve any telecommunication or supervisory control problem.*

The emphasis placed on the staff in this statement underscores the fact that any technical group is a complex team. And, of course, the effectiveness of the team is to a considerable degree dependent upon the ability of its members to communicate clearly with one another, and with "outsiders," concerning the technical problems the team undertakes to solve. Sometimes it is a temptation to decide that this necessary communication can be accomplished entirely through the language of mathematics, without having to depend on words. While contemplating this idea, however, we remembered a highly amusing essay entitled "Mathmanship" which has become famous in the world of science and engineering. It is the last of the four items reprinted below.

The first item, "Training the Professional-Technical Employee," was presented as a paper at a meeting of a division of the American Petroleum Institute and subsequently printed and circulated by the Ethyl Corporation. Only a small portion is presented here.

*Quoted by permission of the Stromberg-Carlson Corporation.

## Training the Professional-Technical Employee*

*T. C. Carron*

### Clear-Writing Program

A research laboratory is unique in that its sole output is ideas. And its only tangible products are the reports and papers that describe these ideas. This photograph (Figure 1) will give you some idea of the magnitude of a research laboratory's writing.

Figure 1. Magnitude of Written Material

The stack at the left is one year of progress reports from just one of the five divisions at our Detroit laboratories. The middle stack is one month's output of similar reports from our two most prolific divisions. And for comparison of sheer amounts of reading matter, the Manhattan telephone directory is shown on the right.

Now some of this writing is just for the record. But most of it is written with action in mind. The author wants a research project continued or expanded; a new product marketed; a new plant built or modified; and so on. A lot of key people must read these reports not only to keep up to date, but in order to make decisions. Not surprisingly, we found that

*T. C. Carron, *Training the Professional-Technical Employee* (Ferndale, Michigan: Research and Development Department, Ethyl Corporation, no date), pp. 1–3. Quoted by permission.

poor organization and complex writing were greatly magnifying the time required to read these reports. Therefore, we decided to try to teach our scientists and engineers to write more clearly and concisely.

Of course writing is especially important in a research group such as the Ethyl Corporation Research Laboratories that issued the material just quoted. In fact, it is common for people in research to remark that the end product of their work is a report. For example, L. R. Buzan, of General Motors Research Laboratories in Warren, Michigan, says simply that " . . . we consider new information our primary product." What about other areas of science and engineering, then? Well, as we said earlier, it depends on all the circumstances; but here is some relevant evidence. Westinghouse Electric Corporation undertook a study of report writing in which 70 managers representing all managerial levels within the corporation were interviewed on the subject of what they most needed and desired in reports. The following is a brief summary of what these managers, representing a cross section of the activities in this large corporation, had to say concerning the best general approach to the whole problem of report writing. This summary is reprinted from the pamphlet *What to Report* (Westinghouse, 1962).

### Approach to Writing

The point of attack in report writing is the analysis of the problem. An effective report, like any good engineered product, must be designed to fill a particular purpose in a specific situation. Effective design depends on a detailed analysis of the problem and an awareness of the factors involved. In analyzing your writing problem, you should ask the following questions:

- What is the purpose of the report?
- Who will read it?
- How will it be used?
- What is wanted?
- When is it wanted?
- What decisions will be based on the report?
- What does the reader need to be told in order to understand the material?

Once this analysis has been completed and the influencing factors defined, you should evaluate your material and select that to be included in the report, using the analysis as a basis for your selection. The material should then be organized functionally within the requirements of the reporting situation just as an engineering design must be functional within its technical and environmental requirements. If the analysis of the problem was thorough, you are reader-oriented, and writing the report becomes a much easier task.

For our next item, we go back to the point of view of a research group—which, as the preceding item makes clear, is actually not significantly different from the point of view of a broad cross section of managers in industry. This next item appeared as an editorial in *The Bell Laboratories Record*, published by the Bell Telephone Laboratories, Inc.

### The Design of Engineering Writing*

Engineers usually handle the language of their craft—mathematics—with supple ease, but often feel clumsy at explaining their work in words. It's true that words are slippery things and an equation may be infinitely more precise than an English sentence. From one point of view, however, the language of an engineer's work and the language of his writing have much in common.

A mathematical statement is more than the product of an engineer's work; it is actually the instrument of his thought—the most concise expression of the relationships between the physical forces he is trying to manipulate. In writing for journals like the RECORD, the engineer substitutes English for mathematics—the English sentence becomes the instrument with which he orders and controls the ideas of his narrative.

To be easily understood a narrative should be concise, with extraneous details pruned away, revealing the essence of the subject. Just as mathematical elegance is achieved, in part, by reducing the number of terms in an equation and the number of steps in an exposition—to utmost simplicity—so narrative elegance is approached by selecting only what is necessary to make the subject meaningful to the reader. "Prose," as Ernest Hemingway said, "is architecture, not interior decoration."

In his mathematics, the creative engineer strives for the simplest yet most useful description of his work. He must apply the same criterion to his writing. Elegance, rather than eloquence, should be his goal. Just as each step in a mathematical exposition is vital to the complete structure, each sentence and each paragraph in an article should support and advance the flow of the final design.

In the face of a difficult technological problem, not many engineers will haphazardly combine many different components into some arbitrary system hoping for a useful result. Most often an engineer starts with a little systems engineering. He decides what he wants to accomplish; he selects from various alternatives, and makes a mathematical model of the circuit or the crucial parts of his design. If the model is a good one he can, by changing its parameters, predict a great deal about the behavior of his final design. This procedure usually produces a viable product.

Writing is also an analytical and deductive process. It has stages of systems engineering, research, of experimental development, of final development. The first step is to define the desired output. And, as in engineering any communications device, the signal-to-noise ratio must be good. Extraneous paragraphs, tortuously complex sentences, eloquence for its own sake all add noise. In his own mind, the engineer-writer can construct a sort of verbal model of his final design and mentally change its parameters in order to predict the impact of his completed article.

After he has defined the parameters of his article by selecting and excluding details, the engineer-writer begins final development. The article's construction is modular; its building blocks are sentences and paragraphs. Redundancy may be added where it is necessary to reinforce a point, thereby improving the reliability of the communications process. And, if the actual product does not fulfill the original design intent, the engineer may have to go back to the writing table.

This process of setting an objective, continual testing and redesigning, reshaping and refining, if done conscientiously, may result in an elegant piece of work that is an efficient and reliable communications device, bridging the gap between the creative engineer and the men who use his work.

The next and last item, referred to earlier, must have touched a sensitive spot, to judge from the amount of interest it provoked. We hope you will enjoy it. The footnotes are reprinted exactly as they were in the original. The article is reprinted here from the June, 1958 issue of *American Scientist.*

## Mathmanship

*by Nicholas Vanserg*

In an article published a few years ago, the writer[1] intimated with befitting subtlety that since most concepts of science are relatively simple (once you understand them) any ambitious scientist must, in self protection, prevent his colleagues from discovering that *his* ideas are simple too. So, if he can write his published contributions obscurely and uninterestingly enough no one will attempt to read them but all will instead genuflect in awe before such erudition.

### What is Mathmanship?

Above and beyond the now-familiar recourse of writing in some language that looks like English but isn't, such as Geologese, Biologese, or, perhaps most successful of all, Educationalese[2], is the further refinement of writing everything possible in mathematical symbols. This has but one disadvantage, namely, that some designing skunk equally proficient in this low form of cunning may be able to follow the reasoning and discover its hidden simplicity. Fortunately, however, any such nefarious design can be thwarted by a modification of the well-known art of gamesmanship[3].

The object of this technique which may, by analogy, be termed "Mathmanship" is to place unsuspected obstacles in the way of the pursuer until he is obliged by a series of delays and frustrations to give up the chase and concede his mental inferiority to the author.

### The Typographical Trick

One of the more rudimentary practices of mathmanship is to slip in the wrong letter, say a $y$ for a $r$. Even placing an exponent on the wrong side of the bracket will also do wonders. This subterfuge, while admittedly an infraction of the ground rules, rarely incurs a penalty as it can always be blamed on the printer. In fact the author need not stoop to it himself as any copyist will gladly enter into the spirit of the occasion and cooperate voluntarily. You need only be trusting and not read the proof.

### Strategy of the Secret Symbol

But, if by some mischance, the equations don't get badly garbled, the mathematics is apt to be all too easy to follow, *provided* the reader knows what the letters stand for. Here then is your firm line of defense: at all cost *prevent him from finding out!*

Thus you may state in fine print in a footnote on page 35 that $V^x$ is the total volume of a phase and then on page 873 introduce $V^x$ out of a clear sky. This, you see, is not actually cheating because after all, or rather before all, you *did* tell what the symbol meant. By surreptitiously introducing one by one all the letters of the English, Greek and German alphabets right side up and upside down, you can make the reader, when he wants to look up any topic, read the book backward in order to find out what they mean. Some of the most impressive books read about as well backward as forward, anyway.

But should reading backward become so normal as to be considered straightforward you can always double back on the hounds. For example, introduce $\mu$ on page 66 and avoid defining $\mu$ until page 86*. This will make the whole book required reading.

### The Pi-throwing Contest or Humpty-Dumpty Dodge

Although your reader may eventually catch up with you, you can throw him off the scent temporarily by making him *think* he knows what the letters mean. For example every schoolboy knows what $\pi$ stands for so you can hold him at bay by heaving some entirely different kind of $\pi$ into the equation. The poor fellow will automatically multiply by 3.1416, then begin wondering how a $\pi$ got into the act anyhow, and

*All these examples are from published literature. Readers desiring specific references may send a self-addressed stamped envelope. I collect uncanceled stamps.—N. Vanserg

finally discover that all the while $\pi$ was osmotic pressure. If you are careful not to warn him, this one is good for a delay of about an hour and a half.

This principle, conveniently termed pi-throwing can, of course, be modified to apply to any other letter. Thus you can state perfectly truthfully on page 141 that F is free energy so if Gentle Reader has read another book that used F for *Helmholtz* free energy, he will waste a lot of his own free energy trying to reconcile your equations before he thinks to look for the footnote tucked away at the bottom of page 50, dutifully explaining that what you are talking about all the time is *Gibbs* free energy which he always thought was G. Meanwhile you can compound his confusion by using G for something else, such as "any extensive property." F, however, is a particularly happy letter as it can be used not only for any unspecified brand of free energy but also for fluorine, force, friction, Faradays, or a function of something or other, thus increasing the degree of randomness, dS. (S, as everyone knows stands for entropy, or maybe sulfur). The context, of course, will make the meaning clear, especially if you can contrive to use several kinds of F's or S's in the same equation.

For all such switching of letters on the reader you can cite unimpeachable authority by paraphrasing the writing of an eminent mathematician[4]: "When I use a letter it means just what I choose it to mean – neither more nor less . . . the question is, which is to be master – that's all."

## The Unconsummated Asterisk

Speaking of footnotes (I was, don't you remember?) a subtle ruse is the "unconsummated asterisk" or "ill-starred letter." You can use P* to represent some pressure difference from P, thus tricking the innocent reader into looking at the bottom of the page for a footnote. There isn't any, of course, but by the time he has decided that P must be some registered trademark as in the magazine advertisements he has lost his place and has to start over again. Sometimes, just for variety, you can use instead of an asterisk a heavy round dot or bar over certain letters. In doing so, it is permissible to give the reader enough veiled hints to make him *think* he can figure out the system but do not at any one place explain the general idea of this mystic notation, which must remain a closely guarded secret known only to the initiated. Do not disclose it under pain of expulsion from the fraternity. Let the Baffled Barbarian beat his head against the wall of mystery. It may be bloodied, but if it is unbowed you lose the round.

The other side of the asterisk gambit is to use a superscript as a key to a *real* footnote. The knowledge-seeker reads that S is – 36.7[14] calories and thinks "Gee what a whale of a lot of calories" until he reads to the bottom of the page, finds footnote 14 and says "oh."

*April fool. See what I mean?

## The "Hence" Gambit

But after all, the most successful device in mathmanship is to leave out one or two pages of calculations and for them substitute the word "hence" followed by a colon. This is guaranteed to hold the reader for a couple of days figuring out how you got hither from hence. Even more effective is to use "obviously" instead of "hence," since no reader is likely to show his ignorance by seeking help in elucidating anything that is obvious. This succeeds not only in frustrating him but also in bringing him down with an inferiority complex, one of the prime desiderata of the art.

These, of course, are only the most common and elementary rules. The writer has in progress a two-volume work on mathmanship complete with examples and exercises. It will contain so many secret symbols, cryptic codes and hence-gambits that no one (but no one) will be able to read it.

## References

1. Vanserg, Nicholas. *How to Write Geologese*, Economic Geology, Vol. 47, pp. 220-223, 1952.
2. Carberry, Josiah. *Psychoceramics*, p. 1167, Brown University Press, 1945.
3. Potter, Stephen. *Theory and Practice of Gamesmanship or the Art of Winning Games without Actually Cheating*. London: R. Hart-Davis, 1947.
4. Carroll, Lewis. *Complete Works*, Modern Library edition p. 214.

# 2

# Five Basic Principles of Good Technical Writing

Chapter 1 outlined the general subject and plan of this book and indicated the importance of writing in the professional work of the engineer and scientist. The present chapter is devoted to a highly condensed preliminary statement of five basic principles that will later be presented in detail. There are many more than five principles involved in good technical writing, but the five stated below are so important that they may be taken as a foundation on which further development rests.

1. Always have in mind a specific reader, real or imaginary, when you are writing a report; and always assume that he is intelligent, but uninformed.
2. Before you start to write, always decide what the exact purpose of your report is; and make sure that every paragraph, every sentence, every word, makes a clear contribution to that purpose, and makes it at the right time.
3. Use language that is simple, concrete, and familiar.
4. At the beginning and end of every section of your report check your writing according to this principle: "First you tell the reader what you're going to tell him, then you tell him, then you tell him what you've told him."
5. Make your report attractive to look at.

You will find that these principles are involved in one way or another with practically everything that is said throughout the rest of this book.

# 3

# Style in Technical Writing

## Introduction

Before we set down any suggestions or "rules" for you to follow in achieving a desirable style in your reports, we need to explore the meaning of the term "technical style," or—to begin with—the meaning of the term "style" itself.

You probably have heard discussions of style in courses in literature, and perhaps have been asked to analyze the style of writers whose work you were reading. Your experience was not unusual if you had difficulty in making such an analysis. Style is hard to describe simply and directly. The whole always seems greater than the sum of the parts. On the other hand, it isn't particularly difficult to describe the impression or effect a style of writing makes. It's like the personality of an acquaintance. We can readily judge the total impression he makes as energetic and cheerful or nervous and excitable, though we may never be able to list all the traits that create this impression.

The difficulty of arriving at an exact definition of style may be responsible for the many colorful aphoristic definitions that writers and literary critics have invented. Jonathan Swift said that style is "proper words in proper places"; Lord Chesterfield, that it is "the dress of thoughts"; and the Comte de Buffon, that "style is the man." Jacques Barzun and Henry F. Graff say in their book, *The Modern Researcher:* "The qualities . . . Clarity, Order, Logic, Ease, Unity, Coherence, Rhythm, Force, Simplicity, Naturalness, Grace, Wit, and Movement . . . are not distinct things; they overlap and can reinforce or obscure one another, being but aspects of the single power called Style."* We would say, simply, that style is the way you write. This means the way you put words together in sentences, the way you arrange sentences into paragraphs, and the way you group paragraphs to make a whole composition.

Technical style, then, is the way you write when dealing with technical subject matter. But because of the specialized nature of technical subject matter, the functions which reports about it serve, and certain well-established conventions or traditions relating to its presentation, we can characterize technical style more explicitly. We can say that technical writing style is distinguished by a calm, restrained tone, by the use of specialized terminology (though every specialized field has its own distinctive terms, of course), and by an accepted convention of the use of abbreviations, numbers, and symbols.

## Objectives of This Chapter

Neither aphoristic comments nor descriptive definitions help much, however, when it comes to the practical problem of how to write a satisfactory style. What you will want to know is what is desirable in report-writing style and how it may be achieved. As Barzun and Graff say, ". . . you cannot aim directly at style, at clarity, precision, and all the rest: you can only remove the many possible obstacles to understanding, while preserving as much as you can of your spontaneous utterance. . . . Clarity comes when others can follow; coherence when thoughts hang together; logic when their sequence is valid."**

Our purpose in this chapter is to consider the most desirable characteristics of an effective style for technical writing, the problems which give inexperienced writers most trouble, and the accepted conventions for using abbreviations, symbols, and numbers, as well as some miscellaneous matters of usage. In carrying out this purpose we have divided the chapter into two parts. In Part One we will discuss, in the following order, the problems of reader adaptation, the scientific attitude, the construction of sentences that say what they are supposed to say, precision in the use of words, and the structure and length of sentences and paragraphs. In Part Two we shall turn to common faults in technical usage and to the mechanics of style.

*Jacques Barzun and H. F. Graff, *The Modern Researcher* (New York: Harcourt, Brace & World, Inc., 1957), pp. 269–70.

**Barzun and Graff, *op. cit.*, p. 270.

# PART 1

## Reader Adaptation

*The Importance of Reader Adaptation.* Perhaps the easiest way to think about this problem of reader adaptation is simply to look back over some of the reading you have done and ask yourself how you felt about it. No doubt you have had the experience of enthusiastically starting to read a technical article about some subject you thought interesting, only to find the article either dull or incomprehensible. It doesn't necessarily follow that the article was no good, only that it was no good for you. As an extreme example of how things can go wrong in this respect, we might imagine the feelings of a small boy who has noticed a halo around the sun, asked his father what it was, and received a reply which began, "My boy, we are here dealing with a phenomenon involving spicules of ice having a uniform refracting angle of 60 degrees and yielding, at minimum deviation, a deflection of 22 degrees of the incident light."

This problem of adapting the writing to the reader is so common that another example is worth a little time. One of the remarkable personalities in the contemporary world of journalism is a man named Roy Morrison, a member of the Society of Automotive Engineers. He conducts a question-and-answer column in which he gives technical advice on recreational vehicles (campers, travel trailers, etc.). The style and content of what he writes can only be called inimitable. For one thing, he obviously takes little interest in the niceties of style or grammar. For another, he has his own opinion about what ought to be said. He has delighted—and, we suspect, sometimes thoroughly puzzled—thousands of readers. We think you'd enjoy what he writes. See what you can make of the following. A reader sent a letter to Roy Morrison asking this simple question: "Does an anti-spin differential increase the wear on the rear tires?" Here is the published answer.

> Dana rear ends are identical to conventional differentials and I defy a mechanic or lube man to look at one and tell the difference from the outside. The carrier cage that holds the spider gears has a small five plate clutch that is actuated by a ramp on the cross that holds the spider gears. This ramp is never in action until the driver applies the foot brake. There has to be a wheel slipping, in other words, standing still and one turning up causing the differential carrier cage to spin. By applying the brake causes the spinning wheel to drag and the carrier to "crawl". This action cause the spider or cross to crawl up the ramp and press tightly against the clutch, thus engaging the "dead" wheel. This will remain in a DRIVE position until the throttle is released and the drive line released and slack comes into the pinion and ring gear. This

in turn allows the spider or cross to return to its normal position releasing the clutch. Sluggish action is caused by using the wrong rear end lube oil and one should warn his lube man that he has an anti-spin rear end and use only the oil as recommended by Dana.*

Now, what is the answer to the question the reader asked? Actually, we believe the answer, or at least a part of the answer, can be found in what Roy Morrison wrote; but what, we wonder, did the reader think when he saw that reply to his simple question?

Let's contemplate still another situation. Let's imagine that in a time of war an American naval vessel, a destroyer, has entered a combat area. While it is on patrol there, a crucial element in its electronic equipment breaks down. The man in charge of maintaining the equipment, a noncommissioned officer, turns to the equipment manual for help in finding the trouble. This manual is the only source of information or assistance he has. Will he find the manual useful? For some thoughtful and informed discussion by naval personnel of precisely this problem, please turn to Appendix D, item 2 near the top of page 526, and to pages 540–542 beginning with the heading "Theory of Operation."

The crucial implication of these examples is that every aspect of an effective piece of technical writing is in some degree designed for a particular reader or group of readers. Another way of looking at this idea is to compare technical writing to other forms of writing, such as poetry, or even to an art form such as music. People sometimes write poetry or play a musical instrument to express their feelings. But who has ever seen a technical report that was written to express the author's feelings? We do not mean to say that no personal feeling is involved in all the activity that leads up to and perhaps even culminates in a technical report. As a matter of fact, the feeling is often intense, as revealed in very different ways in such remarkable books as Michael Polanyi's *Personal Knowledge*, and James Watson's *The Double Helix*. Our point about the purpose of technical writing can be put very simply: (1) poetry may be written and musical instruments may be played just for the fun of doing it, but (2) technical writing of the kind we are concerned with in this book is done primarily for the purpose of helping somebody else do something. This idea is neatly summarized by J. A. Hutcheson, a vice-president at Westinghouse: "Too few engineers realize that everything they write is used at some time by someone to help him make a decision." Mr. Hutcheson's comment happens to be directed to engineers, but his point applies equally well to most of the writing done by scientists of

*From *Trailer Life Magazine* 27, North Hollywood, California (June, 1967), 33 ff.

any kind. For example, should a botanist advise a farmer to plant Siberian pea shrubs in a hedge row? See "Insects and Diseases of Siberian Pea Shrub . . . ," an example in Chapter 8.

You will probably agree that it is wise to design your writing for your particular reader or readers. Nevertheless, you may be wondering if this kind of problem is really of any concern for a young person just starting out on his first job. After all, he will be writing for colleagues who are as well trained as he is, and for supervisors who know a good deal more than he does. True, in a way; but again a look at the facts introduces a surprise.

The supervisor or manager whose opinion of your writing is most important to you may not have been trained in your field at all. A little common-sense reflection makes this point meaningful. Suppose you are a mechanical engineer and you are a member of a team designing and building a prototype of an electric-powered automobile. What field of training will your section chief represent? Mechanical engineering? Electronics? Chemical engineering? Chemistry? Here is what one big corporation has to say to its own people on this subject of how to write for your supervisor. The italics are found in the original.

> *Usually* the management reader has an educational and experience background different from that of the writer. *Never* does the management reader have the same knowledge of and familiarity with the specific problem being reported that the writer has. *Therefore, the writer of a report for management should write at a technical level suitable for a reader whose educational and experience background is in a field different from his own.* For example, if the report writer is an electrical engineer, he should write his reports for a person educated and trained in a field such as chemical engineering, or mechanical engineering, or metallurgical engineering.*

This statement goes a little further than we'd be willing to go in this book. That is, circumstances vary from one group or company to another and we wouldn't want to adopt as a dogmatic rule the conclusions stated in the quotation. Nevertheless, this quotation does reveal the great practical importance of reader adaptation, and we most strongly and urgently endorse the principle that you should never write with the assumption that your reader is informed about your particular subject. The document from which this quotation was taken goes on to make the interesting assertion that whenever possible, highly complex material should be put into an appendix. This idea opens up the whole subject of how organization can be used to help

**What to Report* (Pittsburgh: Westinghouse Electric Corporation, 1962), unpaged.

solve problems of style. Before considering this new subject, however, we shall make some concrete suggestions about analysis of the reader.

*Reader Analysis.* It's easy to say that one reader knows more than another reader about a particular subject. But that doesn't help a great deal in our analysis of how to write for the one or the other. How much more does the one reader know than the other? What we need is clearly a list of kinds or categories of readers that will permit finer distinctions.

Such a list appears below, divided into six categories. The reader in category 1 knows the least, the reader in category 6 the most. But the reader knows the least, or the most, about what? Let's pick out some specific technical subject and assume that we are comparing the categories to one another on the basis of the readiness with which readers in each category could understand a report on the subject we choose. Any reasonably complex technical subject would do for our purpose; let's choose a newly designed electronic circuit. In practice, you will, of course, always be judging your reader according to whatever subject you happen to be writing about.

The following six categories, then, help in judging a reader according to whether he has:

1. Little ability to understand any complex subject
2. Considerable education, but not in science
3. Considerable knowledge of science, but not of electronics
4. Considerable knowledge of electronics, but little interest in the area represented by the new circuit (for example, a power engineer reading a report about sophisticated radar equipment)
5. Considerable knowledge of electronics in the area represented by the new circuit, but no detailed familiarity with the work done in designing the new circuit
6. Considerable knowledge of electronics in the area represented by the new circuit, and detailed familiarity with the work done in designing the circuit.

It is unlikely that a discussion of an electronic circuit would be written for readers in category number 1. Much writing is done for people in this category, however. Usually it is in the form of instructions. (See Chapter 7 for some examples.) What sort of reader might be found in numbers 2 and 3? Members of the Board of Directors. Customers. And in numbers 4 and 5? The manager of your department.

In any event, the list clearly shows the extent of the problem the writer may face in adapting his report to the reader it is intended for, and we should next consider what this means in practical terms. What elements of a report should you work on in adapting it to a given reader?

*Elements Requiring Adaptation.* There are two answers to this question. The first is the most general, and in the long run, the better. This answer is simply, as we said earlier—everything. It is difficult to think of any element in technical writing that is not influenced in some way by the intended reader. This fact will take on added significance as your experience with technical writing continues.

The second answer is that in this chapter on style we want to call attention to the three particular elements of style that should be given *first* attention in adapting your writing to your reader. These three elements are vocabulary, sentence length and structure, and organization. Our purpose at this time will be merely to indicate the importance of these elements and to prepare for later discussions that will be both greater in scope and presented from a different point of view.

So far as vocabulary is concerned, the problem is usually simple. Just don't use words your reader won't know. When you do want to use a word that may not be familiar to him, define it. Definitions are taken up in Chapter 5, and problems of word choice on page 41. For the present, we leave the matter of vocabulary with this thought: in many years of experience we have never heard of a complaint that the vocabulary in a report was too simple.

Sentences, the second element, should be relatively short, and should be simple in construction. What does "relatively" mean? That is, relative to what? Well, here is an example. Suppose you are writing a set of directions on how to assemble a piece of equipment and your reader is a man who dropped out of high school in the tenth grade. You'd better use only simple declarative sentences and restrict their average length to not more than 10 or 12 words. On the other hand, if you are writing about a chemical process and your reader is a Ph.D. in chemical engineering, you can probably get by with sentences that, on the average, are almost twice as long as well as more complex in structure. See page 49 for further comment.

It may seem odd to speak of adapting the organization of the principal parts of a report to a reader. And it may seem even odder to claim, as we did earlier, that organization can be helpful in solving stylistic problems. Probably most people think of the organization of a piece of writing as flowing naturally from the content. For example, the "natural" organization of a report of a scientific experiment would be to explain the purpose of the experiment, to describe the experiment, and to end by presenting the conclusions reached. In technical writing, on the contrary, conclusions are often presented at or near the beginning. One of the advantages of this arbitrary way of organizing the report is that it helps solve a stylistic problem.

The stylistic problem we have in mind is that of writing for a group of readers among whom some have a much greater familiarity with the subject than others. In this situation, if the writer uses technical terminology, some of his readers won't understand what he is saying. On the other hand, if the writer keeps explaining his terms, those readers who are familiar with the subject may feel he is wasting their time. What can the writer do?

One of the best solutions to this problem is to begin the report with a brief summary. In this summary, the principal findings and conclusions are presented in language intelligible to the readers who are least familiar with the subject. These people will read only the summary. Consequently, the writer can use more technical terminology in the main body of the report, which will be read only by people having the necessary technical background. Let's consider a concrete example of the stylistic problem just described, and then return briefly for another look at the kind of solution suggested.

Let's assume a report is to be written that will be of interest to readers who would fall into categories 3, 4, and 5 in the list of categories set up earlier. An oceanographer, let's say, is writing a report and has just completed this sentence:

> Two small commercial propellant-embedment anchors, nominally rated at 5- and 10-kip capacity, were investigated.

This sentence actually appears on page 34 of a 49-page booklet entitled *NCEL Ocean Engineering Program*, issued by the U.S. Naval Civil Engineering Laboratory in Port Hueneme, California. The purpose of the booklet is apparently to acquaint the educated public with the broad outlines of this laboratory's recent accomplishments and with its plans for the future. And for such readers it is a fascinating booklet, containing, among other things, a two-page summary of the laboratory's part in the search for some nuclear bombs accidentally dropped into the ocean off the coast of Spain, as well as brief descriptions of three proposed designs for manned underwater stations in which scientists could live at a depth of 6000 feet in the ocean. The sentence just quoted is concerned with the problem of anchoring down such underwater stations.

Now, in this context, should all intended readers of this booklet be expected to understand the sentence quoted above? One more look at our list of categories indicates that it is quite reasonable to expect readers to be drawn from categories 3, 4, and 5. Actually, even 2 might be represented. You would probably agree, at any rate, that a reader from category 3 would be likely to have trouble with the sentence quoted above because of lack of familiarity with the terms "propellant-embedment anchors" and "kip."

The author of the NCEL booklet evidently felt the same way because he provided this explanatory sentence:

> The propellant-embedment anchor is a self-contained device similar to a large-caliber gun consisting of a barrel, a recoil mechanism, and the projectile which is the anchor.

Fine. All is clear now. But what does "kip" mean? The author never did say. Why didn't he? Perhaps because he was beginning to wonder how his page was going to look if he added still another explanatory sentence.

In brief, here is an illustration of a kind of stylistic difficulty which is sure to result in annoyance for one reader or another. The reader from category 5 won't need the explanation of "propellant-embedment anchor" and will feel the passage is wordy. He might not need an explanation of "kip" either; but see the essay entitled "Mathmanship" at the end of Chapter 1. (As a matter of fact, a kip—kilo pound—is a force of a thousand pounds.) On the other hand, the reader from category 3 won't understand the initial sentence when he first reads it, and won't be interested in how many "kips" there are anyhow. Probably what this reader would like to have is something resembling the following:

> Upon reaching the ocean floor at a depth of as much as 6000 ft, the station will be secured in place by anchors fired like projectiles from the barrel of a specially designed gun.

A brief summary of the main points of the entire booklet, written in simple language like that of the sentence above, would be a real kindness to readers in categories 3 and 4. And thus a stylistic problem would be solved through use of an organizational device.

As you probably know, a routine part of a great many industrial and scientific reports is an initial summary or abstract. Often, although not always, this initial section is intended to help readers who are unfamiliar with the subject. As a matter of fact, many different terms besides "abstract" and "summary" are used for the kind of initial section just described, terms such as "results," "conclusions," "findings," "epitome," "digest." These terms don't necessarily all mean the same thing, but sometimes they do. As a matter of fact, some companies develop a number of elaborate forms for the organization of their reports, each form to be used for certain specified purposes or occasions, and these elaborate forms may begin with a formidable series of epitomes, results, findings, conclusions, and so on. These forms are often called "kinds of reports," and given names, such as "Partial Report." There is no harm in these practices. They are easy to become accustomed to, and are sometimes useful. Further information about kinds of reports and their formats can be found in Sections Four and Five.

We have commented on the reader who desires a summary at the beginning because the detailed text would be difficult for him. It is also worth remembering the advice Westinghouse gave about putting difficult material into an appendix, another helpful device. The company was thinking particularly about readers at the managerial level.

What about the group of readers who are thoroughly familiar with the subject in general but who nevertheless want to read only the abstract? These are people who feel they need to know only the conclusions reached in the report and who want to learn what these conclusions are as quickly as possible. These readers represent categories 5 and 6; of course some of them might be at the managerial level too. Perhaps this situation raises a question. Should a young writer save some of his own time, which is precious, by working hard on only the beginning sections of his report and getting out the rest in a hurry? Well, he could, but it might be risky.

The kind of problem we are getting into here is so important that we want to pause for a very careful comment on it. We have pointed out that some, or even many, of the readers of a report may look at only the introductory portions of it. It would be tremendously misleading, however, to imagine that the full text of a report is seldom read, or is seldom read by people in executive positions. As J. A. Hutcheson observed in the statement quoted earlier, reports are written to help people make decisions; and intelligent people don't make important decisions until they have examined all the facts available. One of the co-authors of this book remembers vividly his own first important (to him) effort at writing a technical report. When the report had been submitted, his boss summoned him to a conference about it. "Your conclusions," his boss began, "do not adequately reflect the content of your report." And he went on to analyze in mortifying detail the fact that the conclusions really did not indicate the full significance of what had been presented in the report.

In later chapters the writing of the initial sections of technical reports will be discussed and examples shown. At present, our purpose is simply to point out ways in which a writer can adapt the style of his report to a given reader by working on vocabulary, sentences, and organization.

*The Communication Situation.* One other aspect of the general problem of adapting the report to the readers who will use it remains to be discussed. This is the need for analysis of what we shall call the communication situation.

So far, what we have said about readers has been confined virtually to consideration of how much they would know about a given subject. Obviously, in analyzing readers' needs and interests, a writer has a lot more to think about than just the state of their knowledge. The readers we are contemplating are, after all, real live people working in a complex situation in which they have responsibilities, ambitions, worries. When looked at in this larger context, the report itself takes on new dimensions. It is the sum of all these additional elements concerning both the reader and the report that we have reference to in the phrase "communication situation."

The scope of what we will have to say on this particular subject will extend beyond what is usually thought of as style, which is our principal subject. We believe you will feel, as we do, that the practical problems of style in technical writing can be understood only in the context of the total situation.

Every communication situation is unique, and adapting to it requires two distinct steps: first, perceiving it accurately, and second, making sensible decisions as to what to do about it. In a general way, it is easy to itemize the elements in terms of which any communication situation in technical writing can be perceived. On the other hand, almost the only means of discussing sensible decisions is to appeal to experience. We will therefore list and comment briefly on the elements of the communication situation, and then present a few examples.

As seen from the writer's point of view, the elements in the communication situation can be broken down into two categories, as follows:

A. Users of the report
   1. Who they are
   2. How much they know about the subject of the report
   3. What their responsibility is concerning any action taken on the basis of the report
   4. What their probable attitude is toward the conclusions and recommendations of the report
B. Uses of the report
   1. As a guide for action

2. As a repository of information
3. As a long-term versus a short-term aid
4. As an aid under specified conditions

Points A1 and A2 have already been discussed in preceding pages and are repeated here only for the sake of the completeness of these categories. Points A3 and A4 are probably clear without explanation, but they do need to be illustrated, and will be in a moment. The elements listed in category B are perhaps less clear.

First, with respect to the elements in category B, let's consider the difference between B1 and B2. This difference may be seen by contrasting two imaginary reports, one recommending that production be started on a new additive for lead-acid storage batteries and the other explaining that efforts to find an effective additive have failed. The latter kind of report is put away in the files as a repository of information, useful as a safeguard against wasteful duplication of the work at some future time. (An abstract of just such a report describing unsuccessful research on a storage battery additive may be found on page 99.) The recommendation report, unlike the one just mentioned, is intended as a guide for action, or at least as a way of initiating action. Of course a recommendation report may also serve as a repository of information, but that is not its primary function.

Some reports have a very short useful life; others, an indefinitely long one. A field report on the repair of a customer's equipment may be of interest only until the repairs have been completed, whereas a report of successful basic research may have permanent value. A report of the latter sort should be written in such a way that it will still be comprehensible many years later. Files of old field reports are often referred to by designers and engineers for clues to design weaknesses, and we don't mean to suggest that clarity is unimportant in such reports. On the other hand, the comprehensiveness of a report on what was done in a basic research project would be only an unnecessary burden in most "short term" reports. This difference between the two kinds of reports is the subject of B3.

The fourth category, B4, the report as an aid under specific conditions, is concerned with the adaptation of the report to the immediate circumstances in which it will be used. For example, "trouble-shooting" manuals designed as an aid in emergency repairs on a piece of equipment are likely to be quite different from manuals designed for routine maintenance. Further illustration may be found on pages 532–533, where differences between naval and civilian use of electronics service manuals are discussed.

As we said, these ideas about the elements of the communication situation are rather simple and obvious, once they have been pointed out. It is only when writers don't stop to think about them at all that foolish things can happen. On the other hand, a good deal of imagination is required to take full advantage of the possibilities afforded by an accurate perception of these elements.

One very common, and often amusing, example of what happens when people simply don't think about these elements is that they apply some "standardized" way of organizing a report to a situation it doesn't fit. One of the authors of this text once examined several dozen proposals written by the technical staff of a large company in which this difficulty appeared. A proposal, as you may know, is a report in which a company attempts to sell its services to some other company or agency for the carrying out of a specific project. (See Chapter 15.) Since proposals are means of obtaining contracts, they are extremely important, and they should be tailor-made for every project sought. In our example, however, as the examination of the proposals written by the technical staff continued, it became evident that in every single proposal the major headings were identical. These headings were "Introduction," "Items and Services to Be Supplied," "Technical Approach," "Management Plan," "Reliability and Quality Assurance," and "Conclusion." For many of the proposals, these headings were in fact appropriate. For some, they most decidedly were not. One proposal, for instance, was concerned with a research study project in which nothing was to be delivered to the contracting company except a report containing the results of the study. In this proposal, nevertheless, the heading "Items and Services to Be Supplied" appeared as always—followed by the single word "None"!

How can imagination be used to adapt a conventional report to an unusual communication situation? The best answer, we believe, is not an abstract explanation but a concrete illustration.

We know an engineer who came to the conclusion that the company he worked for should invest a considerable sum (more than $100,000) in some new environmental test equipment. His study of work loads and of present and probable future commitments on contracts, and his estimate of company growth all led him to believe that money would be saved in the long run by replacement of the present equipment. The problem he then confronted was how to get approval for such an expenditure from his superior, the general manager of the company. This problem was especially forbidding because the general manager was a conservative man, a man who was rather negative about spending large sums of money. He would certainly take a hard look at the facts. On the other hand, the engineer believed the general manager respected his ability and judgment.

With all this in mind, the engineer wrote a 50-page report recommending purchase of the new equipment. In it, he provided detailed support of every aspect of his recommendation. To take the place of the usual introductory summary, however, he wrote a 2½-page section designed especially for his particular communication situation. These 2½ pages were made up principally of a series of questions and answers, each question typed entirely in capital letters. Underneath each question appeared the answer to it. He was thus attempting an adaptation which involved style in the conventional sense as well as organization.

The engineer's recommendation was approved. As a matter of fact, the general manager told us he thought the engineer had saved him a lot of time and energy by concentrating the most important data at the beginning of the report.

We certainly do not suggest that every recommendation be presented in the way this one was. Indeed, the point of our story is that almost any report can benefit from a sensible analysis of the particular circumstances in which it will be received and used. The questions the engineer presented in his first 2½ pages included the following: Why is new environmental test equipment needed? How much will it cost? How soon will it pay for itself? Where can space for it be found? What will be done with present equipment? How long will the proposed equipment serve company needs? What is its probable longevity? As you see, these are really just common-sense questions. They probably first appeared in the form of entries here and there in rough lists the engineer jotted down of things that must be included in his report. Such lists are the natural first step in the process of organizing a body of information. The crucial next step is to arrange the items in such lists into logical groups, and the third is to organize these groups into a formal outline. The success of the report we have just described can be attributed in large part to the skill with which the engineer presented at the beginning of his report a grouping of information which was well adapted to his particular, unique, communication situation.

The subject of organizing a body of information is discussed further in Chapter 4, "Outlines and Abstracts."

As a last example of adapting a report to a particular situation, we have in mind a case somewhat resembling the one just described, but one in which a very different report was written. Again, the following circumstances are actual.

An engineer was assigned to the job of surveying the layout of an assembly line in a manufacturing plant. After making his survey, he came to the conclusion that the entire assembly plant should be redesigned and re-equipped for more efficient operation. The cost of such a change would be very great, not to mention the loss of income occasioned by the considerable amount of "down time." He was convinced, however, that the change-over would ultimately result in a significant saving of time for product assembly, would reduce the number of defective items, and—in short—would result in a marked decrease in unit cost, thus saving the company a great deal of money.

Unfortunately, the management was hostile to any increase in capital outlay, particularly in view of their current high unit cost. He therefore decided that a report which defined the problem and immediately suggested a solution might very well so antagonize the management that they would pay little attention to those parts of the report which proved his case. His solution was to begin with a careful survey of those production problems which were already recognized by the management and were known to be causing them concern. He then proceeded to the presentation of his answers for those problems. In other words, his organization was essentially Socratic; he organized his body of information in such a way that his readers were virtually compelled to accept each new conclusion as they went along. This report, too, was successful.

As we have described this incident, the importance of the style of the report, in the conventional sense, appears much subordinate to that of the organization in achieving an effective adaptation to the situation. The fact is, however, that these two factors must work together. Were you to examine the report, we believe you would agree that its tone of detached and thoughtful analysis was well designed to harmonize with the kind of organization chosen.

The two examples we have just presented are both related primarily to points A3 and A4 in the list of the elements of the communication situation. That is, they are primarily related to the areas of the reader's responsibility and probable attitudes. Perhaps you have already noticed that, altogether, we have spent a good deal of time discussing the problems associated with category A, the users of the report, and relatively little discussing the problems of category B, the uses of the report. We have allotted the time in this way because the problems associated with category B are much simpler and more easily defined. We do not say they are less important than the problems of category A—only that they are less complex. Sometimes it is wise to remind ourselves, especially when the excitement and pleasure of a new achievement or challenge in science or engineering absorbs our interest, that the most complex thing known to exist in the universe is a human being. "Users" are human beings.

In summary, we have discussed adapting the style of a report to, first, the state of the reader's knowledge of the subject, and second, the total situation in which the reader examines and uses the report. We have presented the discussion primarily in terms of vocabulary, sentences, and organization.

We will now leave the subject of reader adaptation and turn to a discussion of the scientific attitude in technical writing. The following quotation will serve as a transition to this new subject. It is taken from a booklet issued by the Whirlpool Corporation for its own employees. It is transitional because it combines the subject we have just been considering, reader adaptation, with concern about point of view—which is one aspect of our next general subject, the scientific attitude.

> In report writing use your own style, not an imitation of anyone else's. Write the report from an objective point of view so as to focus the reader's attention on the subject and not on you; an objective viewpoint does not necessarily mean an impersonal viewpoint. Avoid personalities because this will have little value with the passage of time. Prefer the concrete expression to the abstract, the direct statement to the indirect. Give definite dates and values whenever possible. *KEEP YOUR READER IN MIND WHEN WRITING THE REPORT AND USE LANGUAGE HE CAN UNDERSTAND.**

The capital letters and italics in this last sentence are found in the original.

## The Scientific Attitude

While you must be careful to adapt your writing to your reader, you must also adapt it to yourself. Technical writing is a communication *to* somebody *from* somebody. What sort of person do you want your reader to feel that you are, as he reads your report?

Many years ago the answer to this question was both simple and, as we look back at it, rather strange. In writing about science, said the answer of 30 years ago, the author should appear to be no sort of person at all, simply a pure intelligence. As Albert Einstein wrote, "When a man is talking about scientific subjects, the little word 'I' should play no part in his expositions."** This point of view has been

**Style Manual for Project Reporting* (Benton Harbor, Michigan: Research and Engineering Center, Whirlpool Corporation, 1968), p. 19.

**Albert Einstein, *Essays in Science* (New York: The Philosophical Library, 1934), p. 113.

pretty much abandoned now, and the character of technical writing has been changed accordingly. The reasons for this change in point of view are complex, and there are differences of opinion about them. To avoid becoming involved in the details of a big subject, we shall merely say a few words about what has happened.

Historically, the change in the character of technical writing has been roughly paralleled by certain changes in science, or at least by the emergence of certain kinds of new questions in science. These questions are about the relationship between an experimenter in science and the events he observes. In classical physics it was usually assumed that the experimenter observed certain physical events, and that his experiment—and even his report of his experiment—was concerned with nothing except the "objective" physical events. In modern physics, as no doubt you know, this assumption is often questioned. Now it is often asserted that the experimenter himself is subtly and importantly involved in the content of his observations of physical events. A characteristic statement of this point of view is made by F. S. C. Northrup in his introduction to Werner Heisenberg's *Physics and Philosophy:*

> Quantum mechanics, especially its Heisenberg principle of indeterminacy, has been notable in the change it has brought in the physicist's . . . theory of the relation of the experimenter to the object of his scientific knowledge.*

Accompanying this new point of view there has been a strong tendency to doubt the old idea that science should be considered a monolithic, impersonal method. Instead, emphasis has begun to shift to the importance of the mind and temperament of the experimenter. No criticism of science is intended in this shift, but only a more thoughtful look at what actually happens.

It is this change in emphasis that accounts for the choice of title for a book about science, which we mentioned a few pages back, Michael Polanyi's *Personal Knowledge.* The thesis of Polanyi's book, we might add, has aroused much debate. Another interesting book concerned with this general subject is P. W. Bridgman's *The Way Things Are.*** At one point in his book, Dr. Bridgman, a winner of the Nobel prize in physics, devotes a couple of pages to a rather surprising subject—an explanation of why he decided to use the first-person singular pronoun in his writing.

Dr. Bridgman's unexpected interest in pronouns brings up again

*Werner Heisenberg, *Physics and Philosophy* (New York: Harper & Row, 1958), p. 4.

**P. W. Bridgman, *The Way Things Are* (Cambridge, Mass.: Harvard University Press, 1959), p. 5.

our initial question of how you present yourself to your reader. The use of personal pronouns has a lot to do with the impression of himself a writer conveys. Let's look at a part of what Dr. Bridgman says:

> Insistence on the use of the first person . . . will inevitably focus attention on the individual. This, it seems to me, is all to the good. The philosophical and scientific exposition of our age has been too much obsessed with the ideal of a coldly impersonal generality. This has been especially true of some mathematicians, who in their final publications carefully erase all trace of the scaffolding by which they mounted to their final result, in the delusion that like God Almighty they have built for all the ages.

Now, should you—like Dr. Bridgman—use the first-person pronoun in your technical writing? Would such a stylistic practice be consistent with a properly scientific attitude?

Let's examine the second question first. What is consistent with a properly scientific attitude depends upon how the scientific attitude is defined. As may be inferred from the evidence just looked at, the scientific attitude used to be associated with impersonality, exclusion of emotion, objectivity. Such a view leads to a style of writing in which these qualities are easily maintained, and it is not surprising that writing done in the third person, passive voice, came to be thought of as the scientific style. (We will illustrate this style in a moment.) On the other hand, those people who believe that the scientific attitude does not have to exclude feeling, that it is not completely impersonal, will at least in some degree accept Dr. Bridgman's opinion that there is no harm in using personal pronouns.*

Actually, we think it would be a mistake to attach great significance to the question of whether a personal pronoun should or should not be used. What matters, we think, are honesty about the facts, care in obtaining and evaluating the facts, dignity and restraint in manner. Probably you would agree that these are the kinds of qualities that make up a good scientific attitude. Suppose, for instance, that someone were to write a sentence like this: "There can be no doubt that this product is infinitely superior to all the others on the market; as a matter of fact, the others are worse than useless—they are shoddily made and placed on the market, it would appear, by an unscrupulous group of shysters." This sentence is written in the "scientific style" of third person, passive voice, but surely it does not express an acceptable scientific attitude. It might be effective

*For a further discussion of this point of view, see the following article by the physicist Gerald Holton: "Science and New Styles of Thought," *The Graduate Journal*, 7 (Spring, 1967), 399–422.

journalism, but a scientist would say it lacks an honest concern about the facts, it lacks dignity, it lacks restraint. A scientist might write, "According to the criteria under consideration, test data show this product to be markedly superior to the others tested." Here, then, is one aspect of your writing to keep in mind as you consider what sort of person your reader will think you are.

As to whether or not you should use personal pronouns in your technical reports, the first thing you should do is to learn the opinion of the organization you happen to be associated with. Some scientific and industrial organizations seem to have been influenced by the historical trends we have mentioned; others have not. In the first edition of this textbook, published in 1954, we advised our readers that they would be expected to avoid the personal pronoun. We were sure this was good practical advice because we had examined a great deal of evidence. Now examination of an equally great amount of new evidence reveals a widespread change toward acceptance of a more casual, personal style. But it is by no means a complete change. For one example of this more personal style, see the report quoted on pages 190–195, where the author casually uses the pronouns "I" and "my."

To be as concrete and practical as we can in our advice on this subject of the personal element in style, we shall put it into the form of comments on the six illustrative sentences that follow. These six sentences represent the range of choices you have in deciding how personal to let your style become, so far as use of pronouns is concerned.

A. *First Person Singular, Active Voice*
   I got surprising results from the three tests I made.
B. *First Person Plural, Active Voice*
   We got surprising results from the three tests we made.
C. *Third Person (Singular or Plural), Active Voice*
   1. The laboratory staff got surprising results from the three tests they [or "it"] made.
   2. The three tests gave surprising results.
   3. The writer got surprising results from the three tests he made.
D. *Third Person (Singular or Plural), Passive Voice*
   Surprising results were given by the three tests.

Until recently, the style represented by choice A was traditionally considered unacceptable in formal reports, and in most other technical writing as well. The style of any of the other five sentences was acceptable (with the natural exception that "we" in choice B was acceptable only if it referred to a group, not just a single author). And the style of choice D (third person, passive voice) was widely regarded as *the* scientific style, as we said earlier. Now, on the other hand, you can write in the style of any one of the six sentences unless there are particular reasons for not doing so. There are, however, two kinds of possible objection to one or another of the four choices: (1) personal conviction, and (2) lack of effectiveness for a given purpose.

As an example of an objection based on personal conviction, here is what the Bureau of Mines of the U.S. Department of the Interior has to say to its staff:

> Personal pronouns are acceptable in certain informal manuscripts such as book reviews, journal notes, and letters to editors of trade and technical publications. In published reports, use of the third person, singular and plural, helps the writer escape many awkward constructions. However, such pronouns as "I," "you," "we," "our," and "us" normally should not appear in manuscripts destined for publication in the Bureau's regular series unless they are part of quoted material.*

Clearly, the staff at the Bureau of Mines are expected to avoid personal pronouns.

Here's an example of a contrary opinion, this one from Stromberg-Carlson.

> Personal pronouns should not be used to excess. However, they should be used when writing instructions for actions to be performed by the reader and to avoid awkward or stilted sentence structure.**

Stromberg-Carlson obviously takes a much more casual view than does the Bureau of Mines.

We have quoted these two comments because they effectively represent the two basically different attitudes you will find. With these two attitudes in mind, together with the six illustrative sentences presented previously, we have two specific suggestions about use of personal pronouns.

1. If your organization has no objection to your doing so, don't hesitate to let your style become somewhat "personalized" through the use of personal pronouns; but remember these cautions: (a) Too many uses of "I" or "my" may cause your reader to decide you're conceited. Too many personal pronouns can become irritating. (b) There's a difference between reporting on original research you've carried out entirely by yourself and reporting on rather routine work done as a member of a group. In the latter circumstance the advice

***Style Guide for Bureau of Mines Manuscripts* (Washington, D.C.: Bureau of Mines, U.S. Dept. of the Interior, 1962) p. 6.

****Stromberg-Carlson Style Manual for Commercial Publications* (Rochester, New York: Stromberg-Carlson Corporation, 1968), p. 1.

quoted earlier from the Whirlpool Corporation seems particularly sensible: "Write the report from an objective viewpoint, so as to focus the reader's attention on the subject and not on you."

2. Quite apart from the considerations just noted, the six styles represented by the six sentences listed above are not all equally effective. The style of choice C3 ("The writer . . . ") would seem awkward and artificial in most circumstances. Similarly, the style of choice B ("We . . . ") surely would seem pompous except where this pronoun actually refers to two or more people. Among the other four sentences (A, B, C2, D) the principal distinction, so far as effectiveness is concerned, seems to lie in the choice between the active and the passive voices. It is often argued that the active voice is the more vivid and effective, and comparison of sentences C2 and D undoubtedly lends support to this contention. Here they are again:

> The three tests gave surprising results.
> Surprising results were given by the three tests.

The first sentence has the obvious advantage of saying in six words what the second says in eight. And on the purely subjective grounds of how we happen to like the two sentences, we ourselves much prefer the first. But let's look at another example.

> Discussion at the meeting of the Board of Directors clearly revealed the weakness of plan A. Plan B was adopted.

The second of these two sentences is passive, yet it seems to do very well. You could write the active, "They adopted plan B"—but it's no shorter, and the referent of "they" is disturbingly vague. "The Board adopted plan B" is still one word longer than the orginal, and perhaps you will feel that the repetition of "the Board" is slightly unpleasant. We ourselves will stick to the passive, "Plan B was adopted."

These examples seem to point to two conclusions concerning this rather subjective problem of choosing between the active and passive voices. First, thoughtless use of the passive voice may unquestionably produce writing that is both wordier and less vivid than need be. And second, crisp and effective sentences can certainly be written in the passive. We're sorry we don't have any rules for you to go by, but it really isn't a very difficult problem if you just use your head. The best time to attend to this stylistic problem is probably in the revision rather than the first draft of your writing.

In summary, here are our suggestions. First, make sparing use of the style of choice A ("I . . . ") unless it is simply prohibited. Second, use the style of choice B ("We . . . ") only where the referent is obviously appropriate. Third, with rare exceptions, entirely avoid the style of choice C3 ("The writer . . . "). And, finally, choose thoughtfully between the active and passive voices, on the basis of the immediate context.

We close these comments on the scientific attitude with what strikes us as a curious thought. If you wish to write well, your stylistic problems are going to be just a little more difficult than were those of a previous generation because you have more decisions to make. The previous generation didn't use "I." You now can, pretty often. But "I" is a tricky word. Freedom is wonderful, yet it does demand a lot of decision making.

We ourselves welcome this stylistic freedom.

## Making Sentences Say What You Mean

Besides giving attention to the needs of the reader and maintaining an objective manner, the technical writer must be certain that he is expressing his thought accurately. A great deal of bad writing results from the writer's failure to think carefully enough about what his sentences actually say. Perhaps this fault is a habit of mind as much as anything; usually, when a person is shown that a sentence he has written doesn't make good sense, he recognizes the error at once and complains that he can't understand why he didn't see it before. However this may be, one of the essentials of learning to write well is certainly the development of a habit of critically analyzing one's own sentences.

The kind of bad writing we are concerned with here is illustrated by the following passage:

> A problem usually arising in the minds of laymen considering solar heating for their home is the glare which might result from the use of large areas of glass. Actually, however, *just the reverse* has been found to be true. Large windows, while admitting more usable light, *produce less* than several small openings.

The italicized phrases don't convey the meaning the author intended. Just the reverse of what? Produce less what?

Here is a second illustration of the same sort of bad writing:

> The greatest problem which is found due to using large panes of glass is caused by the fact that glass is an excellent conductor of heat.

There are a lot of unnecessary words in this sentence, but that isn't the worst blunder in it. The worst blunder is the statement that glass is an excellent conductor of heat. The author knew very well that glass is a poor conductor, but he had a picture in his head of a lot of heat coming in through the window pane, and he knew the picture was correct. So he wrote down some words that had to do with the transfer of heat and was satisfied. What he was satisfied with was the picture in his head, which was a good picture. He paid little attention to the words. When he was later shown the words, he saw at once that they were wrong.

To avoid mistakes of this kind, put aside a piece of writing for as long as you can after finishing the first draft. Leave it until you can see the words instead of the pictures in your head. For some people, reading aloud is a help in spotting faulty passages. Ultimately, of course, everything depends on using words which mean precisely what you want to say.

## Precision in the Use of Words

Precision in the use of words requires the technical writer to have an exact knowledge of the meaning of words, to avoid words that—in a given context—are vague, to leave out unnecessary words, to use simple words wherever possible, to avoid overworked or trite words, and to avoid technical jargon.

*Knowing What Words Mean.* Unfortunately, many words are used incorrectly in technical writing. We list below a sampling of those which are commonly confused or misused. Reference to a good dictionary or to books like Fowler's *Modern English Usage* and Evans' *Dictionary of Contemporary American Usage* will help you with them—and many others like them.

ability/capacity
adjacent/contiguous
advise/tell, inform
affect/effect
alternative/choice
among/between
anticipate/expect
apparent/obvious/evident
appreciate/understand
assume/presume
assure/insure
balance/remainder
bimonthly/semimonthly
conclude/decide
continual/continuous
deteriorate/degenerate
effective/effectual
encounter/experience
essentially/basically
few/less
filtrate/filter
indicated/required
infer/imply
liable/likely
maximum/optimum
oral/verbal
percent/percentage
perfect/unique
practical/practicable
preventative/preventive
principal/principle
proportion/part
reaction/opinion
replace/reinstall
respectfully/respectively
target/objective
theory/idea, view, opinion
transpire/occur
universally/generally
waste/wastage

*Avoiding Vague Words.* Most often, however, precision of meaning is lost not through outright error in the use of terms but by the use of words which, although not incorrect, do not convey the exact meaning demanded. For example, words like "connected," "fastened," or "attached" are used instead of terms that more accurately denote the nature of a connection—terms like "welded," "soldered," "bolted," and "spliced." Of course there is nothing wrong with any of the words mentioned; it is in the way the words are used that trouble may develop.

*Leaving Out Unnecessary Words.* Words which serve no useful purpose should be rigorously weeded out of your reports during the process of revision. A comprehensive discussion of the ways in which words can be, and are, unnecessarily used lies far beyond the scope of this chapter. We shall discuss only a few ways. The positive principle we want to establish is probably demonstrated as well by a few examples as it would be by a great many. This principle is simply that you should take a hard look at every word in a sentence to make sure that it is there for a good reason.

To start with, various pitfalls are associated with the use of abstract words like "nature," "character," "condition" or "situation." (Please remember that it is not the word itself but the incorrect use of the word that we are criticizing.) Consider the following sentence: "The device is not one of a satisfactory description." What does the word "description" contribute to the sentence? Nothing at all. It is a useless appendage. The sentence might better have read: "The device is not satisfactory." Here are some other examples:

> The principal reason for this condition is that the areas which were indicated for street purposes were not intelligently proportioned. (*Better:* The principal reason for poor traffic flow is that the streets were not intelligently laid out.)

> An easy example for explanation purposes would be a shunt-type motor. (*Better:* A shunt-wound motor is a good example.)

In both sentences the word "purposes" is used unnecessarily. In the following sentence the word "nature" is ineptly used:

> The soldering proved to be of an unsatisfactory nature. (*Better:* The soldering proved to be unsatisfactory.)

Finally, here is a sentence in which "position" is at fault:

> With this work now completed, the plant is in a position to proceed with work on the new product.

What revision would you make of this sentence?

A second common source of trouble is the use of modifying words that look fine at first but actually mean little or nothing. Examine the list below and then notice in your reading how often they turn out to be meaningless.

| | |
|---|---|
| appreciable | fair |
| approximate | negligible |
| comparative | reasonable |
| considerable | relative |
| definite | sufficient |
| evident | suitable |
| excessive | undue |

These are all good words when they are used with a concrete reference. But consider the following examples:

This newly developed machine proved to be comparatively efficient. (This sentence is not meaningful unless we know the efficiency of the machines with which comparison is made.)

Water-flooding effected a substantial increase in production. (This means little without specific amounts.)

The voltage regulator must definitely be checked at periodic intervals. (*Better:* The voltage regulator must be checked at periodic intervals. *Or:* It is important that the voltage regulator be checked at periodic intervals.)

But what is a periodic interval? The time should be stated. A last illustration follows:

Research personnel have made appreciable progress in solving this problem.

(*Translation:* We haven't found out anything yet, but we have several ideas we're working on.)

A third source of unnecessary words is the use of pointlessly elaborate prepositions and connectives. The sentences below illustrate this problem:

Greater success has been enjoyed this year than last *in the case of* [by] the engineering department.

This problem is *in the nature of* [like] one encountered years ago.

Our reports must be made briefer *with a view to* ["to" is enough] ensure more successful research-production cooperation.

This recorder has been installed *for the purpose of providing* [to provide] a constant check of volume changes.

Many phrases and clauses used in introducing the main idea of a sentence are unnecessary, and they are often pompous-sounding and stilted as well. Study these examples:

> It is perhaps well worth noting that the results of this study show that plant efficiency is low. [If the main idea the author wants to communicate is that "plant efficiency is low," the elaborate introductory clause is a waste of words. The clause can be justified only if the writer wants to emphasize the idea that "it is worth noting" that plant efficiency is low. "Perhaps" surely serves no useful purpose in either case.]

> It will be observed that test specimen A is superior to test specimen B. [If the author wanted to say simply that "test specimen A is superior to test specimen B" he should have done so without the introductory clause. If he really wanted to say that the superiority of A to B will be *observed,* then his sentence was all right.]

There is no inherent fault in the introductory clauses used above, or in others like them (such as "it will be noted," and "consideration should be given to"); but fault does lie in saying more than is meant and in using a great many words to say what could be said more emphatically and clearly with a few.

A comprehensive list of wordy, redundant phrases found in technical writing would make a book by itself. We conclude these remarks on wordiness with a miscellaneous list of frequent offenders.

absolutely essential (essential)
actual experience (experience)
aluminum metal (aluminum)
at the present time (at present, now)
completely eliminated (eliminated)
collaborate together (collaborate; "together" is unnecessary in many phrases, such as "connect together," "cooperate together," and "couple together")
during the time that (while)
few in number (few)
in many cases (often)
in most cases (usually)
in this case (here)
in all cases (always)
involve the necessity of (necessitates, requires)
in connection with (about)
in the event of (if)
in the neighborhood of (about)
make application to (apply)
make contact with (see, meet)
maintain cost control (control costs)
make a purchase (buy)
on the part of (by)
past history (history)
prepare a job analysis (analyze a job)
provide a continuous indication of (continuously indicate)
range all the way from (range from)
red in color (red)
stunted in growth (stunted)
subsequent to (after)
through the use of (by, with)
true facts (facts)
until such time as (until)
with the object of (to)

Words are used unnecessarily in many more ways than those we have pointed out, but the problem of avoiding unnecessary words is always to be solved in basically the same way: by thinking about what each of the words in a sentence is contributing to the meaning.

*Using Simple, Familiar, Concrete Words.* Probably nobody would deny the wisdom of avoiding unnecessary words, but young technical writers are often reluctant to admit that simple and familiar words should be chosen in preference to "big" words. In fact, they may resent such a practice as denying them the free use of the technical vocabulary they have been at such pains to acquire. Furthermore, they may feel that substituting simple words for technical terms will inevitably result in a loss of precision of meaning, or even a loss of dignity and "professionalism."

Many years ago, Thomas O. Richards and Ralph A. Richardson, both of General Motors Research Laboratories, pointed out a curiously interesting fact:

> *We have never had a report submitted by an engineer in our organization in which the explanations and terms were too simple.* [Italics ours.] We avoid highly technical words and phrases and try to make the work understandable, because we know that even the best engineer is not an expert in all lines. . . . Most reports err in being too technical and too formal.*

These men are not talking about writing for people without any technical background, someone like a stockholder or a director, but about writing for other technical people.

A large company of builders and contractors declares that one of the essential qualities of a good report is that it be clear, concise, and convenient, and adds that "the use of technical words should be limited as far as possible to those with which the prospective readers are familiar." The Tennessee Valley Authority manual on reports has as one of the criteria in its report appraisal chart the question, "Is the language adapted to the vocabulary of the reader?" In 1945, E. W. Allen, of the United States Agricultural Research Administration, made a comment which possibly reflects the tremendous sense of pressure of the years of World War II, but which is good advice at any time.

> . . . it is necessary to understand and keep in mind the point of view of those it is desired to reach . . . it is not enough to use language that

*Thomas O. Richards and Ralph A. Richardson, *Technical Writing* (Detroit: General Motors Corporation, 1941), p. 4.

> *may* be understood—it is necessary to use language that can not be misunderstood. . . . The style of the technical paper should be simple, straightforward, and dignified.*

The list below provides a few examples of the problem these men were talking about. Most of the terms in the left-hand column are perfectly good words, and they are the best words in certain contexts. But if you mean "parts" why say "components"? Or if you mean what may be written as either "name" or "appellation," why not take the simpler word? Unless you have a good reason don't substitute

| | | |
|---|---|---|
| initiate | for | begin |
| disutility | for | uselessness |
| compensation | for | pay |
| conflagration | for | fire |
| veracious | for | true |
| activate | for | start |
| ramification | for | branch |
| verbose | for | wordy |

H. W. Fowler writes sensibly and wittily of this problem in *A Dictionary of Modern English Usage,* in such articles as "Love of the Long Word" and "Working and Stylish Words."

On the other hand, don't ever sacrifice precision for simplicity. Some ideas can't be expressed in simple language, and there's no use trying.

*Avoiding Overworked Words and Phrases.* Some words and phrases are used so often that they seem to be second nature to technical writers. Although such trite words and phrases are not necessarily wrong, their frequent use makes them tiresome to discriminating readers. Moreover, such terms are likely to be pretentious and wordy. Since the beginning technical writer may have difficulty in recognizing trite words and phrases, take our word for it that the words and phrases we list below are overused. Keep alert in avoiding them—and dozens of others like them.

activate (begin)
approach (answer, solution)
appropriate (fitting, suitable)
assist (help)
cognizant authority (proper authority)
implement (carry out)
indicate (point out, show)
investigate (study)
maximum (most, largest, greatest)
on the order of (about, nearly)
optimum (best)

*E. W. Allen, *The Publication of Research* (Washington: U.S. Agricultural Research Administration, 1945), p. 4.

communicate (write, tell)
consider (think)
demonstrate (show)
develop (take place)
discontinue (stop)
effort (work)
endeavor (try)
facilitate (ease, simplify)
function (work, act)
personnel (workers, staff)
philosophy (plan, idea)
prior to (before)
subsequent to (later, after)
terminate (end, stop)
transmit (send)
utilize (use)
vital (important)

*Avoiding Technical Jargon.* In writing technical documents for readers who lack a thorough familiarity with the subject matter, you should avoid shoptalk or technical slang. Such terms may be clear to workers in your scientific or technical field, they may be colorful, and they may certainly be natural and unpretentious; but they will not serve your purpose if they are not known to your readers. The list below suggests the kind of terms we mean:

breadboard (preliminary model of a circuit)
call out (refer to, specify)
ceiling (limit)
know-how (knowledge, experience)
mike (micrometer, microphone, microscope)
megs (megacycles)
optimize (put in the best possible working order)
pessimize (deliberately put in poor working order)
pot (potentiometer)
state of the art (present knowledge)
trigger (start, begin)
-wise (added to many terms like budget, production, design)

## Sentence Structure and Length

Good technical writing calls for a natural word order, simple sentence structure, and fairly short sentences.

The normal, natural order of elements in English sentences is (1) subject, (2) verb, and (3) object or complement. Each of these elements may be modified or qualified by adjectives or adverbs. The normal position of adjectives is in front of the terms they modify. Adverbs usually appear before the verb, but often after. This order of parts should generally be followed in your sentences for the sake of clarity and ease of reading. Furthermore, subject and verb should usually be close together. Naturally, departure from these patterns is occasionally desirable to avoid monotony.

The following sentences illustrate some typical word orders:

1. *Natural Order*
The machine was designed for high-speed work.

2. *Natural Order with Modifying Words and Phrases*
This 90-ton, high-speed machine was efficiently designed to provide the motive power for a number of auxiliary devices.

3. *Inverted Order*
Remarkable was the performance of this machine.

4. *Periodic Order*
When these tests have been completed and the data have been analyzed, there will be a staff meeting.

The order of sentences 1 and 2 is usually preferred to that of the other two. In sentences 3 and 4 the principal subject is not clear until near the end of the sentence. Periodic and inverted sentences may certainly be used occasionally, but most of your sentences should be in the natural order.

So far we have been concerned with the effect of word order on the readability of sentences. Closely related is the type of sentence structure employed. In general, simple sentences should outnumber the other kinds: complex, compound, and complex-compound. You will recall from your study of composition that a simple sentence contains only one clause and that a clause is a group of words containing a subject and a predicate. Examples 1, 2, and 3 above are simple sentences. A complex sentence contains an independent clause plus one or more dependent clauses. A compound sentence contains two or more independent clauses. A complex-compound sentence contains two independent and at least one dependent clause.

1. *Complex*
When all other preparations are made, the final step may be taken. (The introductory clause here functions as an adverb and is dependent upon the main clause for its full meaning.)

2. *Compound*
The first stage of this process can be completed under the careful supervision of the shop personnel, but the second stage must be directed by trained engineers. (The compound sentence consists of two statements linked by a conjunction.)

3. *Complex-Compound*
If this process is to succeed, the first stage can be completed under the careful supervision of the shop personnel, but the second stage must be directed by trained engineers. (Here a qualifying dependent clause is added to the first main clause. Additional qualifying phrases and clauses could, of course, be added, further complicating the sentence.)

Reading is slowed by too large a proportion of complex and compound-complex sentences. What is too large a proportion? We wish we could answer that question with a precise figure, but we can't. The writer must have a sense of proportion—and we do intend that word to mean two things: a percentage and a balance or harmony.

You should also be careful about the length of your sentences. The amount of difficulty a person experiences in reading a given text is positively correlated with sentence length and number of syllables per word. Research indicates that the average sentence length should probably not exceed 20 words. Of course this does not mean that every sentence should be limited to no more than 20 words. Nor is it necessary to avoid all words of more than three syllables. Technical subject matter often requires the use of a complex technical vocabulary and the expression of complex ideas. But if you should discover that your sentences are long and your words have many syllables, the chances are that you can simplify. And always keep in mind the range of your reader's intellectual ability and his familiarity with your subject. For an interesting illustration of the practical application of these principles, see Appendix D, page 541. Here you will find that the suggested maximum average sentence length is 25 to 30 words. This length seems to us a little too much, but no certain knowledge exists on this point.

Following is an example of how a difficult job of reading can be made easier by simplifying the sentences. For this example, we are indebted to the Ethyl Corporation Research Laboratories at Detroit, Michigan. It was used in a course in writing provided for their staff. First, here is the original passage:

> Although it is recognized that the question of soap content versus lubricating efficiency is a controversial subject in the grease industry, it is believed that the long record of eminently satisfactory lubrication performance, frequently under adverse conditions where no other grease was adequate, is sufficient evidence that high-soap content is not a detriment insofar as barium greases are concerned. Further, it is felt that a comparison between different types of greases solely upon the basis of soap content is rather a pointless argument unless proper cognizance is taken of the differences in the molecular weight of the bases, of the ultimate effectiveness of the greases and lubricating bearing surfaces under service conditions, and of the various factors of composition that radically modify the oil-thickening action of the different soaps.

Here is the way this material was rewritten in the Ethyl Corporation's course in technical writing:

> The grease industry has long debated the effect of soap content on lubricating efficiency. Still, a long record of highly approved performance

> should show a high soap content is no drawback in barium greases. Often they have succeeded where other greases failed.
>
> Comparing greases by soap content is rather pointless anyway unless such questions as these are answered:
>
> 1. How do molecular weights of their bases differ?
> 2. How well does the grease lubricate bearing surfaces in service?
> 3. What else in the grease might alter the oil-thickening action of the soaps?*

The revised version has 88 words, the original 130; sentences are much shorter; and there are fewer polysyllabic words. We think the revised version is a lot easier to understand.

It is quite as possible to go to extremes in the use of short, simple sentences as in the use of complex sentences. If you go too far in the use of simple sentences, you may find yourself writing something like this:

> He did not do well with the company at first. Later he managed to succeed very well. Finally he became president of the company.

This is bad writing because there is no use of subordination in it. All the ideas are given the same weight. Linking the three sentences together with simple conjunctions – "but later," "and finally" – would eliminate the unpleasant choppy effect, but what is really needed is subordination of one idea, something like that in the following complex sentence:

> Although he did not do well at first, he was later very successful, finally becoming president of the company.

The word "although" subordinates the first clause. Such a word is called a subordinating conjunction. Some other words that will serve this function are: after, because, before, since, in order that, unless, when, where, while, why.

In general, then, the best policy is to make most of your sentences simple in structure and natural in order, but to vary the pattern enough to avoid unpleasant monotony and to provide proper emphasis.

## Paragraph Structure and Length

Typically, a paragraph begins with a sentence (the topic sentence) which states the gist of the idea to be developed. The other sen-

*T. J. Carron, *Training the Professional-Technical Employee* (Detroit: Research and Development Department, Ethyl Corporation), p. 4.

tences of the paragraph develop, support, and clarify this central idea. But, as a matter of fact, you have probably observed that this topic sentence may appear anywhere within the paragraph. It may appear in the middle, or it may appear last, as a summary or generalization based on material already presented. Sometimes it doesn't appear at all, in so many words, but is implied. The requirements of technical style being what they are, we urge you to follow the tried practice of placing the topic statement first in the paragraph, or, at the very latest, just after whatever transitional sentences appear. The technical writer doesn't want his reader to be in suspense as to what he proposes to talk about.

Compare the following two versions of a paragraph from a Shell Oil Company manual. Version B is the original; version A is our revision, for the purpose of illustration.

*Version A*
These instructions are not designed to cope with exposure environment where highly corrosive vapors are encountered, although the paints recommended do have substantially good corrosion-resistant properties for normal plant tank farm conditions. Where such environments are encountered, special coatings may be required, such as vinyls, chlorinated rubber, Epon resin vehicle materials, or standard and other special paint systems applied to sprayed zinc undercoatings. In these cases proprietary brands may be used until open formulations are available. Experience in the field and the use of exposure test panels, pH indicators and other methods will determine whether it will pay to apply the more expensive corrosion-resistant coatings. Special corrosion problems should be referred to the Atmospheric Corrosion Committee for investigation. On the other hand, the instructions, specifications and formulations contained in this manual are designed to cope adequately with exposure environments existing in the general run of tank farms where hydrocarbons and the less corrosive chemicals are stored.

*Version B*
The instructions, specifications and formulations contained in this manual are designed to cope adequately with exposure environments existing in the general run of tank farms where hydrocarbons and the less corrosive chemicals are stored. They are not designed to cope with exposure environment where. . . . [This version continues by completing the first sentence in Version A and concludes at the end of the next-to-last sentence in Version A.]*

The main idea (the topic sentence) in version B is stated at the beginning so that the reader will know without delay just what the

*From *Protective Coating Manual*, p. 2. Reprinted by permission of the Shell Oil Company.

object of the discussion is. It is true that the reader needs to know what will not be covered, but it is more important for him to know what will be covered by the discussion. In A he does not find this out until the very end of the paragraph. Version B is the better of the two.

Two considerations govern paragraph length: unity of thought and eye relief for the reader. Since the paragraph is defined as the compositional unit for the development of a single thought, it may seem to you that length should be governed entirely by requirements of the development of the thought. And in theory, that's right. A simple, obvious idea, for example, might not take much development—perhaps no more than two or three sentences. A complex and highly important idea might, according to this line of reasoning, require a large number of sentences, perhaps covering several pages.

Long paragraphs, however, do not permit easy reading. If there is no break in an entire page, or more than a page, the reader's attention flags and he finds it difficult to keep the central idea in mind. Since long, unbroken sections of print repel most readers, the writer should devise his paragraphs so that such sections will not occur.

Breaking up discussion so that the reader's eye is given some relief does not demand that the writer violate basic principles of paragraph development. But neither does it mean that he should simply indent at will. The writer has a good deal of freedom in deciding what will constitute a unit of his thought. An idea containing several parts or aspects may be broken up, with the sentence which originally stood as a topic sentence for a long paragraph serving as an introductory statement to a series of paragraphs. Let's consider a hypothetical case. Suppose a writer had written:

> For a brief explanation of the meaning of the term "skip distance" in radio communications, we must first turn our attention to the phenomena of the ground wave, the ionosphere, and the sky wave.

Suppose further that this sentence stood as his topic sentence and that he developed a description of the three phenomena, all in the same paragraph. The paragraph would run quite long, too long for comfortable reading. His solution would be simple. Instead of one long paragraph, he could write three shorter ones, one on each phenomenon. The original topic sentence could serve as an introductory, transitional paragraph, perhaps with the addition of another sentence something like this: "Each of these phenomena will now be described in detail." In other words, the writer can arrange his organization so that the material can be divided into conveniently small units.

When you desire an especially forceful effect, try using one or more very short paragraphs.

To sum up, remember that all sentences in a paragraph must be about the same topic, but also remember that paragraphs should not be too long. Try to have one or more breaks on every page of your report.

## Summary

Technical writing style is distinguished by a calm, restrained tone, by the absence of any attempt to arouse emotion, by the use of specialized terminology, and by an accepted convention of the use of abbreviations, numbers, and symbols. Most organizations expect reports to be written in the passive voice, but other possibilities are useful. It is highly desirable to develop a habit of looking critically at sentences to make sure that they exactly express the ideas they were intended to express. Words and phrases must be used with precision. Clarity and ease of reading are improved by moderately short sentences and paragraphs. The organization of both sentences and paragraphs should usually be natural, with main ideas appearing near the beginning. Barzun and Graff write wisely in explaining how good style is achieved: "To the general public 'revise' is a noble word and 'tinker' is a trivial one, but to the writer the difference between them is only the difference between the details of hard work and the effect it achieves. The successful revision of a . . . manuscript is made up of an appalling number of small, local alterations. Rewriting is nothing but able tinkering. Consequently, it is impossible to convey to a nonwriter an abstract idea of where the alterations should come or how to make them. Only an apprenticeship under a vigilant critic will gradually teach a would-be writer how to find and correct all the blunders and obscurities that bespangle every first draft."*

# PART 2

## Introduction: Grammar and Usage

We do not believe that a book on technical writing should try to cover—or even review—the subject of grammar, and shall therefore not attempt it. For one thing, it is a subject deserving and requiring book-length treatment; for another, it is not a subject requiring spe-

*Barzun and Graff, *loc. cit.*, 249.

cial treatment as far as technical writing is concerned. The grammar of technical writing is no different from the grammar of any variety of writing. We do feel an obligation, however, to deal with some matters of acceptable usage that, according to our experience, give technical writers a good deal of trouble. But before we get down to specifics, we feel we should describe some important recent developments in the field of grammatical study and, hopefully, clear up some common misconceptions about grammar and usage.

What *is* grammar anyway? The term suggests a set of rules and regulations prescribing certain practices for agreement of subject and verb, or for agreement of pronoun and antecedent. At the same time, it suggests, doubtless, the proscription or condemnation of certain practices, such as fragmentary sentences, the comma fault, lack of agreement, and faulty sentence construction. Put another way, the term *grammar* probably suggests to you the study of what is "correct" and "incorrect" in the use of English.

But to the professional student of language, or the linguist, grammar denotes the systematic way in which a language functions to convey meaning, the systematic ways in which a complex of structural patterns, governing the forms of words and sentences, operates so that we can communicate with one another. In a strict—and limited—sense, grammar has little to do with "correctness." It does have to do with what is possible and what isn't. To the linguist, a statement like "I seen him when he done it" is grammatical: it functions in accordance with a recognized and accepted organization and pattern of words. This does not mean that the linguist approves the statement, of course; he will prefer "I saw him when he did it," since the latter sentence is in accord with accepted usage among educated users of English. In an important sense, the modern grammarian operates like the pure scientist: he observes facts about the way language functions and the "rules" he sets down stem directly from his observations, not from preconceived notions of how the language *should* function.

During the past several decades, "scientific" observation of the ways in which language operates has brought about the development of several new grammars. These grammars are at odds with much that has traditionally been taught in the schools. The traditional, or school grammar, together with its pronouncements on usage and style, had its origin in the eighteenth century with the publication in England of such books as Bishop Robert Lowth's *A Short Introduction to English Grammar* (1762), and in America of Lindley Murray's *English Grammar* (1802). As classical scholars, Lowth and his followers took as their model for English grammar the Latin grammar with which they were familiar, and they formulated English grammar to correspond with that of Latin, ignoring what present-day students of language take as an essential starting place for the discovery of a language's grammar—the ways in which a particular language actually is spoken and written.

Beginning roughly with the work of Leonard Bloomfield, whose *Language* was published in 1933, linguists have sought to discover how English actually functions. One result of their work has been the development of several new grammars, or methods of describing how the language functions. Among these new grammars, two are most important, the "structuralist" and the "generative-transformational."

The methodology of the structuralist grammar begins with identification of the smallest meaningful unit of speech, the phoneme, and moves on to the morpheme (a meaningful sequence of sound signals or phonemes), and then to the sentence—or, as one well-known text put it in its subtitle, "From Sound to Sentence in English."* The second of the new approaches to the development of English grammar—and the one engaging the support and interest of most present-day students of language—is the generative-transformational. Unlike the structuralists, the transformational grammarian begins with a study of the surface and deep structures of basic patterns of strings of words or sentences, proceeds to an analysis of the constituents and features of those strings, and then goes on to a description of the processes or transformations which may operate on these sentences to produce more complex structures.

We are keenly aware of the superficiality of the foregoing sketch of new developments in grammar. If you are interested in looking into the subject further, we suggest that you read H. A. Gleason's *Linguistics and English Grammar* and Jacobs and Rosenbaum's *English Transformational Grammar*. The first of these contains an extensive bibliography of additional readings.

The important consideration for us here is that the new inquiries into the nature of English grammar of the past several decades have been accompanied by a fresh look at the matter of "correctness" and acceptability in usage. From the point of view of a modern linguist, the criterion for good English is not to be found in the older grammarians' set of prescriptions or rules but in the observed practices of successful writers and speakers. As Robert Pooley has said, "Good English is that form of speech which is appropriate to the purpose of the speaker, true to the language as it is, and comfortable to

*A. A. Hill, *Introduction to Linguistic Structures* (New York: Harcourt, Brace & World, 1958).

speaker and listener. It is the product of custom, neither cramped by rule nor freed from all restraint; it is never fixed, but changes with the organic life of the language."*

## Common Errors in Usage

Before taking up specific problems of usage, we want to point out a rather curious fact, a fact which we discovered while making an extensive and careful survey of errors found in technical writing. Technical writers do not characteristically make a great variety of significant errors in grammar or usage. On the contrary, they make only a rather small number of different kinds of errors. But the errors they do make, they tend to make over and over.

The following discussion of problems in the mechanics of style, as well as in usage, has been designed around what we learned in our survey. Chances are, if you master these problems, that your writing will be almost free of errors. We urge you to remember as you work on these problems that you do not need to "learn English." If it is your native language, you already *know* English. All you need to learn—assuming that you have any difficulties at all—is a tiny fraction of what the whole job of learning English involves. Probably all you need to learn is a few of the principles explained in the following pages.

The particular problems of usage we will discuss include certain troublesome subject-verb relationships, vague or indefinite pronoun reference, coordination and subordination, dangling modifiers, and lack of parallel structure.

*Troublesome Subject-Verb Relationships.* You do not need to be reminded that the subject of a sentence must agree with the verb in number; i.e., that a singular subject demands a singular verb, a plural subject a plural verb. Seeing to it that they do agree is another matter, however. Particularly bothersome are sentences in which the subject is an indefinite term (such as *everybody, anybody, anyone, everyone, nobody, no one, each, one, neither*), a collective, or an amount; and sentences in which the subject involves a pair of correlatives, a relative clause, or a compound subject. These problems are better understood with examples.

*Indefinite Subjects:* When the subject is an indefinite word, commonly a pronoun, the subject is usually identified by a following pre-

*Robert Pooley, *Teaching English Usage* (New York: Appleton-Century-Crofts, 1946), p. 14. By permission of the National Council of Teachers of English.

positional phrase, and the number of the object of the preposition normally determines the number of the verb. That is, the sense of the statement governs agreement. Let's look at some examples.

1. Both of these power supplies *are* satisfactory. (But notice that we would write, "Either of these power supplies *is* satisfactory.")

2. *Everyone* in the organization *makes* a weekly progress report. (Several people are obviously involved, but "everyone"—like everybody, anybody, anyone—takes a singular verb.)

3. *Half* of the units *were* faulty. (But we would write "Half of the trouble is the fault of the drafting department.")

4. *Some* of the units *have* been in service for ten months. (But "Some of the material is no good.")

*Collectives:* Words in this category may take either a singular or a plural verb, depending on the sense of the statement. In other words, if the individuals which comprise the collective term are thought of separately, the verb should be plural; if they are thought of as a group, the verb should be singular. Pronouns referring to such terms must also agree. Study the following sentences:

1. The *number* of reports lost last year *was* large.

2. The *majority were* between 1.5 and 2.5 mm long.

3. A *pair* of workmen *were* taking turns inspecting the units.

4. This *pair is* not as good as that.

5. A *number* of the electrodes *were* burnt.

6. The *data is* available for analysis. (Here "data" is thought of as a body of information; on the other hand, we would write, "The data were plotted, point by point, on the chart." We believe that "data" is most commonly thought of as a collective taking a singular verb, but there are those who cling to the fact that it is the plural form of *datum* and insist therefore on the plural verb. We would caution those who rigorously use a plural verb with "data" that plural pronoun forms must then be used in reference to the term. It is also interesting to observe that some writers who are most adamant about using "are" or "were" with "data" forget themselves when using another verb; hence we see such inconsistencies as "The data were carefully studied," followed a little later by a statement like this: "It [the data] proves the validity of our thesis.")

*Subjects of Amount:* As with indefinite subjects and collectives, subject-verb agreement with terms denoting amounts is governed by the sense of a statement, though terms denoting sums, rates, measurements, and quantities more commonly take a singular verb, despite their plural form. For example:

1. One hundred dollars per hour is high pay.

2. A thousand miles an hour is too fast.

3. Thirty-six inches is a yard.

4. Last year about forty hours was spent on that report. (One writer told us that he would think about each one of those forty hours separately —and painfully—and would therefore use a plural verb!)

5. About eighty pounds of carbon is added to the mix. (Here we would choose a plural verb if the carbon is added pound by pound; for the example we have assumed that an 80-pound *sack* of carbon has been dumped into the mix—hence, the singular verb.)

*Correlatives:* When the parts of a compound subject are joined by such pairs as *whether/or, neither/nor, either/or,* the verb agrees with the nearer part of the subject, as in the sentence: "Either the mainspring or the connections are giving trouble." But note that *not only/but also* and *both/and* take a plural verb since "and" and "also" are clearly plus signs.

*Relative Clauses:* You will have no trouble in choosing the verb form in a relative clause if you simply remember that the verb must agree with the antecedent of the relative pronoun (which, that, who). Consider these examples:

1. This is one of those books that are worth studying. ("That" refers to "books" and thus "are" is required.)

2. This is one of those parts which are always giving trouble. ("Which" refers to "parts.")

3. One of the main errors which were involved was the post-computation check. (Note that "which" refers to "errors" and thus requires the plural verb in the relative clause; note also that the subject of the main clause is "one" and thus requires "was" as its verb.

4. This is the one of those items which is faulty. (Note in this sentence that the presence of the word "the" before "one" leads us to use "is" after "which.")

*Compound Subjects:* Simple compound subjects in which the elements are joined by coordinating conjunctions or the correlatives normally present no problem. But compound subjects in which the initial item is singular take a singular verb form if the additional items which augment the subject are joined to it by *together with, no less than, as well as, along with,* and *in addition to.* Consider this example:

> The chief engineer, as well as the twenty engineers working with him, is of the opinion that the plan will work.

As a matter of fact, we should acknowledge that usage condones use of the singular verb with a compound subject in certain circumstances: (1) when the elements forming the compound subject refer to one person, as in "Our Director—and friend—is sick"; (2) when elements forming a compound subject are arranged in climactic order, as in "Our success, our growth, our survival depends on everyone of us working to capacity"; (3) when the elements of a compound subject follow the verb, as in "There is promotion and money in this new effort of ours."

*Vague or Indefinite Pronoun Reference.* Since a pronoun conveys no information in itself but is meaningful only in reference to the word or phrase for which it stands, the reference should be unmistakably clear. Unfortunately, a good deal of ambiguity is found in technical writing, owing to careless use of "this," "which," and "it." We have found that "this" (and "it") is a particularly frequent offender when it is used as the subject of a follow-on sentence. Notice the lack of clearly defined reference in the following examples:

> Panels should be exposed at more than one test station on exterior racks and regular inspections should be made. This will require trained personnel. (Does "this" refer to exposing the panels, making inspection, or both? As the sentence stands, it is impossible to be sure. If inspections, the second sentence should begin "Inspections will . . .")

> The rotating scanning mirror is larger in effective diameter and must turn faster than the scanner. This will result in increased torque, requiring a more powerful drive motor. (Can the reader be immediately sure what "this" refers to?)

> This input is a prediction of cost, prices, taxes, and success based on history and present knowledge. It includes plans for when, where, and how much money will be devoted to each phase. (Can the reader be sure of what "it" refers to?)

> The appended formulation for aluminum is designed to have fairly satisfactory self-cleaning properties which makes it suitable for decorative purposes but not as good as white. (Here "which" probably refers to the fact that the formulation has self-cleaning properties. If reference is to "properties" the verb "makes" should be "make" to agree in number. A better version of the sentence is, "The self-cleaning properties of the appended formulation for aluminum make it suitable . . . ")

Because these sentences have been taken out of context, their faults may appear so obvious that you would be inclined to say that any careful writer would avoid them. Yet errors like these are made over and over again in technical writing.

*Coordination and Subordination.* Most of us were taught that ideas of equal importance are expressed in independent or coordinate clauses and that ideas of less importance are expressed in subordinate or dependent clauses. Moreover, we were taught that certain conjunctions, like *and, but,* and *for,* are coordinating conjunctions and may be used in linking independent clause structures within a sentence; similarly, we were taught that certain adverbial subordinating words, such as *while, since, because, if,* and *when,* are used to introduce dependent, subordinate clauses which contain the lesser ideas or facts.

Although the validity of the "rule" that the main idea or most important fact should always be contained in the main or independent clause is highly questionable (judging from observation of the practice of quite accomplished writers), we can say that it is inefficient and wordy to express ideas of unequal importance by means of equal or co-ordinate structures. The practice of stringing together a series of facts by the addition of successive clauses joined by *and* and *but,* for example, can lead only to obscurity, monotony, and wordiness. Let's look at a few examples:

> This value is best determined by actual test and it is 50 watts. (*Better*: This value, best determined by actual test, is 50 watts.)
>
> Sand is the other important raw material and it is procured from an outside supplier. (Unless the writer wants to give equal stress to both facts expressed, he would do better to write: Sand, the other important raw material, is procured from an outside supplier.)
>
> This estimate has been plotted in Fig. 3 and shows the likelihood that the meters will all fail at the same time. (*Better*: This estimate, plotted in Fig. 3, shows the likelihood that the meters will all fail at the same time.)

We believe it is important to recognize that co-ordination and subordination are formal, grammatical matters and that the structure of a sentence does not necessarily reveal semantic importance or impact. In other words, the most important idea of a sentence may—and often does—appear in a dependent structure (as in "Although your report is full of the grossest inaccuracies, obscurities of expression, and downright inanities, it is well typed."). Nevertheless, the use of subordinating structures is a useful way of achieving conciseness and of stressing what needs to be stressed. Let's take a look at an example of no subordination along with some examples of the same facts expressed in a variety of subordinated structures.

1. The chief engineer's report was a carefully written, brilliant analysis of the problem. It was about fifty typewritten pages in length. (No subordination.)

2. The chief engineer's report, which was about fifty typed pages, was a carefully written, brilliant analysis of the problem. (Subordination by clause.)

3. The chief engineer's report, covering about fifty typewritten pages, was a carefully written, brilliant analysis of the problem. (Subordination by participial phrase.)

4. The carefully written, brilliant analysis of the problem by the chief engineer covered about fifty typed pages. (Subordination by modifying phrase.)

5. The fifty-page report of the chief engineer was a carefully written, brilliant analysis of the problem. (Subordination by single word modifier.)

6. The chief engineer's report, about fifty typed pages, was a carefully written, brilliant analysis of the problem. (Subordination by apposition.)

Clearly, these examples show opportunities for improving upon the version given in item 1. Which do you prefer?

*Dangling Modifiers.* A dangling modifier is one which has nothing to modify logically or grammatically, or one which seems to modify a word it cannot possibly modify. In technical writing, dangling participial and dangling infinitive phrases are very common, mainly because of the difficulties of describing action in the passive voice. Often—perhaps usually—these dangling phrases cause the reader no trouble, and many writers on the subject of usage take a lenient attitude toward their presence in sentences. Bergen and Cornelia Evans* say that

> The rule against the "dangling participle" is pernicious and no one who takes it as inviolable can write good English. In the first place, there are two types of participial phrases which must immediately be recognized as exceptions. (1) There are a great many participles that are used independently so much of the time that they might be classed as prepositions (or as conjunctions if they are followed by a clause). These include

*From *A Dictionary of Contemporary American Usage*, by Bergen and Cornelia Evans. 

such words as *concerning, regarding, providing, owing to, excepting, failing.* (2) Frequently, an unattached participle is meant to apply indefinitely to anyone or everyone, as in *facing north, there is a large mountain on the right . . .*

And Wilson Follett* says that "Some participles have so far lost their obligation to serve nouns as adjectives that they have in effect become prepositions, parts of prepositional phrases, or adverbs."

Nevertheless in conservative, orthodox, formal English, you will surely escape criticism if you take care to relate action to a specific word that names the actor. Let's examine a few typical sentences:

1. *Dangling Verbal Modifiers*

After connecting this lead to pin 1 of the second tube, the other lead is connected to pin 2. (Who connects the lead to pin 1? It can't very well be "the other lead" that does so! Two correct possibilities suggest themselves. "After this lead has been connected to pin 1 of the second tube, the other lead is connected to pin 2." *Or:* "After connecting this lead to pin 1 of the second tube, the technician connects the other lead to pin 2." In this second sentence, the introductory phrase logically modifies the subject of the main clause, "The technician. . . ." He is the one who did the connecting. In the first sentence, the introductory active participial has been changed to passive to agree with the voice of the main clause.)

When starting the motor from rest in the forward direction, the main coil PEM is de-energized and the IR drop across PFN produces a flux to oppose the residual magnetism left by PFN. (The introductory phrase, "When starting the motor . . ." leads the reader to expect that the subject of the main clause will name the starter, but he is disappointed. "Coil" is the subject of the main clause and it did not start the motor from rest. "When the motor is started from rest . . ." would solve the difficulty.)

In selecting the rectifier, current limiting resistors, and holdout coil, this hazard must be considered. (The participial phrase may be kept if the main clause is made to read "The engineer must consider this hazard." Otherwise the introductory phrase must be changed.)

2. *Dangling Infinitive Modifiers*

To start the motor, the starter button must be depressed. ("To start the motor, the driver must depress the starter button" keeps the infinitive phrase from dangling because we now have "the driver" to relate the action to.)

To achieve a mix of the proper consistency, more sand must be added. (Main clause needs a subject like "you" or "the worker.")

*Wilson Follett, *Modern American Usage* (New York: Hill & Wang, 1966), p. 121.

Ordinarily, as we pointed out earlier, dangling modifiers are no real obstacle to understanding for the reader, but now and then, as in the following sentences, they cause him amusement.

> After drying for three days under hot sun, workers again spray the concrete with water.

> After taking in a constant flow of oil for two days, the supervising engineer will note that the tanks are full.

As the Evans say, the trouble with such sentences as these is not so much that the verbal phrases dangle as it is that they don't: they are firmly attached to the subject of the main clause—and should not be.

*Lack of Parallel Structure.* Parallelism means the use of similar grammatical structure in writing clauses, phrases, or words expressing ideas or facts which are roughly equal in value. A failure to maintain parallelism results in what is called a "shifted construction." Parallelism is made clearer by the following illustrations:

1. *Parallelism of Word Form*
His report was both *accurate* and *readable.* ("Both" introduces two adjectives which describe the report. The parallelism would be lost if the sentence read, "His report was both accurate and it was easy to read.")

The process is completed by sanding, varnishing, and buffing the finish. (*Not:* "The process is completed by sanding, varnishing, and the buffing of the finish." The last item in the series is not parallel with the first two.)

2. *Parallelism of Phrases*
Preparing the soldering iron, making the joint, and applying the solder constitute the main steps in soldering an electrical connection. (All the initial terms of the phrases are participials to make the construction parallel. A failure of parallelism would give us something like this: "Preparation of the soldering iron, making the joint, and application of the solder. . . .")

3. *Parallelism of Clauses*
That this machine is superior to the others and that this superiority has been demonstrated by adequate tests have been made clear in the report. (The introductory "that" of both clauses helps make the parallelism clear. A violation of this parallelism would exist if we had: "That this machine is superior to the others and this superiority is demonstrated by adequate tests have been made clear in the report.")

A shifted construction is sometimes caused by a change in point of view, as shown by the following examples:

*A change from a personal style to an impersonal, objective one:* "First I shall consider the points in favor of this program and second the disadvantages to the program will be considered."

*A change from the indicative mood to the imperative:* "First, the wires should be spliced. Next, take the soldering iron. . . ."

*A change from the active to passive voice in the same sentence:* "The workman wraps insulation around the joint before the repaired joint is replaced by him in the circuit."

## Mechanics of Style

What we mean by the term "mechanics of style" is the use of abbreviations, numbers, symbols, word forms (particularly compounds), capitals, italics, and punctuation. Form, layout, and bibliographical forms are also included in the mechanics of style, but these are discussed in later chapters. Our purpose here is to list some dependable rules for handling problems of usage. Since usage in the mechanics of style is not standardized throughout the country, we can lay no claim to final authority in setting down standards to follow. You may discover, for instance, that some of the suggestions we make are not followed in the organization you work for. If so, you should certainly follow the rules of your own group. The rules below, however, are based on those accepted by the most widely recognized authorities and may be used with confidence.

*Abbreviations.* Abbreviations should be used only when they are certain to be understood by the reader. Otherwise the term should be spelled out. Certain terms, of course, are commonly abbreviated everywhere—Dr., Mr., No., and the like.

The best authority for the use of abbreviations of scientific and engineering terms is the list approved and published by the U.S.A. Standards Institute (formerly the American Standards Association); although not followed everywhere (as you will note in the reports quoted later), this standard is approved by most engineering societies. The following rules are in agreement with this publication (a list of the more common, approved abbreviations may be found in Appendix G):

1. In general, use abbreviations sparingly in the text of reports—never when there is a chance the reader will not be familiar with them.

2. Abbreviations for units of measurement may be used, but only when preceded by an exact number. Thus, write "several

inches," but "12 in." Do not use an abbreviation of a term which is the subject of discussion; thus do not write, "The bp was quickly reached." Write "The boiling point was quickly reached." Abbreviations may be justified in tables, diagrams, maps, and drawings where space needs to be saved.

3. Spell out short words (four letters or less) like ton, mile, day.

4. Do not use periods after abbreviations unless the omission would cause confusion, as where the abbreviation is identical to a word. Thus write "in." rather than "in" because the latter might be mistaken for the preposition. Some exceptions are "cot" for cotangent, "sin" for sine, "log" for logarithm. These abbreviations could scarcely be confused with the words.

5. Do not add an "s" to form the plural of an abbreviation. The number preceding an abbreviation of a unit of measurement sufficiently marks the expression as plural. Thus write "128 bbl" rather than "128 bbls." Exceptions are "Nos." for Numbers, "Figs." for Figures, "Vols" for Volumes. In footnotes, the plural of pages is given as "pp."

6. Write abbreviations in lower-case letters rather than capitals unless the term abbreviated is a proper noun. Thus write "hp" rather than "H.P." or "HP" for horsepower, but write "Btu" for British thermal unit. Exceptions are terms used in illustrations or bibliographical forms, as shown above.

7. Abbreviate titles only when they precede a proper name which is prefaced by initials or given names. Write "Professor Jones" rather than "Prof. Jones." "Prof. J. K. Jones" is acceptable.

8. Do not space between the letters of an alphabetical designation of an organization. Write "USASI" for the U.S.A. Standards Institute, "ASEE" for American Society for Engineering Education, "ASME" for American Society of Mechanical Engineers, and so forth.

9. Use abbreviations which are more readily recognized than the spelled-out form. Thus, in reports, "FM" is as acceptable as "frequency modulation."

10. In reports where a term is used repeatedly, use the accepted abbreviation, but give a spelled-out parenthetical explanation upon first using it. Thus you could write ". . . 1200 Hz (hertz) . . ." and thereafter use "Hz."

*Symbols.* Symbols are generally to be avoided in text. Custom may permit the use of certain symbols in particular organizations, however, and our recommendation is that you observe closely what local practice is and follow it. But while symbols are generally to be avoided in text, they are justifiable in tables, diagrams, and the like because of the need to conserve space. You are probably familiar with most of the commonly accepted symbols, such as ″ for inches, ′ for feet, × for by, # for number, / for per, & for and. A few symbols, like % for per cent and ° for degree are so commonly used in text that most readers are as familiar with the symbol as with the spelled-out term.

*Numbers.* The following rules represent commonly accepted practice in the use of figures:

1. Use figures for exact numbers for ten and above and spell out numbers below ten. Where several numbers, some above and some below ten, appear in the same passage, use figures exclusively. Thus write:

   10 days
   eight resistors
   five tubes
   27 motors
   11 condensers, 8 tubes, and 27 feet of wire

2. Use figures in giving a number of technical units, as with units of measurement, whether below or above ten:

   8 kHz
   2500 hp
   28,000 Btu
   3 bbl

3. Spell out either the shorter or the first number in writing compound number adjectives:

   thirty 12-in. bolts
   8 six-cylinder engines

4. To avoid possible confusion in reading, place a zero before the decimal point in writing numbers with no integer:

   0.789
   0.0002

   Do not place zeros to the right of the last figure greater than zero unless you wish to show that accuracy exists to a certain decimal; thus you might write 6.7000 if accuracy to the fourth decimal exists.

5. Spell out fractions standing alone, as "three-fourths of the staff members." But with technical units, use figures:

   3-1/2 gpm
   5-1/4 sec

   (Note the form used; 3½ and 5¼ are not desirable in typed copy because the fractions tend to blur, especially on carbon copies, and because typewriters do not have all fractions.)

6. Omit the comma in four-digit numbers (practice is not uniform on this point, but the trend is toward omission):

   7865
   98,663

7. Follow conventional usage in writing street addresses, dates, and sums of money:

   4516 Spring Lane
   3600 Fifty-fourth Street
   March 11, 1951
   $8,000,000 or 8 million dollars or $8 million

8. Do not use numerals at the beginning of a sentence; numerals may be used for round-number estimates or approximations:

   Twenty-seven seconds elapsed (*Not:* "27 seconds elapsed").
   about 30 a minute
   nearly 500 arrived

9. Do not use two numerals in succession where confusion may occur:

   On August 12, eleven transformers burned out.

10. Use numerals for the numbers of pages, figures, diagrams, units, and the like:

    Fig. 8, stage 4, page 6, unit No. 5, Circuit Diagram 14.

*Hyphenation of Compounds.* Usage is rather uncertain in the handling of hyphenation—as illustrated in the reports that are quoted later—but the following practices are generally approved:

1. Hyphenate compound adjectives which precede the term they modify:

   alternating-current motor
   ball-and-socket joint
   4-cycle engine
   2-ton trucks

2. In general, hyphenate compound verbs such as "heat-treat," "direct-connect."
3. Do not hyphenate adverb-adjective combinations, such as "newly installed," "readily seen."
4. In general, do not hyphenate compound nouns (such as boiling point, building site, bevel gear, circuit breaker) except those composed of distinct engineering units of measurement (such as foot-candle, gram-calorie, volt-ampere, kilogram-meter). Many compounds are, of course, written as one word (such as setscrew, flywheel, overflow).
5. In specific cases, try to observe and follow the practice of careful writers.

*Capitalization.* In general, technical writing style calls for no departure from the conventional rules for the use of capital letters. You have learned to capitalize proper names, names of cities and states, official titles of organizations, and so on. Any reputable dictionary or handbook of English can guide you as to conventional usage (and most of them contain a prefatory section stating the "rules"). We should like to call attention to two practices common to reports:

1. Capitalize all important words in titles, division headings, side headings, and captions. By "important" is meant all words except articles, prepositions, and conjunctions.
2. Capitalize Figure, Table, Volume, Number as part of titles. Thus reference would be made to Figure 4, Table 2.

When in doubt, do not capitalize.

*Punctuation.* The sole purpose of punctuation is, of course, to clarify thought, to make reading easy. Punctuation which does not contribute to this purpose should be avoided. Most of your difficulties with punctuation are likely to arise in the use of the comma, the semicolon, and the colon. For information on other punctuation marks, see any good handbook of grammar.

The principal uses of the comma are:

1. Between independent clauses connected by a coordinating conjunction (and, but, for, or, nor, yet). But if commas are used within any of the independent clauses constituting a sentence (in accord with one or more of the rules below) a semicolon must be used between the clauses. Study these two sentences:

   The fixed coil is permanently connected across the line; and the movable coil is connected across the motor armature.

The fixed coil, providing a unidirectional magnetic field in which the moving coil acts, is permanently connected across the line; and the movable coil, which operates to close the indicated contact, is connected across the motor armature.

2. After introductory clauses or phrases preceding the main clause of the sentence:

After workers had completed the first part of the job, they immediately began the second.

Jumping on the instant of the explosion, he avoided injury.

3. Between items of a series:

The power supplies, the amplifiers, and the resistors are to be considered now.

The engine was efficient, cheap, and light in weight.

4. Around parenthetical, interrupting expressions, appositives, and nonrestrictive modifiers:

This plan, unless completely misjudged, will bring great success.

This circuit breaker must, obviously, be kept in repair.

He approved, for the most part, of our research plans.

Mr. Jackson, chief technical adviser, returned yesterday.

The chief project engineer, who used to work on the west coast, is responsible for the new procedure.

But not around restrictive modifiers:

The generator which was tested yesterday is the one needed in this installation. (Restrictive modifiers, like "which was tested yesterday," cannot be left out without destroying the meaning of the sentence.)

The semicolon is a stronger mark of separation than the comma, almost as strong as the period. It is chiefly used between independent clauses not connected with one of the coordinating conjunctions and between clauses connected with a coordinating conjunction which are quite long, or unrelated, or contain commas. Study these sentences in which semicolons appear:

The first of these devices failed after one year's use; the second has lasted five years.

One of these instruments has never had to be replaced; however, it is showing signs of wear.

Even after months of study, they failed to solve the problem; but, in some ways at least, they made a great deal of progress.

The colon signals that something is to follow, usually something explanatory, as shown in the following examples:

> A few tools were available: a lathe, a power hack saw, and a drill press.
>
> Operation was becoming uneconomical: both labor and fuel costs were more than had been anticipated.
>
> There are three steps in the process: cutting, grinding, and polishing.

The colon is also used in certain special ways, as in the salutation in a business letter (Dear Sir:), in separating hours and minutes in a statement of time (10:30 A.M.), or in separating volume and pages in a bibliographical entry 17:43–50).

One special comment on punctuation in reports: do not place any mark of punctuation after main or side headings (those which are centered on the page and those which stand on a line alone). If, however, text continues on the same line as a heading, as it does with the heading "Punctuation" (page 68), a period may be used. Some organizations, however, prefer the colon. Follow the style used by the particular organization.

## SENTENCES FOR REVISION I

The following sentences contain errors typical of those commonly made by technical personnel in their letters and reports. Rewrite them, correcting the errors.

1. By specifying standardized commercial equipment, the cost of the proposed system can be substantially reduced.
2. He worked for a chemical firm for over a year and he wants to go into that work again.
3. Using the plan as outlined by the agricultural commission, results will be forthcoming within a reasonable space of time.
4. One of the main errors which was involved was that of the post-computation check.
5. Having evaluated equation (1), the following quantities can be computed.
6. The first locomotive to be operated on an American railway was in 1829.
7. After completing the plan, it was seen not to fulfil all requirements.
8. The well produced a continual flow of oil.
9. Discussing the individual elements that comprise the total test apparatus, it is logical to begin with the light source.

10. To provide for adequate terrain clearance data, two antennas are required.
11. More information per sweep of the oscilloscope is being gathered in the proposed system than in the old system. This will require a wider signal band-width.
12. The airborne recording unit mounts on the drone structure.
13. A complete picture is scanned on one tube; then stored while a picture is being scanned on the second tube.
14. One code disc will be calibrated in 0.2 second steps, another in 0.3 second steps, and the third disc will be designed so as to cover the range of 5 to 50 seconds in 5 second steps.
15. A memorandum explaining this drawing and possible alternate designs was prepared and both forwarded to the main office.
16. The technical discussion of the circuit is the same whichever technical solution is taken insofar as construction of the timing device is concerned.
17. The distortion test involves usage of many of the same or similar elements needed for the other tests, thus the same test unit base and many of the other test components are used for all tests.
18. A seal between these two components of the device permits control of the atmosphere inside the system. This helps to reduce the effects of internal surface fogging.
19. Temperature control is one of the principle factors to watch.
20. Hydrofluoric acid is a liquid, volatile, highly corrosive, and cannot be kept in glass containers.
21. It is desirable to remove edge distortion in the system. It seems impractical to design the device that will remove all such distortion.
22. His statement infers that he had confessed.
23. This mixture, after being thoroughly blended, is transported to the site where it is to be applied in wheelbarrows.
24. All editing chores were divided between the three editors.
25. Each of the foregoing uses are important but the importance varies.

## SENTENCES FOR REVISION II

The sentences below are not as succinct or clear as they might be. Rewrite them for greater conciseness, without omitting essential content.

1. They evidenced a surprisingly uniform communality of attitude to the effect that the most vital area of training was the development of military skills and courtesy.

2. It is seen that there are five output voltages from the analog computer. These voltages are proportional to the yawing velocity of the fighter aircraft.
3. Prior to the conductance of these tests, condensed moisture should be removed from the equipment by either inverting or tilting, whichever is more compatible with its configuration.
4. Numerals are used to identify the various adjustment screws provided on the panel located inside the door of the equipment.
5. Poor living accommodations give promise of incrementing the negative side of the morale balance so far as new personnel are concerned.
6. It is expected to complete the full integration of these new units into the system as a whole by early in the next month.
7. It would seem desirable to terminate the prior process and initiate the new one if optimum results are to be secured.
8. The proposed program is intended for the utilization of foresters who are in the employ of the United States government in seeing to it that fire prevention is carried out with optimum results.
9. This diagram indicates that there are twenty-one instrument servomechanisms in the control room which do the necessary computing for the system.
10. Due to the many and varied applications a system of this type may have in the immediate future, it is felt that techniques should be utilized which will give the system the maximum amount of versatility and reliability.
11. On the basis of past history, it is expected by management that great progress will be made by personnel in providing a solution to these problems in the near future.
12. The first thing that must be done is edit the report.
13. Personnel of the purchasing department must prepare a cost estimate for the purpose of making it possible to make a purchase.
14. In most cases the installation of a monitoring device that provides continuous indication of deviations from the normal will permit the reduction of shutdowns.
15. Whether or not these anticipated operations to correct errors in procedure enable the staff to cooperate together more efficiently, it is intended that they be inaugurated without undue delay by reason of communication difficulties.
16. Enclosed herewith is a list of important essentials that should be subject to coverage in the next conference dealing with the matter of absenteeism.
17. It is to be hoped that work to be scheduled will not involve the

necessity of any undue overtime work in the neighborhood of the holiday period.

18. In this quite unique design, labels have been provided for the purpose of identifying each of the various controls.
19. There is a city-owned pier running out from this land which is used by a marine repair firm.
20. From a cleaning point of view, these valves are relatively good.

## EFFECTIVE WORD CHOICE EXERCISE

In the sentences below, you are offered some word choices. Sometimes the choice is between an acceptable ("correct") word and an unacceptable ("incorrect") word; other times, the choice is between a word with precise and suitable meaning and one which is less precise and suitable. Rewrite these sentences and be prepared to justify your choices. You may need to consult a good dictionary or one of the books on usage, such as the Evans' *Dictionary of Contemporary American Usage*, Partridge's *Usage and Abusage*, or Flesch's *The ABC of Style*. These books contain discussions of words which are closely related in meaning and which are frequently misused. In making your choices, you should assume that the sentence is to appear in a formal, written document (speech allows greater freedom —or lenience).

1. Mr. Brown will (accept/except) the invitation.
2. The man paid all interest and part of the (principal/principle).
3. Government taxing and spending seriously (affect/effect) the economy.
4. The next step is to (filtrate/filter) the fluid.
5. The machine has (degenerated/deteriorated) through overuse.
6. A new design is not (indicated/required).
7. He did not (consider/think) many changes should be made.
8. This successful engineer proved to be most (ingenious/ingenuous).
9. The acid finally (eroded/corroded) the pipes.
10. The project engineer (informed/advised) his men that no overtime work would be required.
11. (Oxidization/oxidation) should be prevented.
12. The chief engineer paid him a fine (complement/compliment).
13. Close supervision seriously (affects/effects) our success.
14. The time (passed/past) quickly.
15. His remarks (infer/imply) that he doesn't believe the report.
16. The geologists spent the (remainder/balance) of the time working.
17. The committee selected a (sight/site/cite) for the meeting.

18. A large (percentage/percent) of this report is useless.
19. Our supervisors are (continually/continuously) trying to help us.
20. We would like to (devise/device) a new (devise/device) for this purpose.
21. The secretary sent the memorandum to (its/it's) destination.
22. The president is worried about the (economic/economical) situation.
23. The task was divided (among/between) five engineers.
24. A few days' work destroyed all his (allusions/illusions).
25. We did not think a holiday was (likely/liable/apt) to be given.
26. He did not believe he needed any (council/counsel).
27. All the subordinates were (aggravated/irritated) by him.
28. His plan was (practical/practicable), but it was not (practical/practicable).
29. The department has (all ready/already) met its responsibilities.
30. He did not know how to (adopt/adapt) the report for his purposes.
31. He would not agree to the plan: he was (disinterested/uninterested).
32. There were (fewer/less) people at this meeting than at the last.
33. His (implicit/explicit) instructions were written in detail.
34. The chief engineer was (enthused/enthusiastic) about the plan.
35. Now employed in Houston, he was (formally/formerly) in Washington.
36. He was chosen to (administrate/administer) the program.
37. He gave us (oral/verbal) instructions.
38. The mixture was said to be (inflammable/flammable).
39. The specifications called for (bimonthly/semimonthly) reports.
40. We must proceed, (irregardless/regardless) of difficulties.
41. The new system was designed to (ensure/assure) success.
42. We thought his decision (equitable/equable).
43. Although he did not say so, we (inferred/implied) that he was pleased.
44. The committee asked him for his (opinion/reaction).
45. We asked for a written (estimation/estimate).
46. Many such (incidence/incidents) have occurred.
47. The plan outlined in the report had (obvious/evident/apparent) merit.
48. Although his design was not greatly different from others proposed, it had some (unique/unusual) features.
49. We shall (utilize/use) this material in our report.
50. (More than/Better than) a million dollars were spent on the project.

51. The worker (fixed/repaired) the damage.
52. He (acquainted/told) us (with) the facts.
53. He described his (approach/solution) to the problem.
54. Plans for this project must be (finalized/completed) soon.
55. The supervisor (concluded/decided) to make some changes.
56. We do not (envision/expect) any difficulty in completing the job.
57. The project engineer did not (consider/think) that any changes should be made.
58. The results were (nowhere near/not nearly as) good.
59. The scientist declared that he (appreciated/understood) the problem.
60. This research project does not need to be carried (further/farther).
61. Our (target/objective) for the quarter was an increase in production.
62. In solving the problem, he (assumed /presumed) nothing.
63. The investigators (encountered/experienced) many difficulties.
64. The device proved to be most (effectual/effective).
65. Should it be (desirous/desirable) to obtain a steady tone rather than the interrupted signal, more work will have to be done.
66. Finally, just for fun, see if the following story makes sense: Two well drillers were injured while at work and sent to the hospital. One was seriously hurt and did not improve very rapidly, but the other was soon ready for discharge. The attending physician, who was new to the hospital, asked the head nurse if she could sign the discharge papers for the one who had recovered. She replied, "I will on your say so do so so that that well well driller can go home." And the doctor said, "Well well well drillers are the last thing we need around this hospital." Can you *say* this so that the meaning is clear?

# 4

# Outlines and Abstracts

## Introduction

Outlines and abstracts are very much alike in one respect—both are highly condensed statements of, or descriptions of, the content of a piece of writing. For this reason they are taken up together in this chapter. In some respects it would be more logical to discuss only outlines at this point and to defer consideration of abstracts until after examination of various types of reports. Such a sequence of study can be easily managed simply by skipping the section on abstracts in this chapter, for the present, and returning to it later. On the other hand, you may find that study of abstracts in direct relationship to the study of outlining will be quite helpful in clarifying the basic concepts of organization in technical writing, and will provide a good background for the later examination of various techniques of writing and various types of reports. This advantage can be greatly augmented by a careful study of Appendix E.

Abstracts are written solely for the convenience of the reader. Outlines, on the other hand, serve both reader and writer. An outline serves the writer by providing a means of analyzing the structure of somebody else's writing, and hence of systematically studying the structure of well-designed reports. It also serves the writer as a guide in designing his own reports. It serves the reader in the form of the table of contents of a report, and then also as the system of subheads within the report. In Appendix E you will find an explanation and illustration of how the subheads scattered throughout the text of a report are derived from the table of contents.

Our interest in the present chapter lies almost entirely in the use of an outline as a guide to writing. If you have never done any outlining, however, it might be wise to write some analytical outlines to learn the basic principles. In that way you can concentrate on the form and logic of the outline, without at the same time worrying about whether you're developing a good organization.

Why write an outline? Well, why follow a road map? Probably you have driven a car in a strange city for which you had no map, and after turning around and retracing your route a few times, and after asking pedestrians for information, you have finally pulled up to your destination. Writing is often like that. The writer runs off first in one direction and then another, while the bewildered reader tries to make sense of his tangled trail. A "road map" would have saved time for both writer and reader.

Of course you do not need a road map to drive from your house to the nearest shopping center, nor do you need an outline for a very short report. The longer the road, the more complex the terrain, and the more unfamiliar the country, the more you need a map. So it is with outlines. The more complex the subject, the more unfamiliar you are with the subject, and the longer the report, the more you need an outline.

We shall discuss outlines first, then abstracts, and finally introductory summaries. An introductory summary is a combined introduction and abstract, as will be explained later.

## Outlines

*Kinds of Outlines.* There are three kinds of outlines: topic, sentence, and paragraph. In a topic outline, each entry is a phrase or a single word; no entry is in the form of a complete sentence. Conversely, in a sentence outline every entry is a complete sentence. If you will now turn to pages 94 and 95, you will find examples of these two kinds of outline. The third kind, the paragraph outline, is of no use to the technical writer and we will not discuss it.

The sentence outline has one important advantage over the topic outline, but it also has at least one important disadvantage. The advantage is that in making a sentence outline, the writer is forced to think out each entry to a much greater degree than for the topical form. In a topical form he might say merely, "Materials"; in the sentence form it would be necessary to say something like, "The materials required are seasoned white pine, glue, and whatever finish is desired." The greater thoroughness of the sentence outline lessens the possibility of ambiguity and vagueness in the thought. It also means, on the other hand, that the sentence outline is more difficult and time consuming to write than the topical. The sentence outline is an excellent analytical device for studying the organization of a given piece of writing. The topic outline, however, is more practical as a guide for writing. It is not a good idea to combine the two forms. There is nothing greatly wrong with such a combination, but it does indicate an inconsistency in the logical process—one part of the subject being developed in detail in sentence form, another being limited to topical development.

*The Logic of Outlines.* The fundamental principle of outlining is division. The subject to be outlined is divided into major parts (Roman-numeral divisions); these major parts are divided into subparts (capital-letter divisions); these subpart divisions are divided into sub-subparts (Arabic-numeral divisions); and so the whole is divided into smaller and smaller units to whatever degree seems desirable.

Since outlining is a method of dividing, it naturally conforms, in a certain degree, to the principles of arithmetic. Let X equal the entire subject to be divided, or outlined. Then $X = I + II + III + \ldots n$. In turn, $I = A + B + C + \ldots n$, and $A = 1 + 2 + 3 + \ldots n$, and so forth. Please understand that this is more than an analogy. It is a principle which not only can be, but also should be, applied to every outline you write, to test its logical soundness. For instance, we might consider the following simple example from the outline on page 94. This outline is taken from a report on the subject of sanitation in isolated construction camps.

I. . . . . . . . . . . . .
  A. . . . . . . . . . . . .
  B. Stopping the spread of these diseases by breaking the cycle of transmission.
    1. Removing or destroying the breeding places of insects and rodents
    2. Killing the adult insects and rodents

This might be rewritten in the following form:

> Stopping the spread of these diseases by breaking the cycle of transmission = Removing or destroying the breeding places of insects and rodents + killing the adult insects and rodents. Therefore

$$B = 1 + 2$$

After thinking about this equation, however, you may object and say that it is not necessarily valid. That is, if absolutely all adult insects and rodents were destroyed, it would be pointless to worry about their breeding places. The equation would therefore be reduced to $B = 2$.

This objection is certainly justified. Nevertheless, it is justified only if we think of our subject matter as actually being arithmetic rather than outlining. But we do not want to think of outlining as being identical with arithmetic.

The two propositions we want to make are these. First, as we said above, outlining conforms, in a certain degree, to the principles of arithmetic. Second, as we also said above, the principle that I = A + B + . . . *n*, (etc.) should be used to test the logical soundness of any outline. What we need to think about, then, is how the second proposition can be true if there is a difference between outlining and arithmetic. What is the difference?

We are not entirely joking when we say that this question will be easy to answer for anyone who has tried to kill all the adult insects and rodents around a construction camp. In the practical world of outlining technical reports, we remember what formidable antagonists insects and rodents have proven to be.

Indeed, to tell the honest, unvarnished truth, perhaps we should rewrite our equation like this:

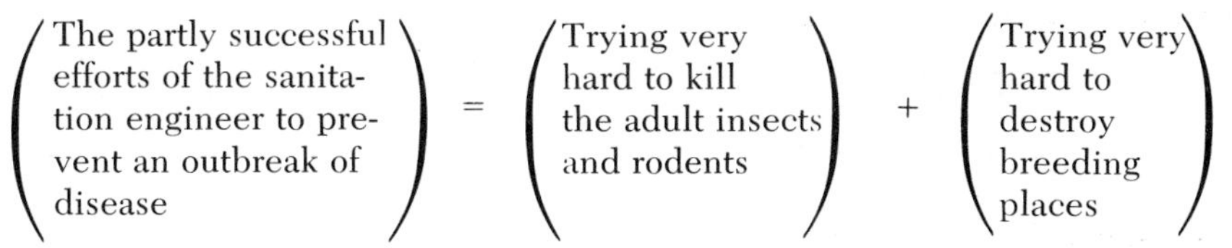


In summary, the principles of division and addition can be used to test the logical soundness of an outline, but they must be used within the practical world of the subject matter of the outline rather than just within the abstract world of arithmetic.

Should this summary leave you with the feeling that there must be some loose philosophical and logical ends lying around, we heartily agree. The whole problem shades off into fascinating but exceedingly puzzling questions. You might enjoy pondering it from the point of view of a little book by Ernest Nagel and James R. Newman entitled *Gödel's Proof.* In practical terms, however, all you have to do is to make sure that the subdivisions of any part of your outline add up in a common-sense way to the point you are trying to make in that specific part.

The discovery that the equation might take the form of B = 2, as noted above, calls attention to a special problem. To have a particularly simple example of this problem to think about, let's turn to a different subject. Let's say we are writing an outline on the subject of power sanders and have got as far as the following entries:

I. Introduction
  A. . . . .

B. . . . .
C. . . . .
D. . . . .
II. Types of Sanders
A. Vibrating
1. Straightline
2. Orbital
3. Combination
B. Belt
1. Straightline

Having got this far, we come to an abrupt halt because we realize there is no further entry to put under "Belt"—no number 2. Since all belt sanders are straightline sanders, there is simply no other subdivision that will here go with "Straightline" in the way "Orbital" did in part A. Consequently, it is pointless to have the heading "Straightline" in part B. So all we can do is cancel this heading and continue with the outline, as follows:

II. Types of Sanders
A. Vibrating
1. Straightline
2. Orbital
3. Combination
B. Belt
C. Disk
D. Drum

This is just another way of expressing the idea that the fundamental principle of outlining is division. And nothing can be divided into fewer than two parts. If there is an A, there must also be at least a B; if there is a 1, there must be at least a 2; and so on.

But, you may be thinking, suppose that for my particular reader I need to do only two things, so far as belt sanders are concerned. First, I need to point out that the belt sander is one of the types to be considered; second, I must explain that there is a smaller range of choices of grits in sanding belts than there is in sanding sheets. Why not simply write the following?

II. Types of Sanders
A. Vibrating
1. Straightline
2. Orbital
3. Combination
B. Belt
1. Limitation of choices of grits
C. Disk
D. Drum

You are wondering, in other words, if there aren't some situations in which it is sensible to have a single subdivision. Yes, if the writing required is brief, and if the outline is just a list of reminders about facts with which the writer is thoroughly familiar, the outline above might be all right. As we further ponder the question of whether it would really make good sense to set up an outline like the one above, however, we come to a helpful insight about how outlines work. We think you will probably decide that the sort of outline shown above would be all right only when the subject was so simple, the amount of writing required so little, and your familiarity with the subject so great that you scarcely needed an outline at all.

Let's imagine we have begun writing a discussion of power sanders, using the outline above as a guide, and have finished with part A and are ready to start on part B.

We sit there staring at the word "Belt." We are supposed to write something. What should we say? Apparently we are supposed to say something we have stored away in our mind about Belt, except that what we say must not be about the choice of grit. As the separate entry on the choice of grit shows, our discussion of grit comes second, after our discussion of Belt.

In short, our outline instructs us to say at least two things about belt sanders, and it reminds us explicitly what one of these things is, but it does not remind us of what the other, or others, may be. When writing about an extremely simple subject, this sort of vagueness may not create much trouble, but you can imagine what happens when the subject becomes so complex that the writer has trouble keeping it all clear in his head. That is a very different situation.

What entries should the outline have had, then? As an absolute minimum, we'd say, something like the following would prove helpful (we're considering only part B).

B. Belt
  1. Special characteristics of belt sanders
  2. Limitation of choices of grit

As we look at these three headings, we may now observe still another fact about how an outline functions. We now have two explicit reminders of what to say about belt sanders. But what about the word "Belt" itself? What do we say concerning it? Absolutely nothing. The fact is, the word "Belt" has now become merely a title.

This idea can be seen more clearly if we divide all the entries in an outline into two kinds: titles, and signals to write. A title is any entry that has subdivisions under it. A signal to write is any entry that is not subdivided. In other words, if we are using the outline above as a guide for writing, we will not write anything about "Belt" but instead will begin by writing about "Special characteristics of belt sanders."

Let's follow this idea one step further, beginning by adding some entries to the outline.

B. Belt
   1. Special characteristics of belt sanders
      a. Speed in removing material
      b. Danger of removing too much material in fine finishing work
   2. Limitation of choices of grit

According to the principle that any entry which is subdivided becomes a title, B.1. is now a title and no longer a signal to write. The first subject to be written about in part B is now B.1.a., "Speed in removing material"—which is one of the special characteristics of belt sanders.

But now we may begin to feel apprehensive about a new problem. Isn't there a danger in starting off a whole new part of our report (part B) by going at once to such a detail as B.1.a.? What if "a" itself had been subdivided, thus becoming a title, and our first signal to write were B.1.a.(1)? Wouldn't following such principles in writing a report result in a text that would appear to be a mass of unorganized detail? It certainly would. And we believe this is in fact one of the commonest trouble spots in the writing of reports.

The fundamental problem we are confronting might be put this way. Since the signal-to-write aspect of outlines tends to lead the writer into beginning a new part of his report by discussing a mere detail, what can the writer do to make sure the details are integrated into a smooth-flowing report? The sudden appearance of a detail which seems unrelated to the preceding text is naturally puzzling.

One answer is that he can put into his outline a special kind of signal-to-write that warns him of the problem and suggests what he should do about it. These special signals-to-write that we are thinking about, as you may already have guessed, are called introductions, transitions, and conclusions or summaries. It is unusual to see the word "transition" in an outline, but of course "introduction" and "conclusion" or "summary" are common. We haven't yet discussed these particular devices, but they are taken up in detail in Chapters 10, 11, and 12 (see also F. Bruce Sanford's example on page 552).

Another answer to the question as to how to integrate the details into a smooth flowing report is this. As a writer acquires skill, he begins to use the "titles" in an outline themselves as signals-to-write—but in the second sense in which we used this term rather than the first. Instead of signaling him to write directly about the subject matter of his report, the titles (like "Belt") signal him to write an introduction or a transition. These ideas will be clarified in the chapters just mentioned.

Another special problem about outlining is that a writer sometimes becomes trapped by his logic. For an illustration of this curious fact, let's go back to one of the earlier versions of the outline we've been playing with.

II. Types of Sanders
  A. Vibrating
    1. Straightline
    2. Orbital
    3. Combination
  B. Belt

The basis of classification used for the sanders within part A is the kind of motion made by the tool's sanding surface. On some vibrating sanders the motion is straight back-and-forth; on others it is orbital; and on still others ("combination" sanders) the user can select either motion by moving a switch. But all this availability of choice does not hold true of a belt sander, in which the sanding surface is literally a belt that runs over rollers. So there are no such subdivisions under "Belt."

A writer who is really trapped by his logic may write page after page under a topic like "Belt" with never a heading to help the reader along. Because the basis of classification used for vibrating sanders doesn't work for belt sanders, he apparently assumes that no subdivision is possible. Incredible as it may seem, we once saw a long proposal, for a contract representing a very large amount of money, in which at one point there was a stretch of over a hundred pages without a single subhead to help the reader. The men who prepared this proposal had been trapped by exactly the same logic you saw in our simple example. Of course there were actually a great many subtopics within those hundred pages, and what the writers should have done was to figure out what these subtopics were and set up subheads for them.

As a matter of fact, examination of the hundred pages soon revealed that one reason there were no subheads was that the writers hadn't thought about their work carefully enough to be sure exactly how the material was being organized. Evidently they just wrote along, trusting to luck. Their luck wasn't especially good.

This bit of reminiscence brings up one other curious fact about the organization of reports—at least we believe it's a fact. We have become convinced through personal experience that very often when people complain that there are a lot of bad sentences in a report, the real trouble lies in the organization. Perhaps their situation resembles that of a man who complains of a pain in his leg but who is told upon examination that the trouble is that his backbone has got a little disorganized. We don't mean to say that poorly written sentences are not a common problem. They certainly are. Nevertheless, our personal experience is that the best place to start, in hunting for the trouble in a report people are complaining about, is not with the sentences but with the organization. This opinion is to some degree corroborated by what Bruce Sanford says in the Introduction to the material in Appendix E. We again urge you to study this Appendix in connection with the whole subject of the logic of outlines.

Previously, we remarked that the outline of a technical report becomes the table of contents and the system of subheads in the finished report. We would like now to take a brief, summary look at outlines from the point of view of some examples of tables of content. The advantages of this point of view are the perspective it gives on the value of informative headings, on the benefits of having neither too few nor too many headings, and on the confusion created by a disregard of the sort of logic we have just been considering.

As you look at the examples that follow, remember that technical reports, even very long ones, almost never contain an index. For assistance in finding a particular passage in the text, the reader must rely on the table of contents.

One difference between the function of an outline and that of a table of contents is that the headings in an outline are primarily reminders to the writer about things he already knows. If the headings in a table of contents are not self-evident sources of information, on the other hand, they aren't much good. A heading like "Vibration Tests," for instance, is not nearly as useful in a table of contents as something like "Procedures for Vibration Tests."

Sometimes the question of whether a heading is sufficiently informative becomes involved with the problem of how to set up the subheads. A common example is found in that kind of report called a "proposal" (see Chapter 15). In a proposal, the most important single section is usually one concerned with how to solve some problem. All too often this section of a proposal will have only some such heading as "Technical Discussion," or "Detailed Description," or "Technical Approach." Such a general heading may be defensible if the discussion is brief; it is not very informative at best, however, and if the discussion is long the table of contents becomes unnecessarily complex. Thus Example 1 below is improved if presented in the form of Example 2.

*Example 1*

III. Technical Discussion
- A. Buoy Electronics Package
  [Nine Arabic-numeral subheads appeared here in the original.]
- B. Shipboard Receiving Station
  [Three subheads.]
- C. Experience in Supplying Related Equipment
  [Three subheads.]
- D. Reliability
- E. Quality Assurance
- F. Testing

*Example 2*

III. Buoy Electronics Package
  [The nine Arabic-numeral subheads now become A through I.]

IV. Shipboard Receiving Station
  [Subheads A, B, and C.]

V. Experience in Supplying Related Equipment
  [Subheads A, B, and C.]

VI. Reliability

VII. Quality Assurance

VIII. Testing

In the next example we leave the problem of whether the heading is sufficiently informative and turn to a consideration of what happens when there are too few headings.

III. Components . . . . . . . . . . . . . . . . . . . . . . . . . . . . . .6
  A. Electronic . . . . . . . . . . . . . . . . . . . . . . . . . . . . . .6
  B. Mechanical . . . . . . . . . . . . . . . . . . . . . . . . . . . . .103

We hope you will agree that the single entry "A" is scarcely enough for 97 pages of discussion! Subentries would have been of great help to a reader wanting to locate descriptions of specific electronic components. This may be an instance of the writer who is trapped by his logic, a phenomenon mentioned earlier.

Here is another example, this one containing too many entries.

A. Experience . . . . . . . . . . . . . . . . . . . . . . . . . . . . .14
  1. Sonar . . . . . . . . . . . . . . . . . . . . . . . . . . . . . . .14
  2. Seismic . . . . . . . . . . . . . . . . . . . . . . . . . . . . .14
B. Personnel . . . . . . . . . . . . . . . . . . . . . . . . . . . . . .14

Why not just write it this way?

A. Experience in Sonar and Seismic Work . . . . . . . . . . . . . . . . .14
B. Personnel . . . . . . . . . . . . . . . . . . . . . . . . . . . . . . .14

So brief a discussion as section A represents is scarcely in need of having its subdivisions labeled by separate headings. (On the other hand, it is very important that the text itself be clearly divided, probably by separate paragraphs for each topic and a transitional phrase or two.)

The following table of contents reveals a defect in logic.

- II. Circuit Redesign
  - A. Preamplifier
  - B. Modulator and Demodulator
  - C. Ripple and Lead Networks
  - D. Servo Amplifier
  - E. Packaging

Clearly, the entry "E" should appear as "III." In terms of our previous discussion of the application of the principles of arithmetic to outlining, the heading "Circuit Redesign" is less than the sum of "A" through "E." Here is another illustration of the same problem.

- V. Data-Transfer Subsystem
  - A. Introduction
  - B. System Operation
  - C. Airborne Unit
    - 1. Data-Processing Unit
    - 2. Transmitter
    - 3. Power Supply
  - D. Ground Unit
    - 1. Translator
    - 2. Telemeter Receiver
    - 3. Subcarrier Discriminator
    - 4. Countermeasures Considerations

Here, obviously, the last item, 4, should appear as "E" rather than 4, since "Countermeasures Considerations" are not a component of the ground unit.

Our last example is difficult to categorize. Does it represent a failure of logic or simply a careless redundancy?

- III. Technical Discussion
  - A. Data-Processing Techniques
    - 1. Time-Shifting Linear Addition
    - 2. Cross-Correlation
  - B. Time Shifting
  - C. Characteristics of Correlator Filter

Something looks very odd here, with one of the subheads of "A" reappearing as "B". Perhaps the general heading "Technical Discussion" could have been deleted and the following setup used, with appropriate subheads under "A" and "B":

- III. Data-Processing Techniques
  - A. Time Shifting
  - B. Cross Correlation

The above excerpts from tables of contents are all drawn from actual reports.

As a final comment on the logic of outlines we should say something about form. Various forms have been worked out, the principle in each case being to devise a physical arrangement that will help reveal the fundamental logical structure of division and subdivision. Tables of contents are often set up in a similar way. The form most commonly used for outlines is the following. The dotted lines represent the text.

```
I. ............................................
   ..........
     A. ........................................
        ......
          1. ....................................
             ........
               a. ...............................
                  ...............................
                    (1) .........................
                        .........................
                          (a) ...................
                              ..........
                          (b) ...................
                    (2) .........................
               b. ...............................
          2. ....................................
     B. ........................................
II. ...........................................
```

Observe the following points: (1) periods are used after symbols (that is, numbers or letters) except when the symbol is in parentheses; (2) in an entry of more than one line, the second line is started directly beneath the beginning of the first; (3) the symbol of a subdivision A, 2, etc.) is placed directly beneath the first letter in the entry of the preceding highest order; (4) periods are placed at the end of sentence entries but not after topic entries; and (5) lines are usually double-spaced.

One other aspect of form calls for attention: the need for parallel grammatical structure in the sentence or topic entries. This may sound like an unimportant matter, but our own opinion is that carelessness in this respect is like the tiny fissure on an exposed slope of earth which, if not attended to, may become a badly eroded gully. Parallelism was previously explained on page 63.

*How to Make an Outline.* Let's distinguish carefully between two things: (1) using the principles of outlining as a help in organizing material about which a report must be written, and (2) writing the report itself, with the outline as a guide. Of these two things, the first is the subject we are now concerned with: how to make an outline.

It is hard to discuss this subject without implying one idea that is actually ridiculous. This ridiculous idea is that the minds of intelligent people work in an orderly process just as an outline is developed. Nothing could be farther from the truth. Of course the purpose of an outline is to establish an orderly relationship among a group of facts or ideas. And intelligence could almost be defined as the ability to perceive relationships. Nevertheless, intelligent thought processes appear to be infinitely more complex and more varied in structure than is even remotely implied in the concept of an outline. Indeed, when you are instructed to prepare an outline for a paper you are to write, you may very well feel a distaste for doing so. You may feel that, logical though it appears to be, an outline doesn't have very much to do with how your mind really works. And you are quite right. It doesn't, except in very limited and explicit ways. These ways do not—apparently—include the most fundamental insights about how to present material.*

Typically, when a person starts making an outline, he has already reached a few conclusions. One is that he has so much material to write about that if he doesn't make an outline he will get mixed up. The second is that he must give some attention to problems of the sort discussed in the section "Reader Adaptation" in Chapter 3. As a matter of fact, people often decide to make an outline only after they have actually started writing and have discovered they really are beginning to get mixed up. We see nothing particularly objectionable about going about the job in this fashion. If the writer doesn't have sense enough to stop writing before getting in too deep, he may waste a lot of time.

At any rate, you will most often begin an outline with a general idea already in mind as to how you want to present your material. Before going very far with the outline you can further clarify this general idea by writing a statement of purpose. That is, write a sentence beginning, "The purpose of this report is . . . ." If you can't finish the sentence, you need to think some more about what you are really trying to do.

The limited and explicit ways in which an outline does reflect how a person's mind works must surely include organizing facts or ideas into orderly groups, once the principle of organization—the purpose—has been decided upon and the logic has been tested.

The first thing to do with respect to the outline itself is to make

*For a discussion of some relevant issues in the philosophy of science, see Michael Polanyi, *Personal Knowledge* (New York: Harper & Row, 1962), especially sections 7 and 8 in Chapter 8.

lists of the topics that must be discussed. The great thing here is not order but inclusiveness. Don't bother about the sequence in which you list the topics you are to discuss; just don't omit anything.

When you have written down all the topics you can think of, the next step is to decide what major groups you will have. These major groups become the Roman numeral divisions in the outline. For example, if you decide you want a division on equipment, "Equipment" becomes the heading for that division; and you then look through your lists to find all the entries that pertain to this topic and jot them down under the heading "Equipment," in whatever sequence you find them in your lists. Later you can arrange them in the best order, and if necessary subdivide them. All of this has to do with organization.

As the outline nears completion you can start testing the logic, according to the principles described and illustrated throughout the preceding discussion and in Appendix E.

In summary, we urge you to think of outlining as simply one of the aids people have found helpful in their writing. Start by listing the topics you need to tell about and then arrange these topics into logical groups and subgroups. And when you have added numbers and letters to differentiate the groups, you have an outline—and a table of contents too.

## Abstracts

An abstract is a short description, or a condensation, of a piece of writing. It is a timesaving device. Naturally, it is a device that is highly popular with executives. The man whose opinion of your report matters most may read only the abstract of it.

We shall identify the two types of abstracts, note the advantages of each, and then make some remarks on how to write an abstract.

*Types of Abstracts.* One type of abstract, the descriptive, tells what topics are taken up in the report itself, but little or nothing about what the report says concerning these topics. This type of abstract is illustrated on page 96. The advantages of a descriptive abstract are that it is easy to write and is usually short; a serious disadvantage is that it contains little information.

The other type of abstract is sometimes called "informational." In this type, illustrated on page 96, there is a statement of the chief points made in the report. Instead of learning merely that such and such topics are taken up in the report, we are told something of what the report has to say about these topics. The advantage of an informational abstract is that it provides much more information than does a descriptive abstract. Of course it is harder to write, and it may be a little longer than the descriptive type. Except where brevity is of special importance, however, there can be no question as to the superiority of the informational type.

How long should an abstract be? A good rule of thumb is to make it as short as you can, and then cut it by half. Some people say it should be about 5 per cent of the length of the report. In industrial reports, an abstract rarely exceeds one page.

It is illuminating to look at these differences from the point of view of the preparation of abstracts for a great published series such as *Chemical Abstracts*. The following comments about such published abstracts are contained in a paper presented by Mr. J. C. Lane at the 138th national meeting of the American Chemical Society*

> . . . let's look briefly at two extremes [in types of abstracts]. The purpose of Chemical Abstracts has been defined as "preparing concise summaries . . . *from the indexing point of view.*" Here the goal is to provide a timeless reference tool for chemists. Speed in reporting must necessarily be sacrificed for the sake of comprehensive coverage of the world's literature and for thorough indexing and cross-indexing. Thus, the time it takes an abstract to appear averages three to four months. But the value of these abstracts continues for years. In fact, from the literature searchers' point of view, their value increases with time—as their use is facilitated by annual and decennial indexes.
>
> At the other extreme, Current Chemical Papers, which has replaced British Abstracts, cuts appearance time to only a matter of weeks, but at a sacrifice in comprehensiveness. It is primarily an indexed listing of titles. This is practical only because our Chemical Abstracts fills the need for a comprehensive abstract journal written in English.

Usually the abstract written to appear as part of a report circulated within a company or other organization resembles the first type Mr. Lane refers to. That is, it is essentially informational. Examples may be found at the end of this chapter.

In concluding these remarks on types of abstracts, we must point out that most abstracts are not exclusively either descriptive or informational, but a combination of both. This is perfectly all right. As a matter of fact, the first sentence in the descriptive abstract on page 264 is more nearly informational than descriptive. Writing an abstract invariably presents a problem in compromising between saying everything you think you ought to and keeping it as short as you think you

*John C. Lane, "Digesting for a Multicompany Management Audience," Sept., 1960. Quoted by permission.

ought to. Descriptive statements here and there in an informational abstract often help solve this problem. Sometimes the term "epitome" is applied to a very short informational abstract in which only the most important facts or ideas are presented, and the term "abstract" is reserved for a longer, more detailed statement. Whatever the terminology you encounter, you have fundamentally two sets of conflicting variables to balance: brevity vs. detail, and description vs. information.

*Suggestions About Writing Abstracts.* The best single suggestion we can make about writing an abstract is to have a well-organized report to begin with. Having that, you simply write a brief summary of each one of the major divisions of the report. To help avoid getting tangled up in the details in the text you may find it a good idea to write the abstract from the outline or table of contents rather than from the text itself. This advice is based on the assumption that you are preparing an abstract of a report you yourself have written and that you are therefore thoroughly familiar with the content. It is often difficult for an author to compress into the brief space available in an abstract all the thinking that has gone into a long report he has written. Working just from the outline or table of contents may benefit his sense of balance and perspective.

If you are preparing an abstract of a report someone else has written, it is helpful to underscore the key sentences in the report before beginning the abstract. You can think of the task of writing the abstract as a problem in bringing together into one brief space the content of these key passages. Of course you might find the same method helpful with a report you had written yourself.

In any event, begin by writing a sentence that focuses the reader's attention on what you think is the principal idea communicated by the report. If you have difficulty deciding what that idea is, you may find some assistance in Chapter 11, which is concerned with the writing of introductions. Examples of good beginning sentences will be found in the illustrative materials at the end of the present chapter.

Once you have your beginning sentence, each additional sentence should constitute a development of the principal idea expressed in that beginning sentence.

As will be seen in the illustrative materials, there is a tendency for abstracts of reports of research work to have a rather specialized form. Emphasis falls on the problem and the scope of the attack on the problem, on findings, on conclusions, and on recommendations. In contrast, reports on other kinds of subjects may lead naturally to abstracts containing virtually none of these subtopics.

In form, the abstract is usually set up as a single paragraph, double-spaced, on a page by itself. It should be written in good English: articles should not be omitted, and no abbreviations should be used which would not be acceptable in the body of the report. A special effort should be made to avoid terminology unfamiliar to an executive or any reader who is not intimately acquainted with the work. With the exception noted in the next section, the abstract should be regarded as a completely independent unit, intelligible without reference to any part of the report itself.

## Introductory Summaries

Abstracts are sometimes called summaries, so it is easy to guess that an introductory summary is a combination of introduction* and abstract. It isn't exactly a combination, however, in the way that $H_2$ and O make water; it is rather a joining together, as a handle and a blade make a knife. It's still easy to identify both parts.

There are really two kinds of introductory summaries. One is an ordinary abstract put at the top of the first page of the text of a report. The only thing introductory about it is the fact it is the first thing the reader sees. Since this is just a matter of what name you want to call an abstract by, we shall say no more about it.

In the second type of introductory summary, special emphasis is given to the introductory portion. The idea back of this is to show clearly at the outset how the project being reported on fits into the whole program of which it is a part. If the report itself is short there may be no further introductory material. In longer reports there is likely to be a formal introduction following the introductory summary. There is always a temptation, however, to let the introductory summary do the whole job, even when a separate formal introduction is definitely needed.

The introductory summary that follows is a fictitious one which the Hercules Powder Company has used as a model for its staff. The Hercules Powder Company calls it a digest.

### Terpene Sulfur Compounds—Preparation**

#### Digest

In previous progress reports under this investigation, terpene sulfides were prepared and tested as flotation reagents with negative results.

*For a discussion of the elements of an introduction, see Chapter 11.

**From *The Preparation of Reports,* 3d ed. (Wilmington, Del.: The Hercules Powder Company, 1945), pp. 32–33. Quoted by permission of the Hercules Powder Company.

From theoretical considerations, there was reason to believe that terpene mercaptans would be satisfactory flotation reagents. However, no method of preparing these compounds was known. It was suggested that terpene hydrocarbons might add hydrogen sulfide directly to form mercaptans. To test the possibility of this reaction, experiments were carried out, during the period covered by the present report, in which hydrogen sulfide was bubbled through separate samples of pinene and also of Dipolymer at atmospheric pressure and room temperature in the presence of catalysts.

Catalysts employed with pinene were 85% phosphoric acid with and without Darco, 90% phosphoric acid, and 32% sulfuric acid. The best results were obtained with the use of a catalyst consisting of 85% phosphoric acid and a small proportion of Darco. The sulfur content of the product indicated that the apparent yield with such a catalyst was 94%. Without Darco, the yield was 68 and 81% with 85 and 90% phosphoric acid, respectively. With 32% sulfuric acid, the yield was 83%. Dipolymer when tested similarly with 85% phosphoric acid and Darco gave a somewhat lower yield.

Further experiments will be carried out with pinene under other reaction conditions. It is planned to carry out the reaction under superatmospheric pressure. The pure mercaptan will be isolated and tested as a collector in ore flotation.

The first paragraph in the example above is obviously the introductory portion. It gives the reader a clear statement of the general situation. The remainder is an informational abstract.

## Additional Examples of Outlines and Abstracts

The illustrative materials in the following pages are divided into two parts.

The first part is made up of five exhibits, including a single chapter, "Insect and Rodent Control," of a report entitled *Sanitation Requirements for an Isolated Construction Project*, which was written by Mr. Jerry Garrett while he was a student at the University of Texas at Austin. The first four exhibits in Part I are all based on this one chapter.

The second part is taken from a booklet on report writing prepared for use in General Motors Research Laboratories at Warren, Michigan.

*Part I.* The five exhibits in this part are:

A. A topic outline
B. A portion of a sentence outline
C. A descriptive abstract
D. An informational abstract
E. "Insect and Rodent Control"

## A. *Topic Outline*

I. Introduction
   A. Flies, mosquitoes, and rats as the vehicles of infection for ten widespread diseases
      1. Flies
         a. Mechanical transmission of disease
         b. Intestinal diseases they transmit
            (1) Typhoid
            (2) Paratyphoid
            (3) Dysentery
            (4) Cholera
            (5) Hookworm
      2. Mosquitoes
         a. Transmission of disease by biting
         b. Diseases they transmit
            (1) Malaria
            (2) Yellow fever
            (3) Dengue
      3. Rats
         a. Transmission of disease through harboring fleas
         b. Diseases they transmit
            (1) Plague
            (2) Typhus
   B. Stopping the spread of these diseases by breaking the cycle of transmission
      1. Removing or destroying the breeding places of insects and rodents
      2. Killing the adult insects and rodents
II. Breeding control
   A. Introduction
   B. Flies
      1. Breeding habits
      2. Control measures
         a. Sewage disposal
         b. Removal of manure
            (1) Time limit
            (2) Storage bins
            (3) Compression
         c. Destruction of all decaying organic matter
   C. Mosquitoes
      1. Differences from flies
         a. Greater difficulty in control of breeding places
         b. Small percentage that carry disease
      2. Disease-transmitting mosquitoes
         a. Female *Aedes aegypti*
            (1) Transmission of yellow fever and dengue
            (2) Breeding in clean water in artificial containers

        b. *Anopheles quadrimaculatus*
            (1) Transmission of malaria in southern United States
            (2) Habit of biting at night
            (3) Breeding in natural places
                (a) Preference for stationary water
                (b) Protection afforded by vegetation and floating matter
    3. Control measures
        a. Removing water
        b. Spreading oil on stationary water
  D. Rats
    1. Lack of direct ways to control breeding of rats or their fleas
    2. Prevention of breeding in specific areas
        a. Building rat-resistant houses
        b. Keeping rats from food
III. Adult control
  A. Flies
    1. Screens
    2. Traps
    3. Baits
        a. Fish scraps
        b. Overripe bananas
        c. Bran and syrup mixture
    4. DDT
  B. Mosquitoes
    1. Screens
    2. Larvae-eating minnows
    3. Poisons
        a. DDT
        b. Pyrethrum
  C. Rats
    1. Importance in property destruction as well as in disease
    2. Poisons
        a. Barium carbonate
        b. Red squill
        c. 1080
        d. Antu
    3. Trapping
    4. Fumigating

## B. *A Portion of a Sentence Outline*

I. The fact that flies, mosquitoes, and rats transmit ten diseases makes it important that these insects and rodents be destroyed by preventing them from breeding or by killing adults.
  A. Flies, mosquitoes, and rats transmit ten widespread diseases.
    1. Flies transmit five intestinal diseases.
        a. Flies are mechanical carriers of diseases.
        b. They transmit typhoid, paratyphoid, dysentery, cholera, and hookworm.

2. Mosquitoes transmit three diseases.
   a. Mosquitoes spread diseases by biting.
   b. They transmit malaria, yellow fever, and dengue.
3. Rats transmit two diseases.
   a. Rats transmit disease through harboring fleas.
   b. They transmit plague and typhus.

B. The spread of the diseases listed above can be stopped by breaking the cycle of transmission.
   1. The breeding places of insects and rodents can be removed or destroyed.
   2. The adult insects and rodents can be killed.

### C. *Descriptive Abstract*

Ten widespread diseases that are hazards in isolated construction camps can be prevented by removing or destroying the breeding places of flies, mosquitoes and rats, and by killing their adult forms.

### D. *Informational Abstract*

Ten widespread diseases that are hazards in isolated construction camps can be prevented by removing or destroying the breeding places of flies, mosquitoes and rats, and by killing their adult forms. The breeding of flies is controlled by proper disposal of decaying organic matter, and of mosquitoes by destroying or draining pools, or spraying them with oil. For rats, only the indirect methods of rat-resistant houses and protected food supplies are valuable. Control of adult forms of both insects and rodents requires uses of poisons. Screens are used for insects. Minnows can be planted to eat mosquito larvae.

## Insect and Rodent Control*

### Introduction

Flies, mosquitoes, and rats are the vehicles of infection for ten widespread diseases. Flies, which are mechanical carriers, are responsible for the transmission of the intestinal diseases; i.e., (1) typhoid, (2) paratyphoid, (3) dysentery, (4) cholera, and (5) hookworms. Mosquitoes spread diseases by biting; they are vectors in the cycle of transmission of (6) malaria, (7) yellow fever, and (8) dengue. Rats are the reservoirs of (9) plague and (10) typhus, but the rat's fleas are the vehicles of transmission.

There is but one way to stop the spread of these diseases, and that is to break the cycle of transmission. The best way to do this is to get rid of the insects and rodents, and the most effective method of getting rid of them is to remove their breeding places by good general sanitation. The only alternative is to kill the adults. Positive steps which may be taken in these operations are discussed below.

*"Insect and Rodent Control" is Section IV of *Sanitation Requirements for an Isolated Construction Project*, by Jerry Garrett.

## Breeding Control

As pointed out above, if there are no insects or rodents the diseases which depend on them for transmission must vanish. It is certainly cheaper and simpler to destroy their breeding places than to try to kill billions of adults only to find more billions waiting to be killed.

*Flies.* One characteristic of the fly makes it particularly susceptible to breeding control. The fly always lays its eggs in decaying organic matter, preferably excreta or manure. Three stages in the life of the fly—the egg, larva, pupa—are spent in the manure. A minimum of eight to ten days is spent here before the adult emerges. Therefore, the measures are relatively simple. First, there should be proper sewage disposal; i.e., the flies are never permitted to come into contact with human excreta. Secondly, all animal manure should be removed within four or five days, or in other words, before pupation takes place. The manure should either be placed in fly-proof storage bins or tightly compressed so that the adult fly cannot emerge after pupation. The final breeding control is to destroy all decaying organic matter such as garbage by either burying it two feet deep or burning it.

*Mosquitoes.* It is not as simple to control the breeding places of the mosquito as it is to control those of the fly. But it can be done! First, it must be realized that there are many kinds of mosquitoes and that only a few are disease vectors. Still they must all be killed to be sure the correct ones are dead, and they are all important as pests anyway. The female *Aedes aegypti* is the vector for yellow fever and dengue; this mosquito breeds only in clean water in artificial containers. In the southern section of the United States (the chief malaria area in the United States), the malaria vector is the *Anopheles quadrimaculatus*, a night biter, which breeds in natural places, particularly where the water is stationary and where there is vegetation and floating matter to protect the eggs, larvae, and pupae.

Therefore, the best way to prevent the breeding of mosquitoes is to remove all water in which they breed by draining or filling pools, and removing or covering artificial containers. However, since the construction project is only temporary, the operators will be interested in the most economical measures rather than the most permanent. Artificial containers must still be covered, but it might be cheaper to spread a film of oil over all the natural, stationary water rather than to try to drain it or fill in the low spots.

*Rats.* There are no direct ways to control the breeding of rats or their fleas, but sufficient control can be exerted to make them take their breeding elsewhere. This is done by building rat-resistant houses and by preventing the rats from reaching food.

## Adult Control

*Flies.* Houses should be screened to keep the flies from getting to food. Then, traps such as the standard conical bait trap should be distributed. The most attractive baits, as established by experiment, are fish scraps, overripe bananas, and a bran and syrup mixture. DDT may be used effectively to leave a residual poison for flies.

*Mosquitoes.* If a house is well screened, the mosquitoes cannot get into the house to bite their victims. Advantage can be taken of the mosquitoes' natural enemies by stocking waterways with minnows which eat the larvae. Poisons which may be used against mosquitoes are DDT and pyrethrum.

*Rats.* Besides carrying diseases, the rat of course destroys much property. Usually, however, the construction project operator need be concerned with rats only to the extent that they endanger his workers' health. Poisons which may be used against rats are barium carbonate, red squill, 1080, and antu. Other effective means of getting rid of rats are by trapping and fumigation.

*Part II.* The material that follows is taken from a rather typical booklet of instructions prepared for the technical staff of General Motors Corporation.* We do have one half-hearted complaint to make about the abstract (Figure 1) that appears as part of this material. We don't think the first sentence is effective. See if you feel that something like the following would be better:

> Tests of ten electrolyte additives revealed that none of them improved battery performance. Ten organic additives . . .

Our complaint is only half-hearted because we are fully sympathetic with any research man's reluctance to emphasize a negative result in his report. Could it be that this reluctance helped determine the form taken by the beginning of this abstract? However that may be, this is a rather typical example of an abstract of a research report. And for the most part it is well done. The marginal comments appear in the original, as does the box within which they are enclosed.

### Abstract

The abstract should present the report in a nutshell. Many readers, including management, may be too busy to read the entire report; therefore, the abstract might be the only chance to tell them what has been accomplished. Another reason for writing meaningful abstracts is that the Library is becoming increasingly dependent on them for cataloging information properly.

To communicate its message effectively, the abstract appears double spaced on a page by itself. It should not exceed 150 words.

Textbooks recommend that an abstract be informative, that it convey the essence of the report rather than merely amplify the title or describe the

*Robert F. Schultz, *Preparing Technical Reports*, Research Publication GMR-427. (Warren, Mich.: Research Laboratories, General Motors Corporation.)

contents. In some cases, however, such as when a report consists of a procedure or a data compilation, a descriptive abstract is the only alternative. Following is a suggested guide for writing an abstract:

- The problem (what you are trying to do, and, if not obvious, why)
- The scope of your work
- The significant findings or results
- Any major conclusions
- Any major recommendations

ABSTRACT

**The "why"**

Chemists have long sought an electrolyte additive that would improve battery performance, especially at low temperatures. Ten organic additives were investigated to learn more about their effects on the rechargeability of an experimental lead-acid cell at 5°F. In each set of trials a different additive (0.001% by weight) was mixed with the sulfuric acid in the cell. The cell was then charged and discharged at eight current intensities. The additives included adipic acid, azelaic acid, p-phenolsulfonic acid, hydroquinone, 1-naphthol 5-sulfonic acid, phthalic acid, sucrose, tartaric acid, alpha-naphthol, and 8-hydroxyquinoline.

**What you are trying to do**

**Scope**

Adipic acid resulted in the best high-rate charge acceptance; however, the level reached was only 96% of that achieved without any additives. Alpha-naphthol, the worst, reached a level of only 20%. In addition to the obvious conclusion that these additives do not improve rechargeability, the following trend appears valid: the higher the recharge voltage, the lower the charge acceptance.

**Findings**

**Conclusions**

*Figure 1. Sample Abstract*

## SUGGESTIONS FOR WRITING

Opportunities for practicing the arts of abstracting and outlining are everywhere, and we suggest that you form a habit of noticing and taking advantage of them. For example, when you go to a public lecture see if you can come away with a mental abstract of the lecture. Each time the speaker starts to discuss a new aspect of his subject—a new Roman numeral division—say over to yourself what seems to you the proper heading for it, and also for any divisions that preceded it. With only a little active concentration you can leave the lecture with a mental outline of its content. With this outline in mind, give a brief resumé of the lecture to one of your friends, and you will have made an abstract, even though it isn't written down.

But practice in writing is essential too. Choose a magazine you enjoy reading, and outline and abstract an article you find interesting. It is particularly instructive to choose a journal whose articles are abstracted in one of the professional journals of abstracts listed in the bibliography in Appendix A of this book, and to compare your abstract with the one published in the journal.

Finally, whenever it is feasible to do so, prepare an abstract of reports and exercises you yourself have written.

*SECTION TWO*

# Special Techniques of Technical Writing

*Five writing techniques are of special importance to technical men: definition, description of a mechanism, description of a process, classification, and interpretation. These techniques will be discussed separately in the five chapters which make up this section.*

*For emphasis, it is worth repeating that these techniques must not be considered as types of reports. Usually, several of them will appear in a single report. It would be exceptional to find an entire report, even a short one, containing only one of these techniques. For example, two or more techniques might be closely interwoven as a writer described the design, construction, and operation of a mechanism. The intermingling of these techniques, however, does not alter the basic principles of their use. And these basic principles can be studied most effectively by taking one technique at a time.*

*The treatment of these techniques will stress the practical rather than the theoretical, particularly in the chapters on definition and classification.*

# 5
# Definition

In this chapter on definition we have three specific objectives: (1) to set down some facts intended to clarify the problem of what should be defined in technical writing; (2) to suggest effective methods for defining what needs to be defined; and (3) to point out where definitions can be most effectively placed in reports.

## What to Define

Before we can tackle the problem of *how* to define, we must think about *what* should be defined. It is not possible, of course, to set up an absolute list of terms and ideas which would require definition, not even for a specific body of readers, but it is possible and desirable to clarify the point of view from which the problem of definition should be attacked.

First of all, let's recall a rather obvious but extremely significant fact about the nature of language: words are labels or symbols for things and ideas. The semanticists—those who study the science of meaning in language—speak of the thing for which a word stands as its "referent." For instance, five letters of the alphabet, l-e-m-o-n, are used as a symbol for a fruit with which we are all familiar. In a sense, it is unimportant that these letters happen to be used, for the lemon would be what it is no matter what combination of letters was used to name it. This fact, however obvious, is an important one to keep in mind, for it often happens that a writer and reader are not in perfect agreement as to the referent for certain words. That is, the same word, or symbol, may call to the reader's mind a referent different from the one the writer had in mind, and thus communication may not be achieved. Or, more importantly for our purposes, a word used by the writer may not call to the reader's mind any referent at all. Thus a reader who is familiar with banana oil may not have it called to mind by the technical term "amyl acetate" because the latter term is unfamiliar to him.

The relationships of words to the ideas and things for which they stand can become very complex, but without going into the problem of semantics any further we can discern a simple and helpful way of classifying words as they will appear to your reader. The words you use will fall into one of the following categories:

1. Familiar words for familiar things
2. Familiar words for unfamiliar things
3. Unfamiliar words for familiar things
4. Unfamiliar words for unfamiliar things

Each of these categories deserves some attention.

*Familiar Words for Familiar Things.* The only observation that need be made about the first category is that familiar words for familiar things are fine; they should be used whenever possible. To the extent that they can be used, definition is unnecessary. This might be dismissed as superfluous advice were it not for the fact that a great many writers often appear to seek unfamiliar words in preference to everyday, simple terms. There is, as a matter of fact, a tendency for some people to be impressed by obscure language, by big words. Thus we find "amelioration" when "improvement" would do as well, "excoriate" for "denounce," "implement" for "carry out" or "fulfill," and the like. It scarcely needs to be pointed out that a "poor appetite" is not really changed by being called "anorexia"; yet there are those who would much prefer the latter term. Nothing is ever gained by using, just for their impressiveness, what you have probably heard called "two-dollar" words; often, much is lost.

*Familiar Words for Unfamiliar Things.* The words in this second category present a rather special problem to the technical writer and one that he needs to be especially alert to. These are the everyday, simple words which have special meanings in science and technology.* Most of them may be classed as "shoptalk," or language characteristic of a given occupation. Because they are a part (often a very

*Our phrase "familiar words for unfamiliar things" does not cover all situations. Sometimes a well-known word is *unfamiliarly applied* to a well-known thing, and hence needs explanation. In anatomy, for example, the word "orbit" (a familiar word) means what most people call "eye socket" (a familiar thing).

colorful part) of the technical man's everyday vocabulary, he is apt to forget that they may not be a part of the vocabulary of his reader, at any rate not in the special sense in which he uses them. Consider a term like "puddle." Everyone knows this word in the familiar sense, but not everyone knows that in the metallurgical sense it means a mass of molten metal. Or take "quench" in the same field. Quenching a metal by immersing it in water or oil bears some relation to quenching one's thirst, but it is a distant relationship.

Every field of engineering and science has a great many of these simple words which have been given specialized meanings. Examine the following list (you could probably add a number from your own experience):

apron: as on a lathe, the vertical place in front of the carriage of a lathe. This term is also used in aeronautics, navigation, furniture, textiles, carpentering, hydraulics, and plumbing, with different meanings in each field.

backlash: play between the teeth of two gears which are in mesh or engaged. Not quite the same thing the word would mean to a fisherman!

blooms: heaving semifinished forms of steel.

chase: iron frame in which a form is imposed and locked up for the press.

cheater: an extension on a pipe wrench.

Christmas Tree: the network of pipe at the mouth of an oil well. Also red and green lights in a submarine control room to show closed and open passages.

diaper: a form of surface decoration used in art and architecture consisting of geometric designs.

dirty: to make ink darker.

dwell: (of a cam) the angular period during which the cam follower is allowed to remain at its maximum lift; and in printing for the slight pause in the motion of a hand press or platen when the impression is being made.

freeze: seizing of metals which are brought into intimate contact.

galling: a characteristic of metals which causes them to seize when brought into intimate contact with each other.

lake: a compound of a dye with a mordant.

This somewhat haphazard list of terms—it could be extended at length—suggests the nature of the terms we have in mind. The reader may not confuse the everyday meaning of such terms with the technical sense they have in a particular report, but there is not much doubt that the first time he sees a term of this sort (unless he is a specialist in the field being discussed), he will think of its common meaning. Almost instantly, he may recognize that the common meaning is not what the term denotes in its present context, and he may then recall its specialized meaning; or, he may not recognize the specialized meaning at all, depending on his familiarity with the subject matter. In any event, the writer must be alert to the need for defining such terms.

*Unfamiliar Words for Familiar Things.* A moment ago we condemned the writer who prefers to use big and pretentious words for referents with which his reader is familiar. Such a practice should always be condemned if a simple, familiar term exists which means the same thing. But an unfamiliar word for a familiar thing may be used if there does not exist any simple, familiar term for it. Both convenience and accuracy justify it. Suppose you were writing on the subject of hydroponics. You can easily imagine addressing readers who know that plants may be grown without soil, in a chemical solution, but who are unfamiliar with the technical term "hydroponics." Since there isn't a simple, familiar word for this process, you would scarcely want to give up the word "hydroponics" for an awkward, rather long phrase. Your solution is simple: you use the convenient term but you define it. Let's take another example. Suppose an electrical engineer were writing about special tactical electronic equipment making use of direct ray transmission. It is not likely that he would be satisfied to use the phrase "short wave" if he were dealing specifically with, say, the 300 to 3000 megacycle band. On the contrary, he would prefer the phrase "Ultra High Frequency" (UHF). Similarly, a medical man might prefer, in the interests of precise accuracy, the term "analgesic" to the simple word "painkiller."

You will have to judge whether your subject matter demands the use of such terms and whether they are familiar to your readers or not. If they are needed, or if they are justifiably convenient, and you decide that your readers do not know them, you should define them.

*Unfamiliar Words for Unfamiliar Things.* This category, unfamiliar words for unfamiliar things, embraces most of those words that are commonly thought of as "technical" terms. They are the specialized terms of professional groups; big and formidable looking (to the nonspecialist), they are more often than not of Greek or Latin origin. Terms like "dielectric," "hydrosol," "impedance," "pyrometer," and "siderite," are typical. We do not want to suggest that a static, precise list could be set up in this group, but since the reader's response determines the category into which a word falls, a great many of the terms which constitute the professional language of any special science or branch of engineering would for the nontechnical reader stand for unfamiliar things. These same words, however, when used by one expert in talking or writing to another expert would be familiar words for familiar things. It is important to remember, on the other hand, that the "nontechnical reader" does not necessarily mean the "lay reader," for even an expert in one branch of science or engineering becomes a nontechnical reader when he reads technical writing in a field other than his own.

So far our interest has been in the problem of what needs to be defined. We can sum it up this way: you need to define (1) terms familiar to your reader in a different sense from that in which you are using them; (2) terms which are unknown to your reader but which name things that actually are familiar to him, or at least things which can be explained simply and briefly in readily understandable, familiar terms; and (3) terms which are unfamiliar to the reader and which name scientific and technical things and processes with which he is also unfamiliar. With these facts in mind about what to define, we can more intelligently consider the problem of how to define.

## Methods of Definition

Before discussing the methods of definition, we want to remind you that insofar as it is possible to use simple, familiar terminology, the problem of definition may be avoided entirely. In other words, the best solution to the problem of definition is to avoid the need for it. When it is necessary, however, there are two methods or techniques which may be employed. The first may be described as informal; the second as formal. The second takes two forms: the sentence definition and the extended or amplified definition. Each of these techniques has its own special usefulness.

*Informal.* Essentially, informal definition is the substituting of a familiar word or phrase for the unfamiliar term used. It is, therefore, a technique that can be employed only when you are reasonably certain that it is the term alone and not the referent which is unfamiliar to the reader. You must feel sure, in other words, that the reader actually knows what you are talking about, but under another name. Thus you might write ". . . normal (perpendicular) to the surface . . ." with the parenthetical substitution accomplishing the definition. Or "dielectric" might under certain circumstances simply be explained as a "nonconductor." Or "eosin" as "dye."

Instead of a single-word substitution, sometimes a phrase, clause, or even a sentence may be used in informal definition. Thus dielectric might be informally explained as "a nonconducting material placed between the plates of a condenser," or eosin as "a beautiful red dye." Or you might use a clause, as "eosin, which is the potassium salt of tetrabromo-fluorescin used in making red printing ink." In very informal, colloquial style, you might prefer a statement like this: "The chemical used in making red ink and in coloring various kinds of cloth is technically known as 'eosin.' " Or, "When you use rubber insulating tape in some home wiring job, you are making use of what the electrical engineer might call a 'dielectric.' "

Several general facts should be noted about such definitions. First, they are partial, not complete, definitions. The illustrations just given, for instance, do not really define dielectric or eosin in a complete sense. But such illustrations are enough in a discussion where thorough understanding of the terms is not necessary and the writer merely wants to identify the term with the reader's experience. Second, informal definitions are particularly adapted for use in the text of a discussion. Because of their informality and brevity, they can be fitted smoothly into a discussion without seriously distorting its continuity and without appearing to be serious interruptions. Third, we should note that when the informal definition reaches sentence length, it may not be greatly different from the formal sentence definition to be discussed in the next section. It lacks the emphasis, and usually the completeness, however, which may be required if a term defines an idea or a thing which is of critical importance in a discussion. In short, if you want to make certain that your reader understands a term, if you think the term is important enough to focus special attention on it, you will find the formal sentence definition, and perhaps the amplified definition, or article of definition, more effective.

*Formal Sentence Definition.* We have seen that informal definition does not require the application of an unchanging, rigid formula; rather, it is an "in other words" technique—the sort of thing we all do frequently in conversation to make ourselves clear. With formal definitions the situation is different. Here a logically dictated, equationlike statement is always called for, a statement composed of three principal parts for which there are universally accepted names. These are the *species*, the *genus*, and the *differentia*. The species is the subject of the definition, or the term to be defined. The genus is the family or class to which the species belongs. And the differentia is that part of the statement in which the particular species' distinguishing traits, qualities, and so forth are pointed out so that it is set apart from the other species which comprise the genus. Note this pattern:

| *Species* = | *Genus* | + | *Differentia* |
|---|---|---|---|
| Brazing is | a welding process | | wherein the filler metal is a nonferrous metal or alloy whose melting point is higher than 1000° F but lower than that of the metals or alloys to be joined. |

Defined as a process, then, formal definition involves two steps: (1) identifying the species as a member of a family or class, and (2) differentiating the species from other members of the same class.

Don't let these Latin terms worry you. Actually the process of working out a formal definition is both logical and natural. It is perfectly natural to try to classify an unfamiliar thing when it is first encountered. In doing so, we simply try to tie the thing in with our experience. Suppose you had never seen or heard of a micrometer caliper. If, when you first saw one, a friend should say—in response to your "What's this?"—that it is a measuring instrument, you would begin to feel a sense of recognition because of your familiarity with other measuring instruments. You still would not know what a micrometer caliper is, in a complete sense, but you would have taken a step in the right direction by having it loosely identified. To understand it fully, you would need to know how it differs from other measuring instruments, like the vernier caliper, the rule, a gauge block, and so on. In all likelihood, therefore, your next question would be, "What kind of measuring instrument?" An accurate answer to this question would constitute the differentia. Assuming that your friend had the answers, he would then probably tell you something about the micrometer caliper's principle of operation, its use, and the degree of accuracy obtainable with it. Were it not for the fact that you had it in hand, he would also undoubtedly describe its shape, for physical appearance is a distinctive feature of the instrument. To be quite realistic about our hypothetical instance, we must admit that he would probably tell you more than is essential to a good sentence definition; and he would probably use more than one sentence. But if he were to sift the essential distinguishing characteristics of the micrometer caliper from what he had said about it and put them into a well-ordered sentence, he would have made a formal sentence definition—something like this, no doubt: "A micrometer caliper is a C-shaped gauge in which the gap between the measuring faces is minutely adjustable by means of a screw whose end forms one face."

Natural as the process of identifying and noting the particular characteristics of something new may be, it must not be done carelessly. Let's take another look at some of the problems of handling the genus and differentia. The first step in the process of formal definition is that of identifying a thing as a member of a genus, or class. It is important to choose a genus that will limit the meaning of the species and give as much information as possible. In other words, the genus should be made to do its share of the work of defining. You wouldn't have been helped much, for instance, had your friend told you that a micrometer caliper is a "thing" or "device." If a ceramic engineer were to begin a definition of an engobe by saying it is a "substance," he wouldn't be making a very good start; after all, there are thousands of substances. He would get a great deal more said if he were to classify it at once as a "thin layer of fluid clay." With this informative beginning, he would have only to go on to say that this thin layer of fluid clay is applied to the body of a piece of defective ceramic ware to cover its blemishes. Generally speaking, the more informative you can make the genus, the less you will have to say in the differentia. Another way of saying this is that the more specific you can be in the genus, the less you have to say in the differentia.

Care must also be taken in carrying out the second step of the process of formulating a sentence definition. Here the important point is to see that the differentia actually differentiates—singles out the specific differences of the species. Each time you compose a statement in which you attempt to differentiate a species, examine it critically to see if what you have said is applicable *solely* to the species you are defining. If what you have said is also true of something else, you may be sure that the differentia is not sufficiently precise. One who says, for instance, that a micrometer caliper is "a measuring instrument used where precision is necessary" will recognize upon reflection that this statement is also true of a vernier caliper, or, for that matter, of a steel rule (depending, of course, upon what is meant by "precision"). One way to test a statement is to turn it around and see whether the species is the only term which is described by the genus and differentia. Consider this example: "A C-shaped length gauge in which the gap between the measuring faces is minutely adjustable by means of a screw whose end forms one face is a__________." "Micrometer caliper" fills the blank, and if the definition is correct, it is the only term which accurately fills the blank.

The foregoing discussion about methods can be reduced to the statement that an accurate limiting genus coupled with a precisely accurate differentia will always ensure a good definition. A few specific suggestions about particularly common difficulties should be added. Do not regard the itemized points that follow as something to be memorized, but as a possible source of help in case of trouble. We have included a few suggestions for solving some particularly common difficulties.

A. *Repetition of Key Terms.* Do not repeat the term to be defined, or any variant form of it, in the genus or differentia. Statements like "A screw driver is an instrument for driving screws" or "A caliper square is a square with attached calipers" merely bring the reader back to the starting point. These examples may be so elementary as to suggest that this advice is unnecessary, but the truth is that such repetition is not at all uncommon.

There are, however, some occasions when it is perfectly permissible to repeat a part of the term to be defined. For instance, it would be perfectly permissible to begin a definition of an anastigmatic lens, "An anastigmatic lens is a lens . . ." if it could be assumed that it is the *anastigmatic* lens and not all lenses that is unfamiliar to the reader. Or a definition of an electric strain gauge might contain the word "gauge" as the genus. In these instances, the repeated word is not an essential one.

B. *Qualifying Phrases.* When a definition is being made for a specific purpose, a common practice in reports, limitations should be clearly stated. For example, an engineer might write, "Dielectric, as used in this report, signifies . . ." and go on to stipulate just what the term means for his present purpose. Unless such limitations are clearly stated (usually as a modifier of the species) the reader may feel—and rightly so—that the definition is inaccurate or incomplete.

C. *Single Instance, or Example, Definitions.* In an amplified definition, as we shall see in the following section, the use of examples, instances, and illustrations is fine; they help as much as anything to clarify the meaning of a term. But the single instance or example is not a definition by itself. "Tempering is what is done to make a metal hard" may be a true statement, but it is not a definition. So it is with "A girder is what stiffens the superstructure of a bridge." In general, guard against following the species with phrases like "is when" and "is what."

D. *Word Choice in Genus and Differentia.* Try as much as possible not to defeat the purpose of a definition by using difficult, unfamiliar terminology in the genus and differentia. The nonbotanist, for instance, might be confused rather than helped by: "A septum is a transverse wall in a fungal hypha, an algal filament, or a spore." And everyone remembers Samuel Johnson's classic: "A network is any thing reticulated or decussated, at equal distances, with interstices between the intersections."

*Amplified Definition.* Although brief informal definitions or sentence definitions are usually adequate explanations of the unfamiliar in technical writing, there are occasions when more than a word, phrase, clause, or sentence is needed in order to ensure a reader's understanding of a thing or idea. If you think that a sentence definition will still leave a number of questions unanswered in the reader's mind—questions that he ought to have answers to—then an amplified or extended definition is required.

A term like "drift meter" provides an example. A formal sentence definition goes like this: "A drift meter is an instrument used in air navigation to measure the angle between the heading of a plane and the track being made good." It is easy to imagine a reader who would be dissatisfied with this as an explanation, especially if it occurred in a report particularly concerned with the subject of aircraft instruments. He might very well ask, How does it work? What does it look like? What are its parts? Answers to questions of this sort would result in an amplified definition. Here is what the author of the above sentence definition said in his discussion of the term:

> The simplest form of drift meter consists of a circular plate of heavy glass set in the floor of the cockpit in front of the pilot. The plate may be rotated within a ring on which degrees of angle are marked to the left and right of a zero mark. This zero point is in the direction of the forward end of the longitudinal axis of the plane. The plate has a series of parallel lines ruled on it. With the plane in level flight the pilot can look down through the plate and rotate it until objects on the ground are moving parallel to the lines. Under these conditions the lines on the plate will be in the direction of the track being made good, and the angle between the heading and this track may be immediately read on the scale.
>
> Many modern and complicated types of drift sights have been devised, but all of them operate on the fundamental principle described above. In some modern drift sights, a gyroscopic stabilizing system holds the grid lines level even though the plane is not flying level. Astigmatizers are frequently incorporated to assist in measuring drift angle, particularly when flying over water.
>
> In some modern drift sights a system is incorporated so that ground speed may be determined. A pair of wires is marked on the grid, perpendicular to those set parallel to the apparent motion of the ground. The time required for an object on the ground to move from one of these wires to the other will be proportional to the ground speed. The dis-

tance of the plane from the ground must be accurately known, and the objects observed must be directly below the plane to obtain an accurate value of ground speed.*

Below is a second example of an amplified definition. This one is more complex in structure. The amplified definition of the principal term, "closed loop," includes a classification ("regulator," "servomechanism"); and each of the terms in this classification is in turn defined. This author also has a sense of humor; but perhaps that is indefinable.

All regulating systems are closed loop systems. Closed loop systems, consequently regulating systems, are common in our everyday lives. For example, while driving an automobile the driver judges the degree of pressure to exert on the steering wheel by considering such factors as road surface and curvature, speed of the automobile, foreign objects on the road, etc. The driver's eyes monitor these factors and, through the human system, control his limbs to allow safe travel. This is a closed loop system. However, if the driver were to close his eyes, the result would be an open loop system. An open loop system is one which does not allow for unpredictable necessary corrections and is not recommended while driving an automobile. This chapter will be concerned with closed loop systems which do not rely on the human for completing the loop.

There are three terms which are used to describe closed loop systems: regulator, servomechanism, and servo. A *regulator* is a closed loop system that holds at a steady level or quantity such elements as voltage, current or temperature, and often it needs no moving parts. A *servomechanism* is a closed loop system that moves or changes the position of the controlled object in accordance with a command signal. It includes some moving parts such as motors, solenoids, etc. The term *servo* has been generally adopted to mean either a regulator or a servomechanism and will be used as such in this chapter.**

There is no single way to go about amplifying a definition. You must use your own judgment in determining how much, and what, to say. Examination of many definitions, however, does indicate that the following techniques often prove useful.

A. *Further Definition.* If you think that some of the words in a definition you have written may not be familiar to your reader, you should go on to explain them (some readers, for example, might like to have the word "astigmatizers" explained in the above definition of drift meter).

**Van Nostrand's Scientific Encyclopedia,* 3d ed. (New York, 1958), p. 541. Quoted by permission of D. Van Nostrand Company, Inc.

***SCR Manual,* 4th ed. (Syracuse, N.Y.: Semiconductor Products Department, General Electric Co., 1967), p. 263. Quoted by permission of the General Electric Co.

B. *Concrete Examples and Instances.* Since sentence definitions are likely to be abstract statements, they do not contain concrete examples of the thing being defined. It helps, therefore, to give the reader some specific examples. As a matter of fact, this technique is probably the best of all.

C. *Comparison and Contrast.* Since we tend to relate—or try to relate—new things and experiences to those we already know, it helps to tell a reader that what you are talking about is like something he already knows. Remember that the relationship must be one of the unfamiliar to the familiar. If you were attempting to explain what a tennis racket is to a South Sea islander, it wouldn't help much to compare it to a snowshoe! On the other hand, it may be better to stress the differences between the things compared. See (E) below.

D. *Word Derivation.* It rarely happens that information about the origin of a word sheds much light on its present meaning, but sometimes it does and the information is nearly always interesting. Take the term "diastrophism" for instance. It comes from the Greek word *diastrophe* meaning "distortion" and ultimately from *dia* meaning "through" and *strephein* meaning "to turn." Thus the word appropriately names the phenomenon of deformation, that is, "turning through" or "distortion" of the earth's crust, which created oceans and mountains. As you know, etymological information may be found in any reputable dictionary. Whether you use it in developing a definition or not, it is worth noting.

E. *Negative Statement.* Negative statement is mentioned in many books as a possible means of developing a definition. Sometimes it is called "obverse iteration," sometimes "negation," and sometimes "elimination." Whatever it is called, you should realize that you will never really get anywhere by telling what something is not. But in some cases you can simplify the problem of telling what something *is* by first clearing up any confusion the term may have in the reader's mind with closely related terms. You might, for instance, say that a suspensoid is not an emulsoid, but a colloid dispersed in a suitable medium only with difficulty, yielding an unstable solution which cannot be re-formed after coagulation. An emulsoid is a colloid readily dispersed in a suitable medium which may be redispersed after coagulation.

F. *Physical Description.* We mentioned earlier that you could scarcely give a reader a very thorough understanding of a micrometer caliper without telling him what it looks like. Note the treatment of drift meter. So it is with virtually all physical objects.

G. *Analysis.* Telling what steps comprise a process, or what functional parts make up a device, or what constituents make up a substance obviously helps a reader. This technique is applicable to many subjects: a breakdown of a thing or idea permits the reader to think of it a little at a time, and this is easier to do than trying to grasp the whole all at once.

H. *Basic Principle.* Explaining a basic principle is particularly applicable to processes and mechanisms. Distillation processes, for instance, make use of the principle that one liquid will vaporize at a different temperature than another.

I. *Cause and Effect.* Magnetism may be defined in terms of its effects. In defining a disease, one might very well include information about its cause.

J. *Location.* Although of minor importance, it is sometimes helpful to tell where a thing may be found. Petalite, for instance, is a mineral found in Sweden and on the island of Elba; and in the United States at Bolton, Massachusetts; and Peru, Maine; vast deposits are located in Southern Rhodesia and South Africa.

The foregoing techniques do not exhaust the possibilities for amplifying a definition. Anything you can say which will help the reader comprehend a concept is legitimate. We have seen mention of authorities' names (in a definition article on the incandescent lamp, it would be natural to find Edison's name mentioned), history of a subject, classification, and even quotations from literature on a subject used to good advantage. Nor should every one of these techniques be employed in any given case, necessarily; often only a few of them would be pertinent. You will have to depend upon your own judgment to decide how much you need to say and what techniques are best suited in a specific situation.

Two organization patterns are possible for amplified definitions. The first pattern begins with the formal sentence definition and proceeds with supporting discussion. A glance at the definition of drift meter given earlier shows that it is organized in this fashion. After the initial sentence definition there follow in combination the simple explanation of the basic operating principle, description of the functional parts of the device, and method of use. These are then followed by mention of more complex types of drift meters and reference to special uses. In a general way, this pattern or organization may be regarded as deductive in that it begins with a statement regarded as true and proceeds to the particulars and details. Altogether, it is a method to be preferred over the second, or inductive, pattern of organization which places the sentence definition last, as the conclusion to the evidence presented. The deductive method is preferred because there is no point in keeping the reader waiting for information he wants. Where the inductive method is used, the issue is in doubt, in a sense, until the last sentence is reached.

## Placing Definitions in Reports

Very often it is difficult to decide where to put definitions in reports. There are three possibilities: (1) in the text, (2) in footnotes, and (3) in a glossary at the end of the report, or in a special section in the introduction.

If the terms requiring definition are not numerous and require brief rather than amplified definition, it is most convenient to place explanatory words or phrases in the text itself as appositives (set off with commas or parentheses). If you are not sure whether your readers know a term, or if you feel that some readers will know it and some will not, it is probably best to put the definition in the form of a footnote with a numeral or some suitable designating mark or symbol after the word itself in the text. If placing definitions in the text would result in too many interruptions, especially for the reader who may know them, it is a good idea to make a separate list to be put in an appendix. If there are a number of terms of highly critical importance to an understanding of your report, they may be defined in a separate subdivision of the introduction of the report. An introduction to a report on, say, a bridge construction may contain a statement like this: "Concrete, in this report, will mean . . ." with the rest of the statement specifying the composition of the mix.

The point of all this is that definitions should be strategically placed to suit your purposes and the convenience of your readers. Once you decide on the importance of the terms you use and the probable knowledge of your readers, you will find it easy to decide where to put the definitions.

## Summary

Definition is needed when familiar words are used in an unfamiliar sense or for unfamiliar things, when unfamiliar words are used for familiar things, and when unfamiliar words are used for unfamiliar things. The question of familiarity or unfamiliarity applies in all cases to the reader, not the writer. Definitions may be either informal (essentially the subsitution of a familiar word or phrase for the unknown term) or formal. Formal definitions always require the use of a "sentence definition," which is comprised of three principal parts: species, genus, and differentia. Sometimes it is necessary to expand a formal definition into an article. An article of definition may be developed by either the deductive or the inductive method, the deductive being generally preferable. Definitions may appear in the text of a report, in footnotes, in a glossary at the end of the report, or in a special section in the introduction. Their proper location depends upon their importance to the text and on the knowledge of the readers.

## SUGGESTIONS FOR WRITING

1. Very often it is necessary in technical and scientific presentations to give the reader an *ad hoc* or working definition of a term. Such definitions do not need to be as full or complete as an amplified definition; they serve to tell the reader what you mean by a term in the particular document you are writing. Try your hand at such a definition for the terms listed below (you can define the context):

| | |
|---|---|
| concrete (noun) | acid |
| radiation | immunity |
| chronometer | rubber |
| rocket | radioactivity |
| catalyst | reforestration |
| gas | conservation |
| feedback | nucleus |

2. Write an extended or amplified article of definition of a key term or concept in your major field of study (perhaps a central concept from one of the courses you are taking will work well for this assignment). We suggest 200 to 300 words.

3. Assume a reader who is unfamiliar with one of the terms listed below and write a definition that is thorough enough to enable him to understand the term:

| | |
|---|---|
| probable error | solid state |
| liquefaction | transistor |
| hydrocarbon | fault (geology) |
| corrosion | binary system |
| dielectric | computer |

4. If you are at work on a research paper, prepare a glossary of the key technical terms that will appear in your report.

5. Write sentence definitions of five of the following terms, taking care to select a suitable genus for each and making sure that the differentia actually distinguishes the term from other members of the genus:

| | |
|---|---|
| resonance | data processing |
| capacitance | quantum |
| accuracy | lift-off |
| reliability | spin-off |
| parameter | noise (in communications) |
| ablation | heat shield |
| orbit | relevance |

# 6 Description of a Mechanism

This chapter brings us to the second of the special techniques of technical writing—the description of a mechanism. What we mean by "mechanism" scarcely requires explanation. For the sake of the record we might say that a mechanism may be either simple or complex, and either large or small. But the principles of the description of a mechanism apply equally well in any case. In fact, the general procedure in the description is quite simple, and in practice the chief difficulty lies in writing sentences that really say what you want them to say. There is no more fertile field for "boners."*

The three fundamental divisions of the description are the introduction, the part-by-part description, and the conclusion. Before discussing these divisions in detail, we should like to remind you of two things. The first is that a description of a mechanism almost never constitutes an entire report by itself. For practice in the technique, it is wise to write papers devoted exclusively to the description of a mechanism, but it should be understood that such papers will not constitute complete reports of the type found in actual use. The second reminder is that what needs to be said in the description always depends on what the reader needs to know. For example, your reader might want to construct a similar device himself. This would require a highly detailed treatment. Or he might be chiefly interested

*The following extract from a student paper suggests the possibilities: "The Dragoon Colts were issued to the army and sold to civilians equipped with shoulder stocks that locked into the butts to make short rifles out of them."

in knowing what the device will do, or can be used for, and desire only a generalized description. Such is the description of a slide rule, which is often included in the manufacturer's directions for its use.

## The Introduction

Because the description of a mechanism seldom constitutes an article or report by itself, the introduction required is usually rather simple. Nevertheless, it is very important that the introduction be done carefully. The two elements in the introduction that need most careful attention are (1) the initial presentation of the mechanism, and (2) the organization of the description.

*The Initial Presentation.* When your reader comes to a discussion of a mechanism unfamiliar to him, he will immediately need three kinds of information about it if he is to understand it easily and well. He will need to know (1) what it is, (2) what its purpose is, and (3) what it looks like.

The problem of identifying a mechanism for the reader is simply a problem of giving a suitable definition. If the reader is already familiar with the name of the mechanism and knows something about the type of mechanism it is, all you need do is write the differentia. For example, if you were about to describe some special type of lawn mower for an American reader, you would not need to define "lawn mower," but you would need to differentiate between the type you were describing and other types of lawn mowers with which the reader was familiar. Or, if the name of the mechanism to be described is unfamiliar to the reader, perhaps a substitute term will do. Suppose we write in a report, ". . . each of these small boats is equipped with a grains." How should we tell the reader what a "grains" is? We can do so very easily by writing, "A grains is a kind of harpoon." As you no doubt recall, both of these two ways of clarifying what the mechanism is, or defining it, were discussed in the preceding chapter on the subject of definition.

The reader must also know the purpose of the mechanism. Often, an indication of purpose will appear as a natural part of the statement of what the mechanism is. For instance, to say that a grains is a harpoon indicates something about its purpose. To take another example, let's suppose we are writing a description of the Golfer's Pal Score-Keeper. Here, the purpose is suggested by the name itself. It is frequently desirable, however, to state the purpose of the mechanism explicitly. In writing about the Golfer's Pal Score-Keeper we might be more certain that its purpose was clear by stating that this score-keeper is a small mechanical device that a golfer can use instead of pencil and paper for recording each stroke and getting a total. The purpose of a mechanism is often clarified by a statement about who uses the device, or about when and where it is used.

Finally, as the mechanism is initially presented to him, the reader needs a clear visual image of it. The most effective way to give a reader a visual image of a mechanism is to let him see a photograph of it—assuming he can't examine the thing itself or a model of it. A drawing would be second best. Our interest, however, is in creating the visual image with words. Photographs and drawings are more effective than words for this purpose, and should be used if possible, but expense, or the need for haste, or the lack of facilities often rule out the use of such visual aids. A very interesting example of the practical importance of the visual image may be found in Appendix D under the heading, "Information Presentation."

In the initial presentation of a mechanism, the visual image created by words should be general, not detailed. There will be time enough for details later on. Fundamentally, there are two ways of creating this general image. One is to describe the general appearance of the device; the other is to compare it with something which is familiar to the reader. You must be careful, of course, not to compare an unfamiliar thing with another unfamiliar thing. Reference to the Score-Keeper again suggests how illuminating a good analogy can be: "This device is very much like a wrist watch in size and general appearance." To this comparison might be added some such direct description as the following: "It consists of a mechanism enclosed in a rectangular metal case—1 5/16 in. long, 7/8 in. wide, and 1/4 in. thick—to which is attached a leather wristband."

We remarked earlier that stating what a mechanism is constitutes a problem in definition. Now, in concluding these comments on the initial presentation of a mechanism, we should acknowledge that references to purpose and appearance are among the methods of making a definition that were discussed in the preceding chapter. It becomes apparent that there is a close similarity between acquainting a reader with a mechanism new to him and acquainting him with a term or concept new to him.

*Organization of the Description.* It is possible to divide almost every mechanism into parts. Such division is an essential part of a detailed description. In the introduction to a description, a statement of the principal parts into which the mechanism can be divided serves two purposes. The first is that it is an additional way, and an important one, of giving the reader a general understanding of what the mechanism is. From this point of view, what we are saying here actually belongs under the preceding heading ("The Initial Presentation"). The second purpose is to indicate the organization of the discussion that is to follow. The reader is always grateful for knowing "where he's at." Since it is logical to describe the principal parts one at a time, a list of the principal parts in the order in which you wish to discuss them is a clear indication of the organization of the remainder of the description. The list of principal parts should be limited to the largest useful divisions possible. The principal parts of a slide rule, for instance, might be listed as the rule, the slide, and the indicator. Later on the rule and the indicator could be broken down into subparts. In practice, engineering assembly prints can be a highly valuable source of lists of principal parts.

The order in which the parts are taken up will normally be determined by either their physical arrangement or their function. From the point of view of physical arrangement, an ordinary circular typewriter eraser with brush attached might be divided as follows: (1) the metal framework which holds the eraser and brush together, (2) the eraser, and (3) the brush. The metal framework comes first because it is on the outside. From the point of view of function the eraser might come first; then the brush, which is used to clean up after the eraser; and last the metal framework.

Finally, you should make sure that the list of principal parts is in parallel form. It is hard to make a mistake in this because the list will almost inevitably be composed of names—the names of the parts; nevertheless it might be well to check your list. The list is usually in normal sentence form, like this: "The principal parts of the slide rule are (1) the rule or 'stock,' (2) the slide, and (3) the indicator." But if the parts are numerous, it may be preferable to abandon the sentence form and make a formal itemized list, like this:

The principal parts of the slide rule are the following:

1. The rule or "stock"
2. The slide
3. The indicator

## The Part-by-Part Description

The introduction being out of the way, and the mechanism logically divided into parts, we are ready to take up the description of the first part. But the fact is that now, so far as method goes, we start all over again, almost as if we hadn't written a line. For what is the "part" but a brand-new mechanism? The reader wants to know what it is. So we must introduce it to him.

We have divided the slide rule—say—into the rule, the slide, and the indicator and are about to describe the rule. The first problem is to tell the reader what the rule is, and then to divide it into subparts. The general procedure will be—as before—to define the part, to state its purpose, to indicate its general appearance (preferably by a comparison with an object with which the reader is familiar, perhaps an ordinary foot ruler), and finally to divide it into subparts.

And what do we do with the subparts? The same thing exactly. In other words, the mechanism as a whole is progressively broken down into smaller and smaller units until common sense says it is time to stop. Then each of these small units is described in detail.

By this time you may have a mental image of a chain of sub- and sub-subparts stretching across the room with a detailed description glimmering faintly at the end. That certainly isn't what we want. Nevertheless, we do want to emphasize the value of breaking the mechanism down into parts before beginning a detailed description. But, if the breaking-down procedure goes very far before you're ready to describe, it probably means that the principal part with which you started was too broad in scope. You need more principal parts. Although we urge the value of this system as a general policy, it is simply not true that all description must be handled in this way. Sometimes, for example, instead of giving a preliminary statement of *all* the subparts that will be described in a given section of the description, it is desirable not to mention a certain minor subpart at all except when you actually describe it.

"Described in detail" means careful attention to the following aspects of the mechanism:

Shape
Size
Relationship to other parts
Methods of attachment
Material
Finish

Each of these matters needn't be labored over mechanically, in the order stated, in every description. Which ones need attention, and what kind of attention, depends—as always—upon the reader and the subject. For instance, let's take the term "material" in the list above. The discussion so far has implied that the material of which a mechanism is constructed is not discussed until the mechanism has been divided into its smallest components. But if you were describing an open-end wrench made of drop-forged steel, it would seem unnatural to wait until you were taking up one of the smaller parts to let the reader in on the fact that the whole wrench was drop-forged steel.

The same line of reasoning can be applied throughout the description. There is no formula which will fit every situation. The important thing is to decide what information the reader needs, and to give it to him in as nearly crystal-clear a form as you can.

## The Conclusion of the Description

The last principal function of the description of a mechanism is to let the reader know how it works, or how it is used, if this hasn't been done in the general introduction. Emphasis should naturally fall upon the action of the parts in relation to one another. This part of the writing constitutes in effect a description of a process, usually highly condensed (see next chapter).

## Summary of the Principles of Organization

The outline below indicates in a general way the organization of the description of a mechanism. As has been explained, the order of some of the topics listed and the inclusion or exclusion of certain topics depend upon the situation. This outline is to be taken as suggestive, not prescriptive.

### Description of a Mechanism

I. Introduction
  A. What the mechanism is
  B. Purpose
  C. General appearance (including a comparison with a familiar object)
  D. Division into principal parts

II. Part-by-part description
  A. Part number one
    1. What the part is
    2. Purpose
    3. Appearance (including comparison)
    4. Division into subparts
      a. Subpart number one
        (1) What the subpart is
        (2) Purpose
        (3) Appearance (including comparison)
        (4) Detailed description
          (a) Shape
          (b) Size
          (c) Relationship to other parts

(d) Methods of attachment
(e) Material
(f) Finish

b, c, etc.—same as "a."

B, C, etc.—same as "A."

III. Brief description of the mechanism in operation

## Some Other Problems

*Style.* By far the most difficult problem in describing a mechanism is simply to tell the truth. The writer is seldom in any doubt as to what the truth is; he wouldn't be writing about a mechanism unfamiliar to him. But it is one thing to understand a mechanism and another to communicate that understanding to somebody else. Only painstaking attention to detail can ensure accuracy.

It is probably a mistake, however, to try to be perfectly accurate in the first draft of a description. Write it as well as you can the first time through, but without laboring the details; then put it away for as long as you can. When you read it over again, keep asking yourself if what the words say is what you actually meant. At especially critical points, try the experiment of putting what you have said into the form of a sketch, being guided only by the words you have written. Sometimes the results are amazing in showing how the words have distorted your intended meaning.

Whenever you see the letters "ing" or "ed" on a word, watch out for a booby trap (specifically, a dangling modifier). And make sure that every pronoun has an easily identified antecedent.

Finally, don't forget to watch the tense. Usually the entire description will be in the present tense. Occasionally it will be past or future. But almost invariably the tense should be the same throughout the description.

*Illustrations.* People who like to draw and do not like to write are often loud in argument as to the waste of writing anything at all when a drawing would do. We ourselves are rather sympathetic toward this attitude; but the trouble consists in deciding when the drawing will do.

First of all is the question of plain facts. Sometimes it is difficult or impossible to show in a drawing how a device functions or how much tension is found on a certain fitting (where a torque wrench might be used). Words are usually much better than drawings for such matters.

There is again a psychological problem. Some people seem to have a greater aptitude for comprehending things in verbal form than in graphic form, and vice versa; just as some people more readily comprehend the language of mathematics than they do the language of words, and vice versa.

Certainly, the wisest course is to use every means of communication at your command if you really want to make yourself understood. The corollary is to use discretion; you don't want to swamp your reader with either text or drawings.

One of the skills that a technical man should possess is that of effectively relating a written discussion to a drawing. In general, two possibilities are open. One is to print the name of each part of the device on the drawing; the other is to use only a symbol. In other words, if you were discussing the indicator on a slide rule, you might write, "The indicator (see Figure 1) is . . . ." Or, if you had used only a symbol on the drawing, instead of the name, you might write, "The indicator (Figure 1-A) is . . . ." If there is only one figure in the report it need not be numbered. You could then write, "The indicator (A) is . . . ."

Information about the form of drawings and other illustrations can be found in Chapter 21.

A problem that comes up in every description is how many dimensions to indicate, both in the text and on the illustration. A decision must be based upon the purpose of the description. If you anticipate that the description may be used as a guide in construction, then all dimensions should be shown on the drawing and a great many stated in the text.

*Problems of Precision and Scope.* Finally, we want to return to two problems that were mentioned briefly in the first two paragraphs of this chapter. One of these is what we called the problem of telling the truth about a mechanism. The other has to do with the amount of detail in a description.

What does telling the truth mean here? First, it means *seeing*. Second, it means *communicating*. We'd like to make a personal suggestion about this problem.

Don't be surprised if you find yourself having something of an emotional experience over your first attempt to describe a mechanism. If, for example, after working hard at your description, you show it to somebody (a friend, an instructor, a supervisor on your job), what happens next may assume the proportions of a disaster. The person to whom you show it may announce that he cannot understand it at all. If the person to whom you show your description is a friend who is too good-natured to risk hurting your feelings, you can challenge fate by asking him to make a drawing of the mechanism, relying on nothing but your written words for guidance. This is almost guaranteed to result in disaster.

We are only partly joking in using the word "disaster." A very real disaster can, in fact, occur. It does occur if, having run into trouble with your description, you decide that you are simply unable to write a good one. Writing a description of a mechanism is, like most other writing, merely a skill. It is not a divine gift. It requires practice. It requires practice in seeing, and in communicating.

It is no doubt especially a shock for a person to discover that he hasn't really been seeing what he is looking at; and yet this discovery is an almost universal experience. Let's consider an example. In the "Illustrative Material" section below, you will find a long, detailed description of a simple mechanism, a patented paring knife. It is accompanied by illustrations so the details of the writing can be checked. In describing the blade, the author wrote, "the sharpening has been done by grinding the convex side of the slit. As a result, the convex side is flattened along the slit." Is this a true statement? Not exactly. It at least implies that if a straightedge were laid across the area of the blade that was ground away, this area would prove to be uniformly flat. But this implication is misleading. Careful observation of the illustrations reveals that the flattened area on one side of the slit is set at an angle to the flattened area on the other side. Does this inaccuracy in the description matter? Yes, it probably matters a lot. First, these angles presumably are important in the way the blade functions, in the "bite" it takes. Second, the width of the slit—also important in the way the blade functions—is partly dependent on the angle at which the grinding is done.

Did the author fail to see this detail, or only fail to communicate it? There is no way of knowing. But it does seem likely that if he had seen it, he would have given at least some evidence of trying to communicate it.

One suggestion that you may find helpful in training yourself to see what you are looking at is to verbalize. Talk to yourself (silently, preferably!). Instead of just looking at the blade of the knife (for example), ask yourself questions, in words, like—"What is the shape of the cross section of this blade?" "Why does the slit down the center of the blade have this particular width?" Keep talking, and it will help you see.

We've been thinking about accuracy of observation. Let's shift attention to the closely related subject of successfully communicating what is seen. Now the reader becomes our chief concern. Here are two ideas you may find especially useful.

The first is to be very careful about the physical point of view or orientation from which you are asking the reader to look at the mechanism. Obviously, if you ask the reader to start by looking at one end of a mechanism and then, without telling him, you suddenly begin to describe the other end, he is likely to become confused. An example of this need for control of orientation can be seen in the description of the handle of the paring knife just mentioned. Further examples are found in the following interesting comments on problems of orientation by John Sterling Harris.

> Some mechanisms have a natural orientation, some have a variety of natural orientations, and some have no natural orientation at all. Automobiles and airplanes have a natural front and back, top and bottom, right and left, but a rifle has one top in the firing position, and another top in the military *order arms* position. Such items as rivets, pliers and sleeve bearings have no recognizable natural orientation. In description, it is nearly always best to use the natural orientation of the device, if the device has a natural orientation. For mechanisms with two or three natural orientations it is necessary to state clearly which is being used: "With the rifle in the firing position, you will see the sights on the top of the barrel. The blade sight is at the front or muzzle end. The V notch is at the rear or breech end." For mechanisms having no natural orientation, it is necessary to establish a temporary or arbitrary orientation: "To understand the parts of the can opener, hold the can opener in front of you with the key to your right and the turning axis of the key running from left to right. Then turn the longer end of the body of the can opener so that it points upward."
>
> For such orientation, the X, Y and Z axes used in mathematics are often useful; however, the writer needs to be sure that the reader understands what an X axis is. It may be preferable to refer to the three axes as vertical, horizontal and transverse. It is still necessary, however, to relate the device to the axes: "In this discussion, the length of the hammer handle will be called the longitudinal axis, the length of the head will be called the vertical axis, and the width of the head will be called the transverse axis." Without such explanation, the reader might assume that the length of the handle was the vertical axis and the length of the head the transverse. The result would be confusion. Incidentally, a review of basic terms of solid geometry can aid the writer who is going to describe a mechanism.
>
> Specialized fields often have their own established set of orientation terms. For example, in anatomy there are such terms as dorsal and ventral, anterior and posterior, proximal and distal. To the specialist, such terms indicate direction as clearly as the sailor's topside and below, forward and aft and amidships and abeam. However, using such terms outside their own field may prove unwieldy or incongruous. A sailor on watch can say, "Periscope, two points off the starboard bow." But an Air Force gunner can more naturally say, "Bandit, one o'clock high." The

gunner conveys the same information as the sailor while adding data on relative altitude.

A related problem arises from confusing two-dimensional and three-dimensional figures. Thus it is not quite accurate to speak of a hockey puck as round (viewed radially it has a rectangular silhouette) but as cylindrical. A similar kind of problem arises in calling a chisel a pointed rather than an edged tool. Such ambiguities are especially likely to occur when you are working from orthographic drawings rather than from the mechanism itself.*

The second suggestion we have to make about communicating accurately, a very brief one, has already been noted by implication. It is actually a means of testing the accuracy of your description after you have completed it. You can do this simply by making a drawing based entirely on what you have written, being careful not to draw anything you haven't put into words. In this way you can not only help yourself to detect omissions and ambiguities but to see the whole description more nearly as your reader will see it.

Finally, we consider the problem of the scope of a description. All descriptions of mechanisms fall somewhere between two extremes. At one extreme is the detailed description of a particular individual mechanism, including consideration of its condition at the time of the description. At the other extreme is a generalized description intended to convey only a reasonable understanding of what the mechanism is. Imagine, for example, that someone is preparing a catalog of antique firearms. One item for the catalog is a rare blunderbuss worth thousands of dollars. It would be described with care, including its condition. Another item is a not-so-rare musket manufactured in the 1850s. The dealer has twenty of them. Here the description would probably not be of an individual weapon, but of the model in general. Between these two extremes, examples could be found of almost any degree of compromise between detail and generalization.

The problem confronted here is only a special form of the familiar problem of adaptating your writing to your reader. All you need do is apply common sense. Ask yourself how much detail the reader wants, and what kind, and then be consistent throughout the description.

## ILLUSTRATIVE MATERIAL

The following pages contain four examples of the description of a mechanism.

*Quoted by permission of the author.

The first, which is concerned with a weight-equalizing hitch for travel trailers, is presented for the specific purpose of illustrating the value of a carefully thought-out analogy in the description of a rather complex mechanism.

The second and third illustrations are both descriptions of the same simple mechanism, a patented paring knife. One is highly detailed, the other generalized.

The fourth illustration is a generalized description of a complex mechanism.

### *Illustration 1—Analogy*

## Engineering Principles of Weight-Equalizing and Sway Resistant Hitches*

*by T. J. Reese*

### What is a Weight-Equalizing Hitch?

A weight-equalizing hitch is a mechanical device usually consisting of three units:

1. A TOW BAR that is either welded or bolted to the rear of the frame of the car and extending forward within an inch or two of the rear axle housing. This tow bar is made of material that is strong enough to take the force imposed on it that would tend to spring or bend it. (Mr. J. R. O'Brien has outlined the typical method of installing to the car frame or body.)
2. REMOVABLE BALL MOUNT. This is the part that has the hole for the trailer hitch ball and also suitable mechanism for attaching the spring bar members.
3. SPRING BAR ASSEMBLY, or other flexible device, that is attached to the ball mount in an articulated manner, the other end being supported by a flexible member such as a chain which is attached to a trailer frame bracket.

. . .

*The passage quoted, which is only a small portion of the paper read at the SAE meeting, opens with a division of the weight-equalizing hitch into its principal parts. One fact which is perhaps not entirely clear within this passage is that the removable ball mount is attached to the tow bar, which in turn is either bolted or welded to the frame of the car.*

*Having itemized the principal parts, the author presents a general idea of how a weight-equalizing hitch distributes the hitch weight of a travel trailer over all the wheels of car and trailer combined. He does this by using the analogy of a wheelbarrow.*

### How the Weight-Equalizing Hitch Works.

We will assume that we have the tow bar attached very securely to the frame of the car

*Printed by permission of T. J. Reese. The material quoted here is excerpted from a paper read at a regional meeting of the Society of Automotive Engineers. Mr. Reese, Chairman, Reese Products, Inc., is the inventor of the Reese Strait-Line Hitch.

. . . so rigidly, in fact, that any movement that the car makes will be reflected in the movement of the tow bar. In other words, if the car sets level the tow bar will be in a relatively level position. If the car sets low in the back and up in front, that is the whole body is slanted to the back, the tow bar naturally would be tilted to the back.

We will now assume that a pair of wheelbarrow handles are rigidly attached to the ball mount that has been inserted in the tow bar on the car and the car has no load, so it is naturally setting level. The wheelbarrow handles will extend back of the car parallel with the ground. We will now place about 400 pounds of weight in the extreme rear of the trunk. The back of the car will drop 4 to 6 inches, and due to the lever action over the rear axle, the front of the car would raise 1 to 3 inches. With the car thus loaded, the ends of the wheelbarrow handles would be slanted down, either resting on the ground or at least very little above it. Now, if a man were strong enough, he could bring the back of the car up to level by lifting up the end of the handles. Not only does this remove weight from the rear axle and wheels, but due to the lever length of the overhang of the rear of the car, plus the length of the wheelbarrow handles, it transfers much of the weight to the front wheels of the car. It can be more easily understood if the lift on the handles were exaggerated to the point that the rear wheels of the car were lifted free of the ground. The entire weight of the car and trunk load would be supported by the front wheels of the car and the one supporting the wheelbarrow handles.

Let us now substitute a trailer connected to a hitch ball for the weight in the trunk, and assume that the hitch weight of the trailer is the same 400 pounds that was removed from the trunk of the car. The wheelbarrow handles will again assume the tipped down position and again could be lifted up so that the car would be level; also the trailer will set level, if the hitch has been properly installed.

Then we will substitute tapered spring bars for the rigidly attached wheelbarrow handles.

These spring bars are attached to the ball mount in such a manner that they have no movement in a vertical plane, but are free to swing from side-to-side in a horizontal plane.

We will substitute chains and frame brackets for the strong man holding the wheelbarrow handles. [The chains are attached to the ends of the spring bars, and then to the brackets mounted on the tongue of the trailer.] In other words, we lift up the end of the spring bars with chain attached and hook the proper link onto the frame bracket. Since the spring bars are approximately 30″ long and are pulling down on the trailer frame 30″ back of the hitch ball, a portion of the hitch weight is transferred through the trailer frame rearward to the trailer wheels.

### *Illustration 2—Detailed Description*

## The A & J Paring Knife

One of the handiest knives to have around a kitchen is a patented paring knife that is identified only by the letters A & J and the patent number, both stamped into the handle. It is much like any small paring knife in size and shape but novel in both design and construction. In spite of its novelty, it is made up of the three parts common to almost any kitchen knife: blade, tang, and handle.

The blade is designed to remove a uniformly thin peeling, as from an apple, without the need of any special care to keep the peeling thin or to prevent the blade from cutting too deeply into the fruit. For this purpose the blade is rounded, as if a piece of 3/8-in. tubing had been cut in halves lengthwise and the blade made out of one of the halves. The cutting edges, the tip of the blade, and the shank of the blade will each be considered separately.

There are two cutting edges, but one of the novelties of this knife is that they are not the outside edges of the blade. These cutting edges can easily be visualized by thinking again of the piece of tubing that has been cut in half lengthwise. In one of these halves, a broad slit has been cut from near the tip to near the shank of the

*This description of a paring knife, and the much shorter one that follows, are characteristic student exercises. The second is also characteristic of a great deal of writing found in routine scientific and industrial reports. This first one, on the other hand, is not as easily defined. It is quite common to find description as detailed as this in reports. It is not common, however, to find so prolonged and exhaustive a treatment of such a simple mechanism.*

*This description should therefore be thought of as essentially an exercise in —as we suggested earlier —seeing and communicating. It is comparable to practicing scales on the piano or tackling a dummy on the football field. For you, it can serve as an exercise in the following ways. First, with-*

blade, and the inner edges of this slit have been sharpened. When the convex side of this blade is moved along the surface of a fruit or vegetable, one or the other of these sharp edges peels the skin away. The peeling comes away through the opening in the blade formed by the slit. The sharpening has been done by grinding the convex side of the blade down to form a cutting edge along each side of the slit. As a result, the convex side is flattened along the slit. The blade is 3 in. long overall and 3/8 in. wide. The slit is 1-7/8 in. long and 1/8 in. wide.

From the tip of the blade to the slit there is a distance of about 5/8 in., and this area has been rounded up toward the end into a shape resembling the tip of a tiny spoon. This shape is apparently intended to be helpful in digging out any small bad spots found in the fruit or vegetable after it is peeled.

The shank serves to fasten the blade to the tang. The shank can be visualized by once more thinking of the piece of tubing cut lengthwise. In effect, about a half-inch of the end of this cut piece of tubing has been rolled further inward around the long axis of the blade to form a cylindrical collar around the tang. The shank is held to the tang by a friction fit.

The tang is a slender steel rod which holds the blade to the handle. It is 3-1/2 in. long and 1/8 in. in diameter. At the end opposite the blade it has a flange, or head—like the head of a nail; and 1/8 in. from this head two small projections or "ears" have been stamped into the tang. The head and projections both help to keep the tang in place in the handle.

The handle actually serves three purposes. In addition to its normal function as a handle, it provides a frame within which the blade and tang assembly can rotate, and also provides stops to limit this rotation to 90 deg. The purpose of this rotation will be explained below. The handle is constructed entirely of a strip of steel 3/64 in. thick and 3/8 in. wide which has been bent lengthwise into the necessary shape. The shape of the handle can be most easily visualized by starting with this strip of steel still in one long, flat piece as it presumably was after

*out trying to comprehend it in detail, examine its organization with respect to the generalized outline of the description of a mechanism presented above. Look for the statement of the principal parts, use of analogy, division into subparts, statement of purpose, and so on. Second, read the description slowly for its content. As you read, keep checking the word description against the illustrations on page 135. And as you read the text and check the illustrations, keep asking yourself if the text is really communicating what the illustrations show. We have already pointed out one place in which the text is inaccurate.*

*You will probably find that reading this description with real comprehension requires intense concentration. Naturally, writing it must also have required intense concentration. Practicing this kind of concentration will help you acquire the ability to write accurately and clearly.*

first being stamped out. At this stage of construction, a hole 5/32 in. in diameter has been punched exactly in the center, and an open-ended slot has been stamped in each end. The inner ends of these slots are semi-circular. These slots are the same size, 5/16 in. long and 1/4 in. wide. The metal left along each side of the slots is only 1/16 in. wide. These slender, prong-like sides ultimately become the stops mentioned previously.

This long, flat strip of steel can next be visualized as bent into a narrow U, about 3/4 in. across, with the punched hole at the bottom. And the following step in the visualization is to close the U by bending the ends of the sides toward one another. Actually, however, two bends are required.

The first might be imagined as the bending of 3/4 in. of each end inward, toward the other, at an angle of about 100 deg. with their respective sides. When this bending has been completed, one end will overlap the other. The amount of this overlapping of the bent pieces should be adjusted in imagination by sliding them together, one over the other, beginning when the ends have just started to overlap. In this position, a rectangular opening (but with semi-circular ends) is formed by the combination of the slots in the ends. As the two pieces are being slid together, one over the other, it is helpful to imagine a nail with the same diameter as the shank of the blade inserted into this rectangular space. The nail should be perpendicular to the bent pieces. When the pieces are moved toward one another this rectangular space decreases in length until the nail is loosely gripped by the two rounded ends of the slots. Together (but one above the other) the two semi-circular ends have now formed an almost perfectly round hole. The nail withdrawn, this is the hole through which the shank itself will be inserted into the handle. The amount of overlap has now been properly adjusted.

The second of the two bends is now to be made, in imagination. It is readily visualized by assuming that a knife-edge is laid across the hole just created, at right angles to the prong-like pieces forming the sides of the upper slot.

Each of these two prongs is then bent upward, away from the handle, to stand parallel to the axis of the blade and tang assembly. A semi-circular hole is thus left at the base of the prongs, its plane at right angles to the blade. This bend might be compared to laying a paper clip on a table, placing a ruler at right angles across about 1/8 in. of the end of the paper clip, and bending the principal length of the paper clip upright. A 1/8 in. semi-circular "hole" would be left at the bottom; the upright sides of the paper clip would correspond to the upright prongs of the knife handle. Each prong is 1/4 in. long.

After the prongs of the uppermost side are bent, the prongs of the other side are to be bent in the same way, again leaving a semi-circular hole. When the inner faces of the prongs of the two sides are then brought together, or mated, an almost perfectly round hole is formed at their base, to accept the shank of the blade. The mated prongs are tack-welded together. Each of the tack-welded pairs serves as a stop to limit the rotation of the blade and tang assembly. The welding gives rigidity to the entire handle.

One other characteristic of the handle is that beginning just back of the end nearest the blade, where the handle is 7/8 in. wide, each side is curved in about 3/16 in. to provide a comfortable grip.

As remarked earlier, the handle constitutes a frame which permits a restricted rotation of the blade and tang assembly. The hole just described, at the base of the prongs, or stops, serves as a bearing for the shank; the smaller hole, at the other end of the handle, serves as a bearing for the tang. That portion of the shank that is faired out to become the blade stands between the two stops, as does the base of the blade itself. The rotation of the blade and tang assembly is limited by the striking of the base of the blade against one or the other of the stops. It is always the concave side of the blade that makes contact.

The blade and tang assembly are fixed in position within the frame by the head and projections on the tang described earlier. The head is on the outside of the handle, the projections on the inside.

For removing the rind, or outer layer, from fruit or vegetables this knife can be used with a rapidity that would be impossible with a conventional knife. Whichever cutting edge is being employed, it is prevented from slicing in too deeply because the opposite side of the blade is passing over the surface ahead of the cutting edge and limiting its angle of attack. The principle of operation is something like that of a reel-type lawn mower, in which the angle of the blade is controlled by the wheels in front and the roller behind. As is now evident, the freedom of the blade to rotate permits an automatic adjustment to irregularities in the surface. The stops inhibit any tendency of the blade to "roll," and permit the application of some twisting force on the blade if that is necessary to help the edge bite in, as at the beginning of a cut.

Of course this knife cannot be used for slicing, at least in any ordinary sense, but only for paring, or for such cutting as making carrot curls, for example.

The handle, which has a decorative "crosshatch" pattern stamped lightly into its outside surface, appears to be nickel plated. The blade and tang have been given no special finish.

### *Illustration 3—General Description*

#### The A & J Paring Knife

One of the handiest knives to have around a kitchen is a patented paring knife that is identified only by the letters A & J and the patent number, both stamped into the handle. As shown in Figure 1 it is much like any small paring knife in size and shape but novel in both design and construction. In spite of its novelty, it is made up of the three parts common to almost any kitchen knife: blade, tang, and handle.

The blade is designed to remove a uniformly thin peeling, as from an apple, without the need of any special care to keep the peeling thin or to prevent the blade from cutting too deeply into the fruit. For this purpose the blade is rounded, as if a piece of 3/8-in. tubing had been cut in halves lengthwise and the blade made out of one of the halves. The cutting edges, the tip of the blade, and the shank of the blade will each be considered separately.

There are two cutting edges, but one of the novelties of this knife is that they are not the outside edges of the blade. Instead, they are the inner edges of the slit (Figure 1-b) that has been cut lengthwise in the blade. When the convex side of the blade is moved along the surface of a fruit or vegetable, one or the other of these sharp edges peels the rind

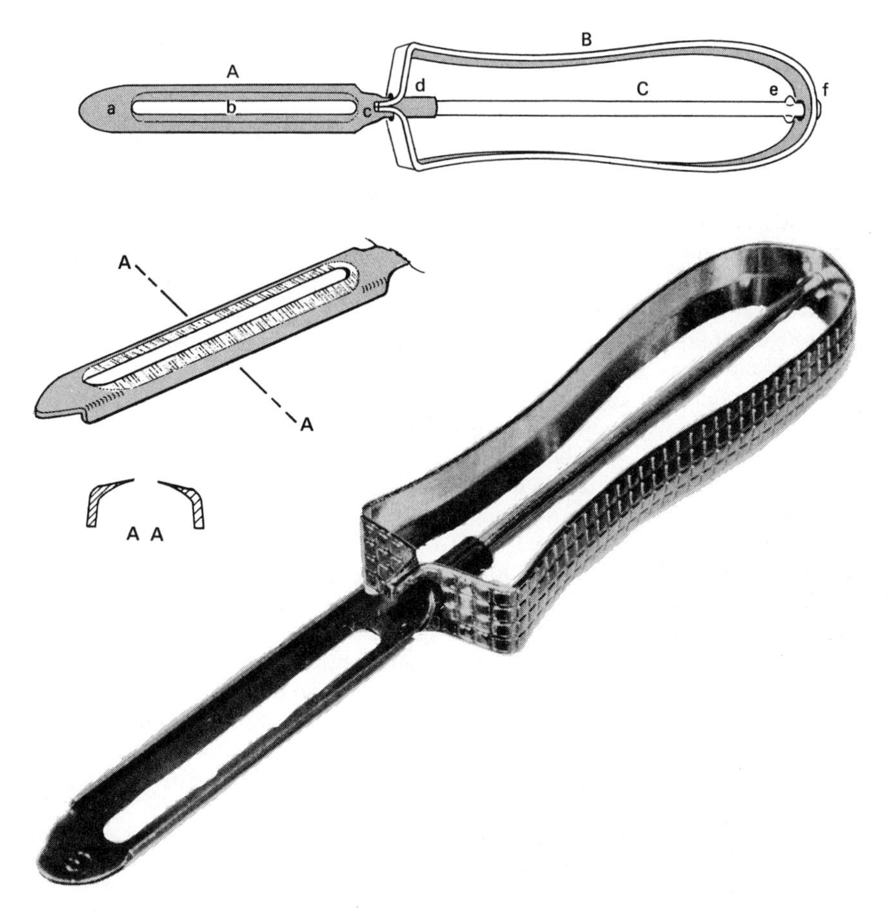


Figure 1. The A & J paring knife. (Top) A: blade; B: handle; C: tang; a: rounded tip; b: slit; c: handle stop attachment; d: shank; e: "ears"; f: outer head. (Middle) Cross-section at AA showing angle at which edges are ground. (Bottom) Photograph showing rounded end of blade and prongs or stops on handle.

away. The peeling comes away through the slit. The sharpening has been done by grinding the convex side of the blade down to form a cutting edge along each side of the slit.

As shown in Figure 1, the end of the blade has been rounded up into a shape resembling the tip of a very small spoon. This shape is apparently intended to be helpful in digging out any small bad spots in the fruit or vegetable after it is peeled.

The shank (Figure 1-d), which serves to fasten the blade to the tang, has been formed, in effect, by rolling the end of the blade around the tang and securing it with a friction fit.

The tang (Figure 1-C) is simply a slender steel rod. It is free to rotate within the handle, and is held in place by a head at the end outside the handle (Figure 1-f) and two "ears" stamped into it just inside the handle (Figure 1-e).

The handle actually serves three purposes. In addition to its normal function as a handle, it provides a frame within which the blade and tang assembly can rotate, and also provides stops to limit this rotation to 90 deg. The purpose of this rotation will be explained below. The handle is constructed entirely of a strip of steel 3/64 in. thick and 3/8 in. wide. As can be seen in Figure 1, holes have been punched in both ends to receive the shank and tang, respectively. At the blade end, the edges of the handle remaining after the hole was stamped out have been bent parallel to the blade and tack-welded together. The prongs thus formed limit the rotation of the blade to 90 deg. by intercepting the movement of the forward portion of the shank, where it is faired out into the blade.

For removing the rind, or outer layer, from fruit or vegetables this knife can be used with a rapidity that would be impossible with a conventional knife. Whichever cutting edge is being employed, it is prevented from slicing in too deeply because the opposite side of the blade is passing over the surface ahead of the cutting edge and limiting the angle of attack. The principle of operation is something like that of a reel-type lawn mower, in which the angle of the blade is controlled by the wheels in front and the roller behind. As is now evident, the freedom of the blade to rotate permits an automatic adjustment to irregularities in the surface. The stops inhibit any tendency of the blade to "roll," and permit the application of some twisting force on the blade if that is necessary to help the edge bite in, as at the beginning of a cut.

Of course this knife cannot be used for slicing, at least in any ordinary sense, but only for paring, or such cutting as making carrot curls, for example.

The handle, which has a decorative "cross-hatch" pattern stamped lightly into its outside surface, appears to be nickel plated. The blade and tang have been given no special finish.

## *Illustration 4—Generalized Description*

### A Program for the Developmental Testing of Turbo-Jet Aircraft Engines*

*by Ephraim M. Howard*
*Allison Division*
*General Motors Corporation*

#### Engine Developmental Test Facilities

*It is unusual to find an entire report devoted to description of a mechanism. The description shown here*

Facilities for engine developmental tests are considered in 2 general areas—ground testing and flight testing (Figure 1). Ground tests are performed in static test cells, altitude test cells,

*Allison Division, General Motors Corporation for *General Motors Engineering Journal*, 3, (Oct.-Nov.-Dec. 1956), 14–17.

and wind tunnels. Flight tests are performed with the engine operating either on an auxiliary mount or serving directly as the aircraft's main power plant.

## Static Test Cell

A static test cell is used for ground developmental tests at zero flight velocity and is composed of an engine mount, sufficient instrumentation to determine test information, a fuel system, and suitable means to control engine operation. A static test cell also includes a filtered air inlet, a plenum chamber, and duct work to carry away the engine's exhaust gases and prevent the possibility of their recirculating back to the engine inlet (Figure 2).

*occupies somewhat less than half of the report in which it appears. It is highly generalized—prepared for a reader who wants information about the general character of the apparatus.*

*The part of the report preceding what appears here was devoted to a discussion of the meaning of developmental testing. The section shown is concerned with the equipment used in developmental testing.*

*In the first paragraph, the major pieces of apparatus to be discussed are listed, and reference is made to Figure 1 (not shown), which is a*

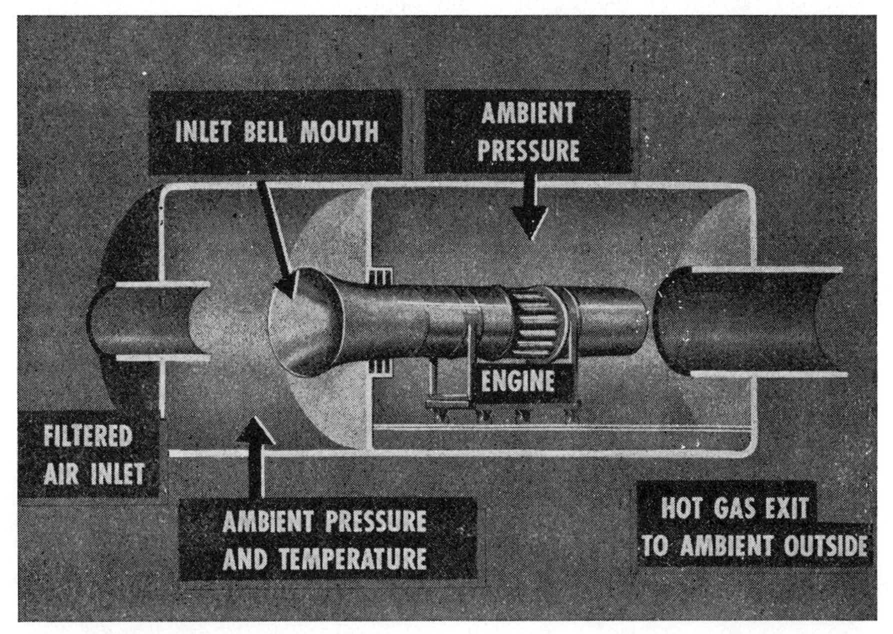


Figure 2. A static test cell is used for ground developmental tests to simulate conditions encountered by a turbo-jet engine in an aircraft at rest. Filtered air at a controlled temperature and pressure is delivered to the engine through a bell-mouth opening. An exhaust duct removes the engine's exhaust gases and prevents their recirculating back to the engine inlet.

Engine thrust is measured by scales or other force-measuring systems, such as strain gages or hydraulic-load capsules. Pressure and temperature pickup probes are provided where needed to determine operating conditions at various locations in the engine. The requisite fuel supply and controls for engine operation also are provided. For safety an engine under test is entirely enclosed by walls; also, sound insulation is provided because of the high noise-level output of turbo-jet engines.

A static test cell for turbo-jet engines can be much simpler in design than the cell shown (Figure 2). Some installations may include only a concrete base on which an engine support is mounted and to which the required services are provided. The extra expense of building a static test cell as shown, however, usually is justified on the basis of safety and efficiency of operation, protection of the engine from damage which might result from ingestion of rocks or other foreign objects, and improvement in test quality by preventing recirculation of exhaust gases to the engine inlet.

Moving the inlet of a static test cell's exit duct (exhaust duct) close to the engine's exhaust nozzle (jet-nozzle exit) extends the range of test conditions obtainable. The hot, engine exhaust gases act as the primary stream of an ejector, and additional air flow is induced into the test cell's exit duct. Air inlets to the aft portion of the test cell are provided so that this air flow can be obtained. This jet pumping, or ejector, action results in a reduction of the static pressure below the ambient value at the engine inlet in the vicinity of the engine jet-nozzle exit. This combination of ambient air inlet and reduced exhaust pressure conditions permits the simulation of some altitude and flight velocity conditions without the use of air exhaust machinery. A static test cell providing such facilities is known as an *ejectorstatic cell.*

As the exhaust duct is moved closer to the engine's jet-nozzle exit, pressure reduction adjacent to the engine jet-exit varies. In the extreme limit the engine jet-nozzle exit may be coupled directly to the exhaust duct with a frictionless slip joint. For this test condition the engine exhaust serves as the primary stream in an ejector with no secondary flow.

*diagram of the relationships of these major components. It would obviously be impractical to try to give a realistic visual image of all this equipment together at the beginning of the description. As you see, however, care has been taken to illustrate the individual major components.*

*Each principal division in the discussion is begun with a statement of definition and purpose and is followed shortly by reference to one of the illustrations. (Figure 3 of the report is not shown here.) To some extent, the material is then further broken down into subparts. For example, under "Static Test Cell" such a list of subparts is given.*

*Incidentally, two common errors appear in the first two sentences under "Static Test Cell." The first sentence is not parallel in structure (here is an improved version: "A static test cell is used for ground developmental tests at zero flight velocity. It is composed of an engine mount. . . ."). The second common error is that the first sentence is not a true statement of fact. As the second sentence makes clear, the static test cell is composed of more parts than the first sentence says it is.*

*As a whole, however, this is a well-written description.*

### Attitude Test Cell

An attitude test cell, which is a variation of a static test cell, permits an engine to be tested in any desired flight attitude and also allows the attitude of an engine to be varied during a test.

A special attitude test cell (Figure 3 top) was used in developing turbo-prop engines for the world's first vertical take-off (VTO) aircraft (Figure 3 bottom). VTO aircraft require engines which can operate for long periods of time in vertical and horizontal positions and also provide reliable operation during transition from the vertical to the horizontal and vice versa. It was through the use of the specially designed attitude test cell that turbo-prop engines providing reliable operation under the various specifications were developed.

### Altitude Test Cell

An altitude test cell is used for engine developmental tests in which the altitude range to be simulated is greater than that provided by the ejectorstatic cell. Where positive flight velocity simulation is required, either ducted-inlet or free-jet test cells may be used. The ducted-inlet type of altitude test cell simulates inlet stagnation conditions while the free-jet type of test cell simulates actual air-stream velocity conditions.

In the ducted-inlet type of altitude test cell (Figure 4) air at the requisite temperature and pressure is provided at the inlet-air plenum chamber. The exhaust chamber is evacuated as required for altitude exhaust simulation. The engine, therefore, is surrounded with an atmospheric environment at conditions corresponding to the required altitude. The choked-nozzle technique may be used to extend the range of simulated altitudes without increasing power or equipment requirements.

In an altitude test cell setup for a free-jet test (Figure 5) air is provided to the plenum chamber at the requisite stagnation temperature and pressure and is then expanded through the free-jet nozzle to the desired stream velocity. The stream of air from the free-jet nozzle is directed

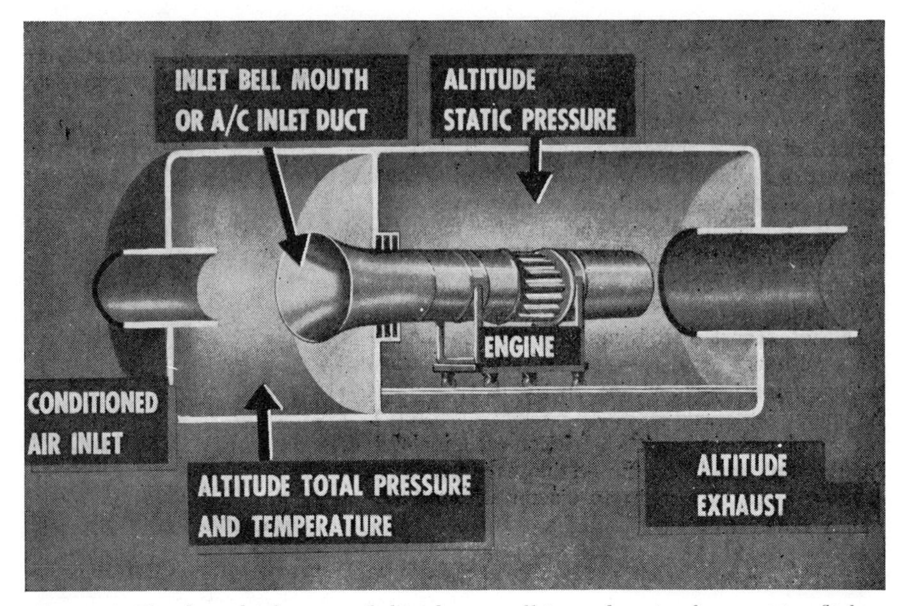


Figure 4. The ducted-inlet type of altitude test cell is used to simulate positive flight velocity conditions encountered by a turbo-jet engine. In the ducted-inlet setup, air is supplied to the engine at the required temperature and pressure corresponding to a specific altitude. The test cell's exhaust chamber, in turn, is evacuated as required to simulate altitude exhaust conditions. The engine, therefore, is surrounded with atmospheric conditions corresponding to a specific altitude.

at the engine-inlet diffuser so that free flight conditions are effectively simulated. For these tests the inlet diffuser is coupled to the engine in the same manner as it would be in the flight vehicle. Performance of the actual configuration of engine-inlet diffuser can then be determined.

With the free-jet test cell setup, engine operation at a specific angle of attack with relation to the air stream also may be studied. The effect of non-symmetrical air flow into the engine due to such operation is very important for an understanding of engine behavior during aircraft maneuvering. Angle-of-attack developmental testing is done by relative misalignment of the free-jet axis and the engine axis. To reduce mechanical complexity it is usually desirable to maintain the engine in a fixed position and vary the position of the free-jet nozzle. In the ultimate setup for this type of altitude test cell the

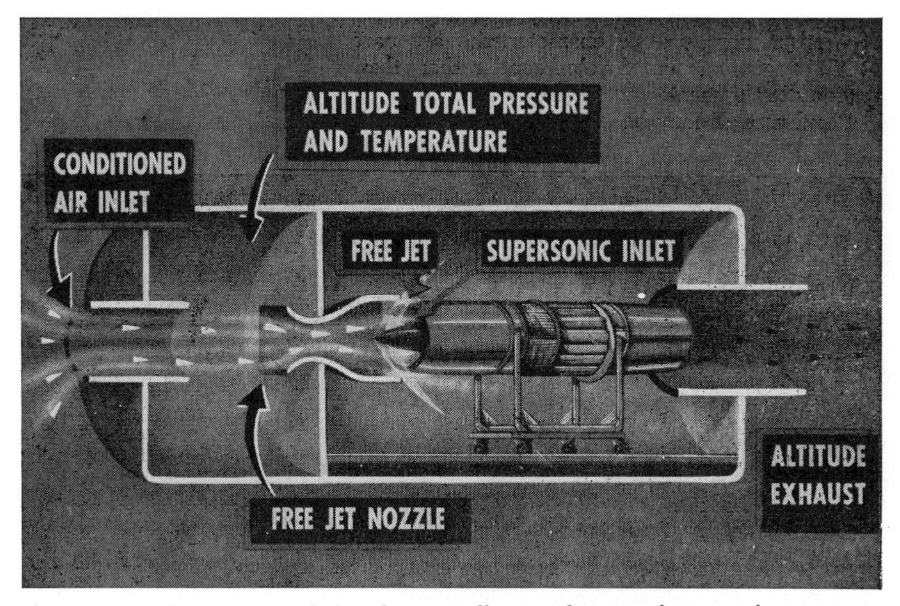


Figure 5. A free-jet type of altitude test cell is used to simulate actual air-stream velocity conditions. Air, supplied through a conditioned air inlet at the required temperature and pressure, is expanded through the free-jet nozzle to the desired stream velocity. The air leaving the free-jet nozzle is directed against the engine-inlet diffuser, which permits effective simulation of free flight conditions. Nonsymmetrical air flow to the engine can be simulated effectively by relative misalignment of the free-jet axis and the engine axis.

simulated altitude, flight speed, and angle of attack might all be varied. In some cases all of the conditions encountered during an actual flight mission could be simulated.

## Wind Tunnel Test Cell

Wind tunnels are needed for tests in which the external aerodynamic characteristics of an engine installation are required. Strictly speaking, tests of external aerodynamic characteristics are part of an engine-nacelle development rather than engine development.

Wind tunnels may be of either the connected-duct or free-stream inlet type. In the test section of a connected-duct type of wind tunnel installation (Figure 6) a duct is coupled to the engine inlet with a frictionless slip joint. Air at the requisite stagnation conditions is supplied through this duct directly to the engine inlet. The entire

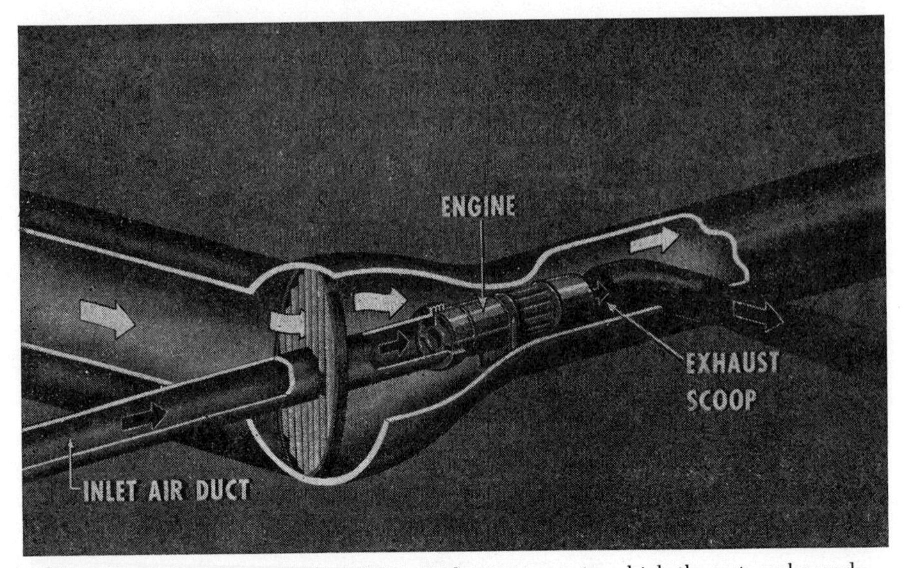


Figure 6. Wind tunnels are used for conducting tests in which the external aerodynamic characteristics of an engine are to be determined. Shown here is the connected-duct type of wind tunnel. Air at the required temperature and pressure is supplied directly to the engine inlet through an air-inlet duct. This duct is connected to the engine inlet by a frictionless slip joint. An exhaust duct carries away the engine's exhaust gases and prevents their recirculating back to the engine inlet.

engine installation is immersed in the wind tunnel air stream. An exhaust-duct scoop downstream from the engine exhaust intercepts the engine exhaust gases and ducts them from the wind tunnel. The wind tunnel air stream, therefore, is not contaminated by the engine exhaust gases. External aerodynamics of the engine installation may be studied in these tests, and the drag force of the engine can be determined.

To simulate actual flight conditions closely a free-stream inlet type of wind tunnel configuration is used. For such tests the inlet diffuser is coupled to the engine as it would be in the flight vehicle. In some cases all or part of an aircraft may be mounted in the wind tunnel to simulate more closely the actual flight operating conditions.

### Flight Testing

In flight testing the engine is operated in an aircraft in flight at the desired test conditions of altitude, velocity, or flight attitude. The engine may be mounted on an auxiliary mount or may be used directly as the aircraft's power plant. As an example of auxiliary mounting, a turbo-prop engine was installed in the nose of a U. S. Air Force B-17 bomber-type aircraft (Figure 7) [not shown here.] During tests the engine furnished part or all of the propulsive power while the aircraft was in flight.

In some instances of engine flight testing for the military, the engines have been positioned on bomb-bay mounts or on supports under aircraft wings, such as bomb shackles. Frequently an engine is flight tested on an auxiliary mount before it is used directly as the aircraft's power plant.

### Advantages and Disadvantages of Ground and Flight Testing

The major advantage of ground developmental tests is that test conditions are constantly under the control of test engineers. If re-runs are necessary, it is possible to duplicate approximately the conditions of a particular test. Extensive instrumentation can be used for each type of test since there are no weight limitations. Similarly, any special auxiliary unit requirements, such as loading or power take-off units, can be readily accomplished. Any changes or adjustments needed on test instrumentation or the engine can be made and installed.

Operational maintenance and observation during a test run are more simple for ground testing than for flight testing. Any test conditions which are within the capacity of the test facility to produce may be obtained at any time. If an engine fails or burns, there is no danger to test personnel.

Disadvantages of ground testing include the extensive array of test equipment required to test an engine completely. Ground testing is costly due to the large quantities of power required, facility amortization, and the multiplicity of special test equipment used. A major disadvantage for ground testing is that all flight conditions cannot be simulated exactly—for example, maneuver loads encountered in flight may not be possible to duplicate in ground tests.

Advantages usually cited for flight testing are the simplicity in providing many test conditions at low cost and the ability to apply actual maneuver loads to the engine while it is in operation. Flight testing also provides actual data on engine operation under conditions closer to those encountered in service at the same time that test data are being obtained.

The disadvantage of flight testing is the resulting high cost and loss in developmental time if failure occurs in either engine or aircraft. Dollar cost of the test flight vehicle is high, and its useful life may be relatively short. The flight testing time is limited by the comparatively short duration of most flight tests and may be still further restricted by weather conditions. Although the cost of flight testing sometimes may be relatively low, potentially it is always quite high.

## SUGGESTIONS FOR WRITING

The things to think about in selecting a mechanism to describe are (1) your personal interests, (2) the information you already possess or can readily obtain, (3) the suitability of the mechanism for a description conforming to the length desired, (4) the suitability of the mechanism for a description conforming to the kind of description desired, and (5) the selection of an imaginary reader.

The first two things are simple and obvious: if you yourself are choosing a mechanism to describe, you will naturally want one you are interested in and know something about, although for a beginning exercise, such as the description of the paring knife quoted earlier in this chapter, the mechanism may be so simple that the choice may not matter very much. The last three things mentioned above, on the other hand, are all dependent on one another. For example, a mechanism that would be well suited for a highly generalized description limited to a few hundred words in length might prove impossibly complex for a detailed description of the same length. Or a mechanism you could easily describe for a reader having a good deal of technical knowledge might become extremely difficult and long for a reader who was unfamiliar with scientific concepts and terminology.

Among the five elements listed, your instructor will probably specify some, such as the length of the description, the degree to which it should be generalized or detailed, and the sort of imaginary reader to whom it should be directed. Any that are left unspecified you can think over and arrange to suit your own preferences.

The following is a list of possible subjects. The subjects designated by Arabic numerals are very broad areas which have been chosen at random; under each of these subjects are three topics, beginning with one suitable for a detailed description and terminating with one suitable only for a generalized description. These topics will no doubt suggest others, some of which you might prefer.

1. Aircraft

   A nonfeathering propeller on a light plane
   Nonretractible landing gear on a light plane
   A sailplane

2. Boats

   The running light mounted on the stern of a small motorboat
   The standing rigging of a small sloop
   A small catamaran

3. Automobiles

   A wheel
   A shock-absorber
   An air-conditioner

4. Farm machinery

   A manual post-hole digger
   A hammer mill
   A hay-bailer

5. Tree-care

   A bow-saw
   A manually-operated sprayer
   A chain-saw

6. Electricity

   A small transformer
   A simple speaker or earphone
   A walkie-talkie

7. Carpentry

   A handsaw
   A plane
   A radial-arm saw

8. Metal-working

   Tinsnips
   A power hacksaw
   An electric arc welder

9. Hydraulics

   A garden hose
   A hydraulic ram
   A hydraulic lift in a service station

10. Optics

    A small hand lens
    Eye-glasses
    Binoculars

11. Earth-moving equipment

    A spade
    A scraper-blade attachment for a tractor
    A back-hoe attachment for a tractor

12. Laboratory

    A ringstand
    A balance
    A centrifuge

# 7 Description of a Process

A process is a series of actions, and fundamentally the description of a process is the description of action. The action may be either one of two types. One type is that in which attention is focused on the performance of a human being, or possibly a group of human beings. A simple example is planing a board by hand; in a description of this process, emphasis would fall naturally upon the human skills required. The other type involves action in which a human operator either is not directly concerned at all, or is inconspicuous. An instance is the functioning of an electric relay. Large-scale processes, when considered as a whole, are also usually of this second type, even though human operators may take a conspicuous part in some of the steps. The manufacture of paper is an example.

This chapter is divided into two main parts, according to these two types of processes. Before taking up the first type of process, however, we shall consider three problems that arise in describing almost any process, regardless of type. These problems are (1) the adaptation of the description to the reader, (2) the over-all organization, and (3) the use of illustrations.

Adapting the description to the reader depends, as always, upon an analysis of the reader's needs. As in the description of a mechanism, if the reader wishes to use the description as a practical guide, it becomes necessary for the writer to give careful attention to every detail. If the reader is interested only in acquiring a general knowledge of the principles involved and has no intention of trying to perform the process himself, or to direct its performance, the writer should avoid many of the details and emphasize the broad outlines of the process.

The fundamental organization of a process description is simple, consisting merely of an introduction followed by a description of each of the steps in the process in the order in which they occur. But this simplicity is usually marred by the necessity of discussing the equipment and the materials used. In building a wooden boat, for instance, the equipment would include hand and power saws, miter boxes, and planes; the materials would include lumber, screws, paint, and others. It is not always necessary to mention every item of equipment or every bit of material (it might be taken for granted that a hammer would be useful in building a boat), but no helpful reference or explanation should be omitted through negligence. Sometimes it is necessary to explain certain special conditions under which the process must be carried out, like that for developing photographic film in a darkroom.

There are basically two ways of incorporating the discussion of equipment and materials into the description as a whole. One is to lump it all together in a section near the beginning; the other is to introduce each piece of equipment and each bit of material as it happens to come up in the explanation of the steps in the process. The advantage of confining the description of equipment and materials to a single section near the beginning is that such discussion does not then interrupt the steps in the action itself. This method is usually practical if the equipment and materials are not numerous. If they happen to be so numerous or so complex that the reader might have difficulty in remembering them, the other method of taking them up as they appear in the process is preferable. The second method is by far the more common.

In summary, we can say that a process description is organized as follows (except that the discussion of equipment and materials may be distributed throughout the description instead of being confined to one section):

Introduction
Equipment and materials
Step-by-step description of the action
Conclusion (if necessary)

The use of illustrations, the last of the three general problems, needs little comment. Certainly, as many illustrations as can be managed conveniently should be introduced. It is difficult to represent action graphically, but sometimes a sketch of how a tool is held or of how two moving elements in a device fit together can add greatly to the clarity of the text. The general problem of the use of illustrations is much the same as in the description of a mechanism.

## PART ONE: PROCESSES IN WHICH AN OPERATOR TAKES A CONSPICUOUS PART

In this part of the chapter we consider three subjects: the introduction to a process description, the step-by-step description of the action, and the conclusion.

### The Introduction to the Description

The introduction to the description of a process is a comprehensive answer to the question, "What are you doing?" (The rest of the description is largely an answer to the question, "How do you do it?") An answer to the question, "What are you doing?" can be given by answering still other questions, principally the following:

1. What is this process?
2. Who performs this process?
3. Why is this process performed?
4. What are the chief steps in this process?
5. From what point of view is this process going to be considered in this discussion?
6. Why is this process being described?

It is not always necessary to answer all six questions, and it is not necessary to answer them in the order in which they happen to be listed. It will be helpful to consider each question in turn to get some notion of what needs to be done.

*What Is This Process?* Very early in the report the reader must be told enough about what the process is so that he can grasp the general idea. The way in which he is told depends upon how much he is presumed to know about the process, as well as upon the nature of the process itself. As in the description of a mechanism, we have come up against the whole problem of definition of the subject of the description. Again we must refer to the chapter on definition for a full treatment of the problem; here we give some particular attention to the use of comparison and generalized description.

A report written for sophomore engineering students on how to solder electrical connections might start by saying merely, "It is the purpose of this report to explain how to solder electrical connections." This simple statement of the subject would be sufficient. If, however, a report on the same subject were being prepared for a reader who had no real understanding of even the word "solder," an entirely different approach would be needed. Let's consider a reader who is very different from the sophomore engineer. Suppose a description of how to solder electrical connections was being prepared for a class of high school girls in home economics. Here, great efforts at clarification of the fundamental concept would be required. For these readers, it would be wise to write a formal definition accompanied by a comparison to soldered articles which most of them had probably seen, and to similar processes which they would know about. Such a report might begin in the following manner:

> It is the purpose of this report to explain how to solder electrical connections. Soldering is the joining of metal surfaces by a melted metal or metallic alloy. This process may be compared roughly with the gluing together of two pieces of wood. Instead of wood the solderer joins pieces of metal, and instead of glue he uses a melted alloy of lead and tin which, like the glue, hardens and forms a bond. Soldering is a very widely used technique; one evidence of its use which probably almost everyone has noticed is the streak of hardened solder along the joint, or seam, of a tin can of food.

The third and fourth sentences above constitute a comparison with a process with which the reader (the high school girl) would probably be familiar. The last sentence in the example is a reference to a familiar device in which the process has been employed.

The preceding introduction might continue:

> The process of soldering consists essentially of heating the joint to a degree sufficient to melt solder held against it, allowing the melted solder to flow over the joint, and, after the source of heat has been removed, holding the joint immovable until the solder has hardened.

This example gives a general idea of the whole process. You will probably have noticed also that it looks much like a definition of the process, and at the same time like a statement of the chief steps (a subject to which we will come in a few moments). As a matter of fact, it is an acceptable definition; and although the list of steps is actually incomplete, the missing steps could easily be added. It is evident, then, that it would be possible to define a process and indicate its purpose, to give a generalized description of it, and to list the chief steps, all in one sentence. Would such compression be advisable? Sometimes; it depends upon the reader. For the high school girl it probably would not be, since we are assuming that she knows nothing of the process. The more leisurely manner in the example above, including both parts, would provide her a little more time to get used to the idea. In short, the question "What is this process?" is simply a problem of definition; and the use of comparison and of generalized description is often particularly helpful.

*Who Performs This Process?* There is not a great deal to say about this matter of explaining who performs the process, except to emphasize the fact that it is sometimes a most helpful statement to make. For example, if a description of the process of developing color film was written for the general public, it might be rather misleading unless the writer explained that most amateur photographers do not care to attempt this complicated process, the bulk of such work being done commercially. Very often the statement about who performs the process will appear as a natural or necessary element in some other part of the introduction. Often no statement is required.

*Why Is This Process Performed?* It is, of course, absolutely necessary that the reader know why the process is performed—what its purpose is. Sometimes simply explaining what the process is, or defining it, makes the purpose clear. Often the purpose of the process is a matter of common knowledge. There would be no point in explaining *why* one paddles a canoe, although not very many people may know *how* to paddle a canoe efficiently. Sometimes, however, the purpose of a process may not be clear from a statement of what it is, or how it is performed. Then it is necessary to be quite explicit in stating its complete purpose. To take a simple instance—one might explain clearly and accurately how to water tomato plants, how much and in what manner, and still do the reader a disservice by not informing him that if the supply of moisture is not sufficiently regular there will be a tendency for circular cracks to appear around the stem end of the ripening tomatoes.

*What are the Chief Steps in This Process?* The listing of the chief steps in the process is an important part of the introduction. It is important because it helps the reader understand the process before the details of its execution are presented. Even more important is its function in telling the reader what to expect in the material that follows. It is a transitional device. It prepares the reader for what lies ahead of him. Naturally, it serves the purpose of a transitional forecast best when it appears at the end of the introduction.

The list of steps may appear as a formal list, with a number or letter standing beside each step. If this method seems too mechanical, the steps can be stated in ordinary sentence form, with or without numbers or letters. Care should be taken with punctuation to avoid any possibility of ambiguity or overlapping of steps. The statement of the major steps in the process of soldering an electrical connection might be written as follows:

> The chief steps in this process are (1) securing the materials and equipment, (2) preparing the soldering iron (or copper), (3) preparing the joint to be soldered, (4) applying the solder, and (5) taping the joint.

Observe that itemized parts of the sentence are grammatically parallel, as they should be. The steps should be discussed in the order in which they are listed.

*From What Point of View Is This Process to Be Discussed? Why Is This Process Being Described?* These two questions, which are the last two, can be discussed together conveniently. Neither of them is properly concerned with the question with which we started this section on the introduction, "What are you doing?" Nevertheless, each of them represents an important aspect of the introduction. Each is concerned in its own way with the purpose of the report.

The latter, "Why is this process being described?" calls for a specific statement of purpose: the purpose of including the description of this process in the report of which it is a part. In other words, the reader will want to know why you are asking him to take time to read your description of the process. More often than not your reason for including the description will be perfectly obvious. If so, there is no need to mention it. Sometimes, however, it may not at once be clear. Perhaps, for example, a later part of a discussion will be incomprehensible without a preliminary understanding of a certain process, and the reader will need to have this fact explained. Be careful to keep in mind the distinction between the purpose of the process itself and the purpose you have in writing about it. These are very different matters.

Perhaps we should add that if your instructor asks you to write a paper devoted to describing a process, we don't think he will need to be told why you are writing the paper.

The first of the two questions above is likewise related to the matter of purpose, but here the interest is not in why the process is being described; rather it is in why it is being described in a particular way or from a given point of view. One illustration of this fact is contained in the different ways that were suggested earlier for the writing of the introduction to the report on soldering. There would be no difficulty in seeing at once that the report written for the high school girl was designed to explain the simple process of soldering so fully that a completely uninitiated reader could successfully use the explanation as a guide. However, it is often wise to state the point of view explicitly, as in the following example:

> The explanation of how to correct the instability of this oscillator will be given in terms of physical changes in the circuit rather than as a mathematical analysis.

One concludes from this statement that the point of view in the report is going to be practical, the treatment simple. The point of view will perhaps be that of a radio repair man rather than that of an electrical engineer.

So much for the introduction to a description of a process. In this discussion we have pointed out what facts the reader of a process description should be aware of when he has finished the introduction. Sometimes almost all problems will be met by the writer in a single introduction; sometimes only a few. But probably they should all be considered. Much depends upon who the reader of the report will be and upon the general circumstances which cause the report to be written.

Of course the writing of introductions may involve many problems not mentioned here at all. In this section we have discussed only those elements which are likely to be involved in the "machinery" of starting off a process description. For a discussion of other aspects of the writing of introductions see Chapter 11.

## The Chief Steps

*Organization.* With the possible exception of the discussion of equipment and materials, the introduction to a description of a process is followed directly by a description of the chief steps in the process. Two problems appear in organizing the description of the chief steps. One problem is how to organize the steps; the other is how to organize the material within each individual step.

The organization of the steps can be dismissed at once. It is chronological, the order of the performance of the steps. Although there are processes in which two or more steps are, or can be, performed simultaneously, you can usually manage fairly well by explaining the situation plainly and then taking one step at a time.

The organization *within* the description of the individual steps requires more comment. For both content and organization of the description of each individual step, there is one idea that is so useful that it cannot easily be overemphasized. That idea is that each individual step constitutes a process in itself. The individual step should, therefore, be properly introduced, and, if necessary, divided into substeps. Its description is essentially a miniature of the description of the process as a whole. Furthermore, if a given individual step can be broken down into substeps, each substep is treated according to the same general principles applied to the whole process.

Of course it would be easy to go too far with this idea. What we just said should be taken with a little salt. In the introduction to the whole report, for instance, it is often desirable to say something about who performs the process, about the point of view from which the process will be described, and about why the description is being written. Usually, when you introduce an individual step, nothing of this sort need be said. Definition, statement of purpose, and division into parts, on the other hand, require the same attention in introducing the individual step that they do in introducing the whole report. The great importance of making the purpose of each step clear may be seen from another point of view in the discussion of a block diagram in Appendix D. Read the second paragraph under "Information Arrangement."

What is to be said in describing the action itself constitutes an entirely new problem. It is perhaps surprising to reflect that of all that has been said so far in this chapter about how to describe a process, which was originally defined as an action or series of actions, nothing has as yet been said about how to describe the action itself. Everything has been concerned with how to get the action in focus, together with all its necessary relationships. The only point in the whole report at which action is really described is in the individual step. And if there are substeps, the description of the action drops down to them.

*The Description of the Action.* In describing the action, the writer must say everything the reader needs to know to understand, perhaps even to visualize, the process. The omission of a slight detail may be enough to spoil everything. Moreover, care should be taken not only in connection with the details of *what* is done, but also of *how* it is done. For example, in telling a reader about heating his soldering iron, it would surely be wise to tell him that if the tip of it begins to show rainbow colors it is getting *too* hot. And in an explanation of how to calibrate a wide-range mercury thermometer in an oil bath, it would be advisable to point out that the oil should not be allowed to get too hot because the thermometer may then blow its top off. Keep the reader in control of the action.

A further illustration of the importance of details and of analyzing the needs of the reader can be taken from the following incident. A lecturer in physics was speaking to a class of college freshmen and sophomores about the fundamental principles of the electronic tube. He pointed out that three basic elements in the tube are the cathode, the grid, and the plate. The cathode, he said, has a negative charge, the plate a positive charge, and the electrons flow from the cathode to the plate, passing through the grid, which is between them. He pointed out that the grid usually has a negative charge, and went on to other matters. A goodly percentage of the class left the room wondering how the electrons got past the negative grid. Perhaps you will feel that these people were not very alert, and perhaps they weren't. On the other hand, the lecturer was speaking *to them.*

We started out in this section by saying that the content of the description of a process is governed by the reader's need to comprehend every step in the action. There is little more that can be said about the description of the action in the various steps of the process, with one important exception: that is, the style.

*Style.* A general discussion of style in technical reports is given in Chapter 3, and what is said there applies to the description of a process. One problem peculiar to the description of a process is not taken up in that chapter, however. This problem is the choice of the mood and voice of the predicate, and of the noun or pronoun used as the subject. A good many possibilities exist, but (neglecting the noun or pronoun for the moment) three are of special importance: the active voice and indicative mood, the passive voice and indicative mood, and the active voice and imperative mood. We shall illustrate each of these and then comment on them.

> A. *Active Voice, Indicative Mood*
> The next step is the application of the solder to the joint. This step requires the use of only the heated iron (or copper), and a length of the rosin-core solder. The solderer takes the iron in one hand and the solder in the other, and holds the iron steadily against the wire joint for a moment to heat the wire. Then he presses the solder lightly against the joint, letting enough of it melt and flow over the wire to form a coating about the entire joint.
>
> B. *Passive Voice, Indicative Mood*
> The next step is the application of the solder to the joint. This step requires the use of only the heated iron, and a length of the rosin-core solder. The iron is held steadily against the wire joint for a moment to heat the wire. Then the solder is pressed lightly against the joint, until enough of it has melted and flowed over the wire to form a coating about the entire joint.
>
> C. *Active Voice, Imperative Mood*
> The next step is the application of the solder to the joint. This step requires the use of only the heated iron, and a length of the rosin-core solder. Take the iron in one hand and the solder in the other, and hold

the iron steadily against the wire joint for a moment to heat the wire. Now press the solder lightly against the joint. Let enough of it melt and flow over the wire to form a coating about the entire joint.

The essential differences among these three ways can be expressed as the differences in the following three statements: (1) The solderer holds the iron. (2) The iron is held. (3) Hold the iron.

Which one of the three ways is best? It depends upon several factors.

The advantage of the first way, the active voice and indicative mood, is that it gives the reader the greatest possible assistance in visualizing the action. It is the most dramatic. It comes as close as it is possible to come in words to the actual observation of someone performing the action. The presence of the person carrying out the process is kept steadily in the mind of the reader. This technique is without question a very effective one, and its possibilities should not be overlooked. Probably its best use occurs when the following three conditions prevail: (1) the process being described is one which is performed by one person; (2) the description of the process is intended as general information rather than as a guide for immediate action; and (3) the description is directed to a reader who knows little about the process. If a guide for immediate action is desired, the terse imperative mood may be preferable—although this is a debatable point. And if the reader of the report already knows about the process in general, he will have little need of aid in visualization.

The disadvantage of using the active voice is that it is likely to become monotonous unless handled with considerable skill. The monotony arises from the repetition of such terms as "the solderer," "the operator," or whatever the person performing the action is called, even though pronouns can be used to vary the pattern a little. Finally, for some curious reason, perhaps because the active voice is not the customary way of describing a process, the writer may feel reluctant and slightly embarrassed to continue saying "The operator does this, the operator does that," and so on. There is no logical reason why he should give in to this feeling.

The advantage of the passive voice is that there is no problem about handling this hypothetical operator. The disadvantage is that the positiveness and aid to visualization of the active voice are missing. For a process performed by one person, or perhaps even a few persons, a combination of the active and the passive voices is possibly a good compromise. We do not care to be dogmatic about this.

The advantages of the third way, the active voice and the imperative mood, are that it is concise, easy to write, and a reasonably satisfactory guide for immediate action, so long as the process is not too complex. It is, however, not really a description at all; it is a set of directions. And, because it is a set of directions, there is likely to be a slighting of emphasis upon purpose, and a consequent weakness of the report as an explanation of the process. The imperative mood promotes action better than it promotes understanding.

There are numerous possibilities in addition to the three just illustrated. In fact, all the practical possibilities can be listed as follows:

> *Active Voice, Indicative Mood*
> The solderer (or "I," "we," "you," or "one") takes the iron. . . .
>
> *Active Voice, Subjunctive Mood*
> The solderer (or "I," "we," "you," or "one") should (or "must," or "ought to") take the iron. . . .
>
> *Passive Voice, Indicative Mood*
> The iron is taken. . . .
>
> *Passive Voice, Subjunctive Mood*
> The iron should (or "must," or "ought to") be taken. . . .
>
> *Active Voice, Imperative Mood*
> Take the iron. . . .

Almost all these forms may be found in use occasionally. We will comment on special problems related to a few of them. (1) We don't advise the use of "one," but still less do we advise the use of "you" as a substitute for "one" (for example, in "You take the iron. . . ." or worse, "You take your iron. . . ."). On the other hand, there can be no objection to "you" when its referent is the reader (for example, "You should take the iron. . . ."). But even if there are no objections to the latter use of "you," there is not much to be said in favor of it, and we do not advise its frequent use. (You will have noticed that we use it often in this text, but we are not describing a technical process. The style of this book is more colloquial than that of technical reports.) (2) The subjunctive mood should be used sparingly. It is a fine form in which to give advice—as we just did. But don't forget the distinction between *describing* and giving advice. (3) It is all right to use different forms within the same process description, but discretion is necessary. It is probably best to use only one of the forms throughout if you can make it sound natural and easy. Please note that we did not say, "if you can *do* it easily." Good writing is easily read; it is not usually easily written. All in all, the three forms illustrated at paragraph length above (active indicative, passive indicative, and active imperative) are by far the most useful, with the active imperative running a poor third. These remarks refer only to the type of process in which there is a conspicuous operator.

## The Conclusion

The last of the major parts of the description of a process is naturally the conclusion. It is not always necessary to write a formal conclusion. Whether one is desirable depends, of course, on whether it will help the reader. Sometimes the reader needs help in matters like the following:

1. Fixing the chief steps in mind (listing them again might help)
2. Recalling special points about equipment or materials
3. Analyzing the advantages and disadvantages of the process
4. Noting how this process is related to other processes, or other work that is being done, or reported on

The writer must analyze his own report and identify his intended reader to decide whether a conclusion is necessary.

# PART TWO: PROCESSES IN WHICH AN OPERATOR DOES NOT TAKE A CONSPICUOUS PART

We turn now to that kind of process in which the human agent is less conspicuous. Such processes may be of great magnitude, like the building of a large dam, or relatively simple, like the functioning of a tire pump. They are distinguished by the fact that little emphasis falls directly upon the performance of a human being or beings. How does a tire pump work? An answer to this question would be the description of a process; but in that description there would be little need to mention the quality of the performance of the operator.

The fact is that the kind of process description requiring little attention to the operator turns up in technical writing more frequently than does the other kind. The technical man is more likely to be called on for an answer to the question, "How does this work?" than he is to the question, "How do you do this?" So our subject here is an important one.

All that need be considered here is how the description of a process in which the operator does not take a conspicuous part differs from one in which he does. The essential differences are three:

1. Emphasis is altogether on the action—on what happens—and not on the operator and how he performs certain actions.
2. The presentation is usually (not always) in the active indicative, the passive indicative, or a combination of the two. The imperative mood never appears.
3. The terms "equipment" and "material" take on a somewhat different meaning and significance.

Point 1 is fairly obvious. Once a train of events has been set in motion, as in a chemical process, interest in the operator who set the events in motion fades. From then on, interest lies in what occurs next. In a process of great magnitude like the manufacture of rubber, where hundreds of operators are engaged and where it is obviously impossible to keep an eye on an individual operator, or even a group, the emphasis must be on the action itself. The reader is simply not interested in the *who* involved.

In view of this emphasis, it is easy to see that either the active indicative or passive indicative (or both) will likely be used and that the imperative cannot be used. The passages quoted below illustrate these styles, the first principally in the passive indicative, the second in the active indicative.

> Following the work of Faraday, Ferdinand Carré developed and patented the first practical continuous refrigerating machine in France in 1860. Carré's idea was to use the affinity of water for ammonia by absorbing in water the gas from the evaporator, then using a suction pump to transfer the liquid to another vessel where the application of heat caused the liberation of ammonia gas at a higher pressure and temperature.
>
> Carré's machine is illustrated by the flow diagram in Fig. VI.* In this ammonia-water system, high-pressure liquid ammonia from the condenser is allowed to expand through an expansion valve and the low-pressure liquid then vaporizes in the surrounding refrigerated space. In these two steps the ammonia absorption system is exactly like the compression system. However, the gas from the evaporator, instead of being passed through a compressor, is absorbed in a weak solution of ammonia in water ("weak aqua"). The resulting strong solution ("strong aqua") is then pumped to the generator, which is maintained at high pressure. Here the strong aqua is heated and the ammonia gas driven off. The weak aqua which results flows back to the absorber through a pressure-reducing valve, the highly compressed ammonia gas from the generator is condensed, and the cycle is repeated.

*See Appendix B.

Except for two in the last sentence, the verbs in the preceding paragraph are in the passive indicative. Now consider another account of substantially the same process, this time in the active indicative:

> The process is shown diagrammatically and much simplified in Fig. X.* Beginning with the ammonia-hydrogen loop, the ammonia gas enters the evaporator from the condenser through the liquid trap which confines the hydrogen to its own conduit. In the evaporator it takes up heat from the surrounding space and vaporizes, its gaseous molecules mixing with those of the hydrogen. The addition of the heavier ammonia molecules increases the specific gravity of the vapor, and it sinks down the tube leading to the absorber.
>
> In the absorber, the ammonia dissolves in the countercurrent stream of weak aqua, while the practically insoluble hydrogen, lightened of its burden of heavy ammonia molecules, ascends to the evaporator to perform again its task of mixing with and decreasing the partial vapor pressure of the ammonia.
>
> Taking up the ammonia-water loop, the strong aqua in the absorber flows by gravity to the generator, where the application of heat drives the ammonia out of solution. A vertical tube, the inside diameter of which is equal to that of the bubbles of gaseous ammonia generated, projects below the surface of the boiling liquid. This "liquid lift" empties into the separator, where the ammonia vapor is separated from the weak aqua. The weak aqua then returns by gravity to the absorber to pick up another load of ammonia.
>
> Finally, the ammonia loop, which has been traced as far as the separator, next involves the "condenser," an air-cooled heat exchanger which removes the latent heat from the ammonia gas, converting it into a cool liquid. Here it passes through the liquid trap that marks its re-entry into the evaporator to serve its purpose of cooling the refrigerated space.

In the account of closed-cycle refrigeration given above, there is an operator, of course, but after he lights the gas flame which starts the process, he takes a back seat, for the process completes itself without any further assistance from him.

Which of the two versions of the process is the better? They are both good. Pay your money and take your choice. And don't fail to ponder the value of the consistency of point of view illustrated in both versions.

The third of the differences listed above has to do with a change in the meaning of the terms "equipment" and "material," as these terms are used in the list of the major parts of a description of a process. Where an operator is conspicuously involved, their meaning is clear; he *uses* equipment and materials in carrying out the process. But in the description of a process like those quoted above, there is

*See Appendix B.

no operator, and—curiously enough—what in the other kind of process would be simply equipment and materials may now be said to be performing the process! For instance, in a description of how a tire pump works, there would be no operator, and instead of being merely the equipment, the pump might be granted the active voice—as in the statement, "The plunger compresses the air in the cylinder. . . ."

Once this fact is understood, you will have no difficulty. And now we can go on to point out an important fact about process descriptions in actual industrial and research reports. More often than not in such reports, the description of the device or devices involved (as discussed in Chapter 6) and the description of the process (as discussed in the present chapter) are inextricably intermingled, and often other elements (analysis, classification, and the like) are involved as well. As we have said several times, process description is one of the special techniques of technical writing; it is not a type of report. The complexity found in actual reports does not invalidate the principles discussed in these two chapters, although it naturally makes them more difficult to apply. But that is not the fault of the principles; it is, indeed, only through the principles that the complexity can be ruled and order created.

Aside from the three differences just discussed, the description of a process in which an operator is conspicuous and of one in which he is not rests upon the same principles.

If, at this point, you have the feeling that there are too many things to be kept in mind in solving the problem of writing a description of a process, don't let the feeling discourage you. To become quite expert in quickly solving these writing problems will naturally take a good deal of practice and experience. But remember that essentially you have just two important things to do: (1) introduce your subject carefully so that your reader will be able to follow you easily when you (2) describe accurately and in the most effective manner the steps of the process.

## ILLUSTRATIVE MATERIAL

The following pages contain three examples of process description. The first, which originally appeared in *General Motors Engineering Journal*, is a highly generalized description. The second is a mixture of detailed instruction and generalized description. Although it is usefully informative in content, it is nevertheless a careless job of writing. The third example, which is well done, is set up in the familiar form of directions for students on how to perform a laboratory experiment. An additional example, illustrating the use of itemized directions in a report designed for professional scientists, may be found on page 265. It is important to recognize that in both of these latter two examples the use of itemized directions unaccompanied by explanations of purpose is justified by the character of the particular situation. The students read their directions as part of a total communication situation involving assigned reading and lectures on theory. The professional scientists read theirs in a total communication situation involving access to other reports on the same subject, as well as long personal familiarity with the laboratory techniques required.

The concept of the "communication situation" is discussed on page 29 and those following.

## An Improved Method for the Economic Reconditioning of Aircraft Spark Plugs

*by Alfred Candelise*

The severe heat stress and wear resulting from extreme temperatures and pressures generated within the combustion chamber of an aircraft engine require that spark plugs be reconditioned after established periods of operation to insure that proper ignition, so vital to the safe performance of aircraft, will be maintained.

When plugs are removed from an engine they are examined thoroughly and, if found in suitable condition, are put through a reconditioning process which consists of several individual operations performed by trained personnel using specialized tools and equipment usually mounted on individual stands, with the plugs carried from one operation to another in trays.

AC Spark Plug Division's engineers, recognizing the need for improving aircraft spark plug reconditioning practices, recently designed and built a unit known as a Servicing Facility to be used for reconditioning commercial aircraft spark plugs (Figure 1). The Facility concentrates all reconditioning operations on a convenient bench having a length and width of 12 ft. by 2 ft., respectively, with the fixtures arranged in a progressive operational sequence.

*Here is a nicely organized, well-written description of a process.*

*The first six paragraphs, which constitute the introduction, give fairly direct answers to all but two of the questions appropriate to such an introduction (see page 149). The apparently unanswered questions are those concerning point of view and the purpose in describing the process, respectively. However, both point of view and purpose are actually indicated clearly by the character of the journal in which the description originally appeared. Readers of the* General Motors Engineering Journal *are well aware of its policy of printing general technical information for people having some background in science and engi-*

Also, new fixtures and instruments have been designed for the purpose of obtaining safety, economy, and accuracy.

### Sequence of Reconditioning Operations

Fig. 2 shows a cutaway view of a typical commercial aircraft shielded spark plug and its construction details.

The seat gaskets are first removed from the plugs which are then degreased, dried, and carried to the Facility in a specially designed "buggy" which prevents damage to critical portions of the plug. The plugs then are inspected and those found suitable for reconditioning are placed on the Facility's extreme left-hand shelf.

Eight operations are performed on each plug in the following order as it passes through the Facility: (a) buffing of shield and shell threads, (b) cleaning of spark plug firing end, (c) cleaning of shielding barrel insulator, (d) resetting gap clearance between electrodes, (e) electric breakdown and gas leakage test, (f) center electrode wire resistance check, (g) final inspection, and (h) rustproofing of threads and identification painting.

### Buffing of Shield and Shell Threads

The shield and shell threads are cleaned and buffed simultaneously by placing the plug in a sliding fixture and pushing it against a revolving wire brush. This reconditioning operation prevents faulty installation by removing hard deposits of lead and carbon from the threads.

### Cleaning of Spark Plug Firing End

The center and ground electrodes as well as the insulator tip are cleaned of lead and carbon formation and dirt by an abrasive blasting compound. The plug is first placed into an adapter which is designed specially to fit each type of plug to be cleaned. A foot pedal is then depressed which activates the blasting compound. When the foot pedal is depressed 1/3 of its travel, a blast of air is applied to the firing end for the purpose of cleaning away blasting compound remnants. The complete cleaning of the firing end enables the energy of the ignition

*neering. The purpose, therefore, is to provide general information about this process, and the point of view is moderately technical.*

*The illustrations mentioned, Figures 1 and 2, are not shown here, but in the original they do add helpful visual images of the plug and the apparatus, or Servicing Facility, in the way that was suggested in the chapter on description of a mechanism. (We'll admit, incidentally, that we would find it easier to be serious about this apparatus if it were not known by the name of the Facility.)*

*Since this description is, in one sense, concerned with the equipment throughout, only a few descriptive comments on the equipment are given in the introduction; others appear in the text as needed.*

*The introduction closes with a statement of the chief steps, or operations.*

*The purpose of each of the first four steps is carefully stated. In the first two steps, however, the purpose is stated at the end rather than in the more common position at the beginning. This variation in organization seems to work out all right here.*

system to be discharged through the gap without any losses caused by leakage along the dirty insulator surfaces.

### Cleaning of Shielding Barrel Insulator

The shielding barrel insulator must be cleaned of any accumulated dirt which, if not removed, serves as a conductor of electricity and creates a possible leakage path. The cleaning of the shielding barrel insulator is accomplished rapidly with a specially designed rotating fixture incorporating a rotating rubber plunger and a special cleaning compound. The rubber plunger is inserted into the plug with a slight hand pressure. The plug is cleaned, rinsed with a spray of water, and finally dried with a blast of air. The water spray and air blast are actuated by suitable controls within easy reach of the operator.

### Resetting Gap Clearance Between Electrodes

Electrode wear causes a widening of the gap between the electrodes. As the gap increases, the resistance the spark must overcome to jump it also increases. This may result in erratic engine operation and requires a resetting of the gap to the specified value.

Resetting of the gap is performed with the aid of a specially designed gapping fixture having a handle and a pressing tool which make it easy to move the prongs of the ground electrodes until the correct clearance is obtained.

### Electric Breakdown and Gas Leakage Test

An electric breakdown test is conducted to ascertain whether the ceramic core is cracked or broken internally. A gas leakage test is conducted to insure that high-pressure gases developed within the combustion chamber will not leak past the plug's seals.

The electric breakdown and gas leakage tests are accomplished through the use of a fixture referred to as a "test bomb." The bomb is an airtight container having a glass window and a hole for the insertion of the plug. After the plug is inserted into the bomb a locking nut is turned which causes a microswitch to activate automatically an air valve and a high tension circuit. Dry air or other suitable gas is admitted into the bomb until a suitable pressure is attained. High-tension current is then applied to the plug's terminal and sparking is observed through a mirror against which the spark is reflected. The air pressure reading at the spark's point of suppression, shown by a dial located on the Facility's instrument panel, indicates the electrical characteristics of the plug.

### Center Electrode Wire Resistance Check

In order to minimize electrode erosion as much as possible the majority of aircraft spark plugs have a resistance of 1,000 ohms nominal value built into the center electrode of the plug as part of the insulator assembly. To check this resistance a specially designed 1-kilovolt resistance meter is used which is not affected by small contact resistances along the path of measurement. The spark plug is first inserted vertically into a fixture with the center electrode in the upward position. A switch is then turned on to energize electrically a needle-like probe which is pressed against the center electrode. The value of the electrode's resistance is indicated on the dial of the resistance meter mounted in a convenient position on the panel facing the operator.

The resistance meter has three scales: (a) 0 ohms to 300 ohms, (b) 300 ohms to 3,000 ohms, and (c) 3,000 ohms to 30,000 ohms. A scale of the correct range can be selected by turning a multiplier dial to an X1, X10, or X100 position. Center electrode resistance checks for aircraft spark plugs can usually be made with the X10 dial. If a warning lamp is lit, however, during a resistance check the next higher scale is used.

### Final Inspection

Final inspection of the spark plugs is accomplished with the aid of a lighted magnifier conveniently mounted on the Servicing Facility. The magnifier is used to inspect each plug for (a) cracked or chipped insulator barrels, (b) cracked, broken, or chipped insulator nose, (c) worn out electrodes, (d) damaged threads, and

*The purpose of the last two steps (g and h) is surely self-evident, as the author assumes. The definition of the last step is also self-evident (actually there are two steps, or two substeps, depending on how you think*

(e) damaged shell hexagon. When final inspection is completed, the operator places the plug in a tray located to the right of the Facility's inspection area where it is then ready to have the final reconditioning operation performed.

### Rustproofing of Threads and Inspection Painting

Rustproofing of threads is accomplished with a motor-operated fixture having a double set of rollers located inside a trough-like box which is filled with the desired rustproofing compound. While the plug is being rotated in this fixture, a paint band is applied to the body of the plug for identification purposes. After the paint has dried, the plugs are packed and stored until ready for use.

### Summary

The capacity of the Servicing Facility is approximately five-hundred spark plugs per day per operator. It is well suited for commercial airline requirements and greatly reduces the amount of floor space usually required for reconditioning. Also, economic advantages are realized by conveniently locating the reconditioning equipment in such a manner that all unnecessary handling is kept to a minimum.

*of it), but the definition of final inspection is clearly presented through the list of substeps making up the operation.*

*Finally, we should mention the style. It is a good style in general, but what we particularly want to emphasize is the absence of the imperative mood. This writer is not giving orders or instructions; he is not concerned with getting something done, at least not immediately. He is describing a process for a reader who likes to be informed about such things. The next illustration gives examples of the use of the imperative mood.*

## General Instructions for Storing, Mixing, and Installation of Castable Refractories*

The storing, mixing, and installation procedures for all castable refractories are quite important and the recommended instructions should be followed carefully. In general, with all castable refractories the following procedures and precautions should be observed.

### Storage

All castables contain a hydraulic-setting binder and if exposed to dampness can pick up

*Here is a description of a process that we thought you might have some fun practicing on. The over-all organization is good, and the content is clear; the writing was pretty obviously done by a man who knew the subject thoroughly. But as a whole this description is still in a rather crude state.*

*Courtesy, A. P. Green Fire Brick Company, Mexico, Mo.

moisture from the air resulting in the formation of hard lumps in the bag. When this occurs, usually the castable has partially set and ordinarily will not be suitable for use.

While castables are shipped in moisture-resistant bags, they still should be stored in a dry place. Special attention should be paid to rotating warehouse stocks of castables in order to ship out the oldest material first.

### Mixing

When mixing by hand, the entire bag should be dry mixed before adding water. This is most important where only a portion of the bag is to be used because some segregation of the ingredients may have taken place during shipment. Dry mixing is not required where the entire bag is to be mixed in a mechanical mixer.

Castables may be mixed in a mortar box, bucket, wheelbarrow, or paddle-type mechanical mixer. Mixing on the floor is not recommended because of the tendency to wash out part of the binder. A paddle-type mixer is recommended rather than a drum-type mixer as it gives more positive and thorough mixing. Also, many castables are too sticky to mix in a drum-type mixer and will adhere to the sides of the drum.

Add only clean, cool water and be sure the container for the water and the mixing container are clean. Even a small amount of foreign material may lower the strength or prevent setting.

Do not use too much water. An excess of water over that recommended for the particular castable will reduce the strength. The term "pour" or "pouring" of castables has been used for many years, but those experienced with this type of material know that this term is not to be taken literally. Castables are not poured like water. When mixed to the proper consistency, a castable can be formed into a ball with the hands and will hold its own shape. Of course, too little water and not sufficient puddling can result in a honeycombed structure; however, most people not experienced with castables are inclined to put in too much water rather than

*Evidently it is a first draft. We'll make three specific criticisms to illustrate what we mean in saying the description is crude, and then you may want to make others for yourself.*

*Our first criticism is that the major divisions are not broken down into substeps. For example, under "Installation" several operations are discussed, but no indication is given at the beginning as to what they will be. For all the reader can tell as he starts to read this section, he may be in for a detailed description of every step in the installation process. But such an anticipation would be quite incorrect, of course. Only certain aspects of the installation are discussed. What are they? Why have they been chosen?*

*Our second criticism is that the language is careless. For instance, look at the second sentence under "Storage": "When this occurs. . . ." What is the referent of "this"? Of course the reader can figure it out. What is objectionable is the fact that he is forced to figure it out. Similarly, look at "resulting" in the preceding sentence. What does it modify?*

too little. Follow the directions on the bag for proper amount of water.

A convenient method for checking the proper puddling consistency is the "Ball-in-Hand" test. To determine the proper consistency, form a compact ball of the mix in the hands and toss it upward about one foot and catch it in one hand, as illustrated on Page 2. [Figure 1 in this text.] This method is prescribed by A.S.T.M C268 for checking consistency of castables for molding testing specimens for modulus of rupture tests.

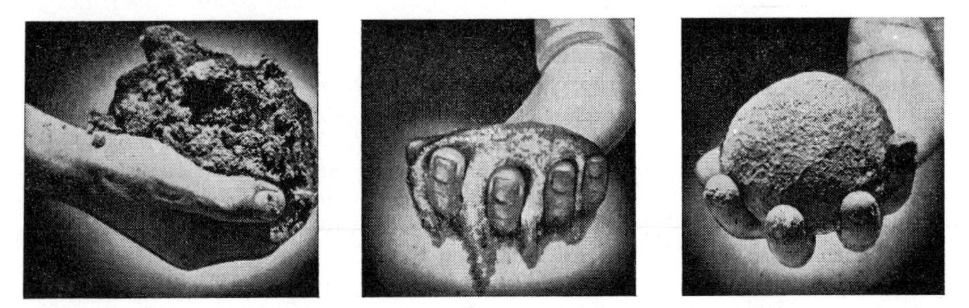

Figure 1. The ball-in-hand test. (Left) Too dry. (Center) Too wet. (Right) Just right.

## Installation

Wooden forms or porous backing-up material should be waterproofed to prevent absorption of the water from the castable which would result in lower strength of the castable because it needs water for its proper hydraulic set.

Most castables set up fairly rapidly and should be placed soon after mixing—usually within 20 minutes after water has been added. If a castable has started to set up in the mortar box or mixer, the addition of more water merely lowers the strength. It is better to discard the whole batch than to try to salvage with additional water and mixing.

Continuous reinforcing rods, such as used with structural concrete, should not be used with refractory castables. Steel reinforcing rods have twice the thermal expansion of castables, and as the castable is heated the rods may disrupt and break the castable lining.

All castables during installation should be thoroughly puddled or tamped to work out all voids. A blade or spade-type tool for puddling is

*Finally, we have here an illustration of an undisciplined use of the imperative. If it is good to say, in the seventh paragraph, "Do not add too much water," why not change the first sentence of the eleventh paragraph to the following: "Do not use continuous reinforcing rods,* preferable to a 2×4 or blunt end board except in the case of castables that are very sticky (A. P. Green GREENCAST-12, for example) and the castable insulations containing vermiculite. In these cases a blade or spade-type tool tends to form voids or planes of weakness and a blunt end board is a better puddling tool. Overpuddling of the lightweight vermiculite castables may break down the aggregate and produce a denser, heavier material with loss in insulating efficiency.

The top surface should only be screeded off level and not troweled to a smooth, slick finish as a slick surface will retard the escape of moisture. If the screeding operation tends to "tear" the surface, as it may do with drier mixes, a wood float is satisfactory to rework the surface.

The length of time for proper curing varies with individual castables. However, for best results, at least 24 hours should be allowed before application of heat and then the temperature should be raised gradually to prevent formation of steam within the castable during the initial firing. With certain extremely dense castables, the initial firing schedule is very important. The strength of castables depends on the proper hydration of the binder and if the material is dried before a strong hydraulic set has developed, loss of strength and possible cracking may occur. Under normal temperature conditions, castables may be allowed to cure naturally, but when exposed to excessive temperatures in hot furnace rooms or to the direct heat of the sun, they should be sprayed with water or covered with wet bags to keep them moist so as to avoid drying for at least 24 hours. All castables should be protected from freezing after pouring until they are dry.

All the above rules are general in nature. The specific instructions for a particular brand of castable should be studied carefully before using.

*such as are used in structural concrete, with refractory castables." Surely the latter would be an improvement (but compare the language of this description with that of the spark plug cleaning, which is designed for a different situation). Our point is not to urge an abundant use of the imperative mood, or of any other stylistic or grammatical form, but to recommend an intelligent awareness in the use of all of them. If the imperative mood is well suited to the first of the examples above, then it must also be well suited to the second. Are there other places where it should or should not have been used?*

*And finally, you might like to make that title grammatically parallel.*

The third example of the description of a process is set up in the familiar form of directions for a laboratory project. The reader for whom it is intended is a student in a beginning course in chemistry.

Because this description is brief, and was prepared for use by students in a laboratory, the introduction is virtually limited to an answer to the question, "Why is this process performed?" There is no need to explain why the description was written, or to answer the other questions likely to come up in the writing of an introduction to a description of a process. The equipment is itemized; the steps are numbered. It's a fine, clear presentation, and fundamentally a good way to describe a process—provided the circumstances of the reader permit so limited a discussion. What "limited" means here is interestingly illustrated by a comparison of the content of the "Procedure" section with the section labeled "Mixing" in the preceding report on castable refractories. In the latter, much attention is given to explanations of why certain substeps are taken—in contrast to the blunt directions in the "Procedures" section of this description of the experiment on color in oranges. The reason for this difference no doubt lies in the fact that the author of the description for the students assumed they were being given lectures and reading assignments on the theoretical (explanatory) aspects of what they were doing.

We have omitted a brief "Discussion" section and a bibliography that appeared in the original.

## Identity of Artificial Color on Oranges*

*presented by the*
*Educational Services Staff*
*Food and Drug Administration*

Sometimes oranges get ripe as far as taste is concerned before their skins turn completely orange. The development of the orange color in the peel is sometimes hastened artificially by exposing the oranges to ethylene gas in a special chamber designed for the purpose. The ethylene gas reacts with the pigmenting material in the peel to develop the color very much as it would naturally develop if the orange were left on the tree longer. Nothing is added to these oranges, and no special labeling is required.

But not all oranges respond satisfactorily to the ethylene treatment, and growers prefer to use artificial color. You may have noticed that some oranges are individually stamped "Color Added" or "Artificially Colored," while others do not bear this stamp. The uncolored oranges may be streaked with green, although they are ripe to the taste.

The color is permitted for use in making the oranges more attractive

*FDA Publication No. 54. Science Project Series. Food and Drug Administration, U.S. Department of Health, Education, and Welfare. Adapted from *Journal of the Association of Official Agricultural Chemists, 43* (1960).

provided they are in fact ripe and otherwise of acceptable quality, and provided the consumer is notified by the stamp that artificial color has been used. The Federal Food, Drug, and Cosmetic Act (FD&C Act) prohibits the use of the color, however, when it would serve to conceal inferiority or damage, or to make the product appear better than it really is.

The color most often used on oranges is called Citrus Red No. 2. It was developed specifically for the purpose after it was shown that a previously permitted orange color could produce injury to test animals when it was consumed in large quantities.

Like all other colors used in foods, drugs, and cosmetics, the colors used on oranges must be specifically authorized for this use, and the amount used must be within the safe tolerance limit set by the Food and Drug Administration. In addition, a sample from every batch of the color produced by the manufacturer must be submitted to the Food and Drug Administration for tests. If the color is found to be pure and suitable for use, it is "Certified."

The color previously permitted on oranges, but now banned as not having been proved safe, is called Oil Red XO. It is possible—although unlikely—that some growers or shippers might use this nonpermitted color by mistake.

Perhaps it is more likely that Citrus Red No. 2 would be used, but the oranges not stamped. Such undeclared use of color would be a violation of the Federal Food, Drug, and Cosmetic Act and of State laws.

The *spectrophotometric absorption curve* of a color is preferred as a means of identification. However, in the absence of a suitable spectrophotometer, the probable identity of a color can be determined by other means, such as chromatographic properties.

In the following experiment, you can use a paper chromatographic technique to test colored oranges to determine that the proper color has been used, and to determine whether undeclared color was used on oranges not stamped "Color Added."

## Experiment

*Problem:*

To determine by ascending paper chromatography which colors are added to artificially-colored oranges, and whether colors have been used on oranges without proper labeling.

*Equipment needed:*

2 funnels, 125 mm.
6 glass rods
2 pipettes, 25 ml.
2 Erlenmeyer flasks, 125 ml.
2 beakers, 100 ml.
1 micro pipet; a capillary or melting point tube drawn to a fine point
1 steam bath

1 funnel support
2 sheets Whatman No. 3 MM filter paper, 7″ × 9″
1 chromatographic tank, inside dimensions 8″ × 9″ × 4″. Any glass or stainless steel tank of the approximate size will do. Small aquariums are useful and inexpensive.

*Reagents:*

cotton
oranges, at least 3 stamped "Color Added"
and 3 not so stamped
chloroform, 300 ml.
light mineral oil, 5 grams
ethyl ether, 95 ml.
acetone, 200 ml.
distilled or deionized water, 100 ml.
Citrus Red No. 2, 1-(2, 5-dimethoxphenylazo)-2 naphthol, about 0.1 gram
Oil Red XO, 1-xylylazo-2-naphthol, about 0.1 gram

## Advance Preparation

Place the Whatman filter paper sheet so that the 9″ dimension is vertical and the 7″ dimension is horizontal. Using a soft lead pencil, draw a horizontal line across the sheet 1 inch from the bottom edge. Then mark off on the line three 1-½″ segments about ½″ apart. Label the 1-½″ segments as follows:

*First sheet*
1. Citrus Red No. 2
2. "Color Added" oranges
3. Oil Red XO

*Second sheet*
1. Citrus Red No. 2
2. Oranges not stamped "Color Added"
3. Oil Red XO

Make a solution of mineral oil in ethyl ether by dissolving 5 grams of the oil in 95 ml. of the ether. Stir until well mixed. Transfer to a 100 ml. graduated cylinder. Immerse one rolled sheet of the paper in the mineral oil solution for a few minutes. Remove and dry it by suspending it in air. Treat the second sheet in the same way. (Note—Ether is highly inflammable and should be kept away from flames or electric heating elements.)

Make a mixture of 130 ml. of acetone and 70 ml. of distilled water. Stir until mixed and store until ready for use in a glass-stoppered bottle.

Place a small piece of cotton in the bottom of a 125 ml. filtering funnel supported on a stand. Position three glass rods in the funnel in such a manner that they will support an orange so it will not touch the sides of the funnel.

Prepare standard solutions of the two dyes as follows: dissolve 10 mg. of Citrus Red No. 2 in 50 ml. of chloroform. Store in a glass-stoppered, labeled bottle, away from light.

Dissolve 10 mg. of Oil Red XO in 50 ml. of chloroform. Store in a glass-stoppered, labeled bottle, away from light.

### Procedure

1) Set water to boil if steam bath is not available (for step 4).

2) Place a 125 ml. Erlenmeyer flask beneath the funnel and support an orange on the glass rods. Wash the color off a "Color Added" orange by spraying it with 25 ml. of chloroform in the form of a fine stream from a pipet. (Note: Use rubber tube on end of pipet. Avoid getting chloroform in your mouth.) Surface oils, waxes, and natural pigments, as well as the artificial color will be washed off.

3) Repeat step 2 with two more of the "Color Added" oranges, combining the washings in the same flask.

4) Transfer a portion of the solution in the flask to a 100 ml. beaker. Allow the solution in the beaker to evaporate by placing the beaker on a steam bath in the hood, or over a suitable container of hot water, if a steam bath is not available. When the solvent has evaporated, add another small portion of the solution. Continue the evaporation in this manner until all of the solution has been transferred and all of the solvent has evaporated.

5) Using another 125 ml. flask and a clean funnel, glass rods, pipet, and beaker, repeat steps 2, 3, and 4 above, using oranges not stamped as artificially colored.

While waiting for the chloroform in the two beakers to evaporate, proceed with step 6.

6) Pour the prepared solution of acetone and water into the bottom of the chromatographic tank. Transfer a 50 microliter portion (need not be measured exactly) of each of the prepared solutions of Citrus Red No. 2 and Oil Red XO to the marked sheets of filter paper prepared as indicated above, by dipping a pointed capillary tube or melting point tube into the solutions and drawing the liquid in a band along the appropriate 1-½″ lines on the filter papers.

Permit the chloroform to evaporate from the paper for about 5 seconds, and retrace the line with the tip of the pipet twice more. Permit drying between applications. Finally, suspend the papers in air to dry.

7) Allow the beakers to cool after removing them from the steam bath. Dissolve the residue in each beaker in 3 ml. of chloroform. Transfer a 50 microliter portion of the "Color Added" extract to the first sheet of prepared filter paper in the same manner as you did with the authentic dye solutions; repeat on the second sheet with the extract from the presumably uncolored oranges.

8) Lower the two sheets of filter paper in the chromatographic tank so that the 9″ dimension is vertical and the 7″ dimension is horizontal, and suspend from a rod in such a way that the bottom edge of the paper (with the 1-1/2″ marked segments) is immersed about 1/4″ in the solvent. Do not allow the papers to touch each other or the sides of the tank. Cover the tank and allow the papers to remain undisturbed for about 1 hour.

9) Remove the papers from the tank and suspend in air to dry. For each sheet of paper, measure the distance the colors have traveled. By comparing the distance traveled by the known color solutions with that of the solutions washed from the oranges, you can determine which one of these colors was used on the "Color Added" oranges, and whether either of them was used on the other oranges without the required declaration. Usually Citrus Red No. 2 will travel a distance of about 2-1/4″ from the point of origin, and Oil Red XO will travel about 3/4″ from the origin. The natural coloring materials present will remain at the origin.

Special cautions

Preparation of ether-mineral oil solution for treatment of the chromatogram papers should be carried out in the hood.

Evaporation of chloroform solutions should also be carried out in the hood, over a steam bath if available.

## SUGGESTIONS FOR WRITING

The preliminary remarks in the "Suggestions for Writing" in the preceding chapter, "Descriptions of a Mechanism," apply equally well to the selection of a process to describe, so we will not repeat them here. Perhaps we should remind you, however, of the importance of deciding whether you are describing a process or giving directions about performing a process.

Many of the same general areas of subject matter that were considered as sources of topics for the description of a mechanism are appropriate here. As before, the first of the three topics listed under each general subject will be simple, and suitable for a detailed treatment, and the second and third will be progressively greater in scope. And, again as before, we assume that these topics will suggest others to you, some of which you may prefer.

1. Aircraft

   Refinishing the fabric of a fabric-covered airplane
   Landing a single-engine, propeller-driven, light airplane
   Landing a twin-engine propeller-driven airplane

2. Boats

    Tacking a small sloop
    Installing a steering-wheel, including cables and fittings, on an outboard motorboat
    Constructing a wooden or fiberglass boat of 16 to 20 feet in length in your backyard

3. Automobiles

    Changing a tire
    Installing a separate radiator to cool the oil of an automatic transmission
    Rebuilding a worn engine

4. Agriculture

    Attaching a plow to a Ford tractor
    Making an earth dam to create a pond or tank
    Planting, cultivating or weeding, and harvesting a crop (specify the crop)

5. Tree-care

    Making a graft
    Taking down a large tree in a congested area in a city
    Taking proper care of a farm wood-lot

6. Electricity

    Making a simple toy motor
    Installing electric heating in the living room of a house
    Installing a ham station in an automobile

7. Carpentry

    Making a small article like a tie-rack or a foot-stool
    Adjusting a radial arm saw
    Constructing the framework of a summer cottage

8. Metal working

    Making a silver-solder joint
    Making a green sand mold
    Making a utility trailer

9. Optics

    Checking the quality of binoculars on sale in a store
    Grinding a lens for a telescope
    Reconditioning a camera

10. Laboratory

    Making a thermocouple
    Carrying out a laboratory procedure such as determining the flash and fire point of an oil
    Making an annealing oven

11. Masonry

    Making a concrete bird bath
    Making a field-stone and mortar wall
    Making a fireplace

12. Gardening

    Measuring the pH of a soil
    Raising a crop of tomatoes from seed
    Constructing a small greenhouse with provision for control of temperature, humidity and light

# 8
# Classification and Partition

## Introduction

If you were to list, just as they occur to you, all the terms you could think of which name kinds of engines, you might write down a list something like the following: steam, internal-combustion, in-line, aircraft, radial, diesel, gasoline, marine, automobile, two-cycle, four-cycle, rocket, jet, eight-cylinder, six-cylinder, and so on. Such a list, quite apart from its incompleteness, obviously makes little sense as it stands; it has no order or system. If you were then to experiment with the list further in an effort to bring order and meaning to it, you would probably rearrange the items in the list into groupings, each grouping in accord with a certain way of thinking about engines. In other words, you would list kinds according to a point of view. Thus the term "internal combustion" might suggest a grouping according to where the power-producing combustion occurs and would give you two kinds of engines: internal-combustion engines and external-combustion engines (steam engines suggest the latter type.) Other terms of the list would naturally suggest other ways of grouping engines: according to cylinder arrangement, use, number of cylinders, and so on. You would, in fact, be on the way to making a classification of engines, for classification is the orderly, systematic arrangement of related things in accordance with a governing principle or basis. The classifier notes the structural and functional relationships among things which comprise a class.

In recording these relationships, the classifier employs certain conventional terms. Acquaintance with these convenient terms will make the rest of what we have to say easy to follow.

*Genus and Species.* A genus is a class; a species is a subdivision within a class. If "engineering subjects in college" is the genus, then mathematics is a species; if mathematics is the genus, then algebra, geometry, and calculus are species; if calculus is the genus, then differential, integral, and infinitesimal are species. These two terms, genus and species, are very commonly used, but many others can be used if a more complex classification is needed. Recent classifications of animal life, for instance, give as many as 21 categories, from subspecies through species, subgenus, genus, subtribe, tribe, subfamily, family, superfamily, infraorder, suborder, order, superorder, cohort, infraclass, subclass, class, superclass, subphylum, phylum, and finally kingdom, the broadest group of all. Elaborate classifications like this are designed to tell all that is known by man about the structural and functional relationships among the individuals of the classifications.

*Classification.* The term "classification" has a loose popular meaning and a more precise technical one. Popularly, classification is almost any act of noting relationships. Technically, classification is the act of locating a specimen of all the different kinds of objects which possess a given characteristic or characteristics. Initially, of course, classification must begin with the recognition that different things possess similar characteristics. Suppose that one day you happened to see a strange creature swimming around in the water, a creature with the body of a horse, feet like a duck, and a tail like a whale (we're thinking of some local statuary). You'd probably only stare; but if you presently saw a second creature just like the first except that it had a tail like a salmon, you'd possibly say, "There's another of *those things!*" And if, soon after, you saw a third, slightly different from the first two, you might be moved to think up a name (like Equipiscofuligulinae) for the whole family and to spend many years thereafter hunting for new species, and giving them names. You would be classifying.

*Logical Division.* When you got around to sending off some papers to the learned journals on the discovery described above, you would find yourself engaged in logical division. By this time, you would have found all, or at least all you could, of the existing species, and so would have completed your classification. You would write, "The genus Equipiscofuligulinae is made up of 17 species. . . ." In thus dividing, into 17 parts, the collection that had previously been made, you would be doing what is technically called "logical division." In short, classification and logical division are the same gun seen from opposite ends of the barrel.

In report writing it is usually logical division, not classification, that you will be concerned with; nevertheless, the term "classification" is the one you are likely to use even where "logical division" is technically correct. You might write, "Tractor engines can in general be classified as full diesel, modified diesel, and gas." This is logical division simply because it is a division, into three groups, of information already known. But it is likely that a scientist or engineer would say "classified" rather than "logically divided" —to judge from our own acquaintance with technical literature. For that reason, and for convenience in general, we shall use the term "classification" to mean either logical division or classification. And we shall be chiefly concerned with logical division, since the report writer is almost always concerned with arranging a collection of facts or ideas, in order to discuss them in turn, rather than with hunting down new species. After all, the hunting will necessarily have been done before the writing starts.

If we're going to use the term "classification," you may wonder why we bothered to distinguish between it and logical division at all. The reason is twofold: first, just to get down all the facts; and second, to avoid confusion when we go on to the next term on our list.

*Partition.* Partitioning is the act of dividing a unit into its components. The parts do not necessarily have anything in common beyond the fact that they belong to the same unit. A hammer may be partitioned into head and handle. *Hammers* may be logically divided according to the physical characteristics of their heads as claw, ball peen, and so forth. Classification, or logical division, always deals with several (at least two) units. Partition deals with the parts of only one unit. A hammer is a single unit. A hammer head without a handle is not a hammer. The head and the handle are parts of a single unit. You have probably become familiar with a variety of partitioning in a chemistry course when you determined the components of a chemical compound. Partitioning is further discussed later on in this chapter.

*Basis.* Suppose we go back for a moment to the strange creatures with a body like a horse, feet like a duck, and a tail like a fish or a whale. And suppose we imagine two men discussing them after only a very brief glance as the creatures swam by. The first man, A, catches only a glimpse of the body, and he immediately says, "Why, these animals are clearly related to the horse." But B, who is a little slower, sees only the tail, and he protests, "Not at all! They're obviously fish!" Now, what these men need is a basis of classification. Classified on the basis of their main body structure, the creatures are horses. Classified according to their tail structure, they are fish or whales.

Or we might go back to an illustration used at the beginning of this chapter. How should we classify engines? On the basis of power, use, the kind of fuel they use, or where the combustion occurs? The manufacturer who is looking for a gas engine to use in the power lawn mower he wants to make and sell would not thank you for a classification of small engines according to the place where the combustion occurs. What basis would interest him? Power? Weight? Cost?

These terms, then, are the ones to remember: genus and species, classification, logical division, partition, and basis. The rest of this chapter is devoted to a consideration of the times when the techniques of classification and partition may be found useful, what rules govern the use of classification, what rules govern the use of partition, and what particular writing procedures may be involved.

## When Is Classification a Useful Expository Technique?

The foregoing discussion has suggested why classification is a useful technique of exposition: it permits a clear, systematic presentation of facts. When to use this technique depends on whether a writer is dealing with classifiable subject matter and whether his writing can be made more effective by means of the technique.

To get an idea of when the technique may be usefully employed, let us consider a specific writing problem. Let us suppose a writer has to write about a large number of different vat dyes. Let us also suppose that he needs to inform his readers about their properties so that their applications will be fully understood. He could, of course, discuss each dye in turn, giving all the pertinent information about the properties and characteristics of dye No. 1, dye No. 2, and so forth. But let us suppose that the writer knows that there are several points of similarity among the forty or fifty different dyes he has to discuss. He knows, for instance, that they all fall into two groups so far as their derivation is concerned. He knows, further, that some of them are alike in dyeing behavior. If these relationships among the dyes are important to an understanding of the use of the dyes, he may decide that the information he has to present will be more meaningful and more readily usable if he classifies the dyes rather than discusses them one at a time. His reader will be helped if he points out the relationships which exist among them instead of leaving these relationships for the reader to discover.

Classification, then, is useful when you have a number of like things to discuss, among which there are points of similarity and difference which it is important for the reader to understand. Obviously, however, the relationship among the things classified must be a significant one. Consider these items:

Lead pencil
Lead, tetraethyl
Lead, Kindly Light

There is a point of similarity among the items in this list, but it is difficult to imagine that it could be very significant!

## Suggestions or "Rules" to Follow in Presenting a Classification

If a writer decides upon classification as an effective way of presenting related facts, he needs to follow a number of "rules," all of them simply common-sense suggestions for clarity and meaningfulness. There are seven rules altogether.

A. *Make Clear What Is Being Classified.* Making clear what is being classified requires a definition of the subject if there is a question as to whether the reader will be familiar with it. For instance, "colloids" would need definition for some readers before a classification should be begun. Although a formal definition of a classifiable subject is rarely necessary in reports—the nature of the discussion will already have made it clear—remember that grouping the related members of a class will mean little to a reader who does not know what you are talking about in the first place.

B. *Choose (and State) a Significant, Useful Basis, or Guiding Principle, for the Classification.* The basis of a classification governs the groupings of members of a class. If we were to classify roses according to color, each species in our listing would necessarily name a color. Color would be our basis. Thus we would list *red* roses, *pink* roses, *white* roses, and so on until we had named every color found in rose blossoms.

It is possible to classify most subjects according to a number of different bases, some of them informatively significant, and some of them unimportant or of limited importance. Let's consider another example. A classification of draftsmen's pencils according to the color they are painted would be of no value at all, except perhaps to the aesthete who prefers a yellow to a blue pencil. Disregarding personal tastes about color, the draftsman would choose a pencil with lead of a desired hardness or softness. In short, a significant, informative classification could be made according to a basis of hardness or softness of lead, but not according to the color of the encasing wood. The basis should point to a fundamental distinction among the members of a class.

A word or two about a commonly chosen basis for classification: *use*. Everyone is familiar with numerous, practical classifications of objects according to the use to which they are put. A common example may be seen in the terms "sewing machine oil" and "motor oil." What we want to call your attention to is this: Classifications according to a *use* are of limited value except for those who understand what the qualities are which make an object particularly suitable for a special use. In other words, the *real* basis for such a classification is not use at all; the real basis is the qualities or properties that make the various uses possible. To a person with any technical knowledge at all, the terms "sewing machine oil" and "motor oil" automatically suggest a possible real basis of the classification—viscosity. Before employing use as a basis for any classification, ask yourself two things: (1) Is *use* really what I want my reader to understand? Or (2) is the quality or property which distinguishes an object for a special use what I want my reader to understand? We do not mean to suggest that classifications should never be made according to a basis of use; as a matter of fact, they may be very helpful. But do not confuse a use with a quality or distinguishing characteristic.

Finally, we advise stating the basis, clearly and definitely, as a preliminary to naming members of a class. It is true, of course, that this rule need not always be followed; sometimes the basis is clearly implicit, as for example, in the color classification of roses mentioned earlier. But in general, it is a good plan to put the basis in words for the reader to see. The actual statement helps guarantee that your reader will understand you and also helps you stick to the basis chosen.

C. *Limit Yourself to One Basis at a Time in Listing Members of a Class.* Limiting yourself to one basis at a time is simply common sense. Failure to do so results in a mixed classification. This error results from carelessness, either in thinking or in choice of words. The student who wrote, for instance, that engineers could be classified according to the kind of work they do as mechanical, civil, electrical, petroleum, chemical, and research was simply careless in thinking. A little thought would have suggested to him that research is not limited to any special branch of engineering. This error is obvious, especially once it is pointed out, but the other kind mentioned—improper choice of terms—is not as obvious. Just remember that the *names* of the members of the class should themselves make clear their logical relationship to the basis which suggests them. An author, in illogically listing fuels as "solid, gaseous, and automotive," may actually have been thinking correctly of "solid, gaseous, and liquid"; but, no matter what he was thinking, the term "automotive" was illogical. Still another practice to avoid is the listing of a specific variety instead of a proper species name, as listing fuels as gas, liquid, and *coal* (instead of "solid").

D. *Name All the Species According to a Given Basis.* In making this suggestion that every species be listed, we are simply advising you not to be guilty of an oversight. As we pointed out earlier in this chapter, a writer would scarcely consider using classification as a method of presenting facts unless he has the facts to present. Just as important is the need for telling the reader what the limitations are upon the classification you are presenting, so that he will not expect more than it is your intention to give. A complete classification, or one without any limitations placed upon it, theoretically requires the listing of every known species; and sometimes species exist which it is not practical to list. A classification of steels according to method of manufacture, for instance, would not need to contain mention of obsolete methods. Limiting a classification means making clear what is being classified and for what purpose. Thus a classification of steels might begin, "Steels commonly in use today in the United States are made by . . .," with the rest of the statement naming the methods of production. In this statement three limitations are made: steels made by uncommon methods of production are neglected, steels made by methods of the past are omitted, and finally steels made in other parts of the world are ignored (though these latter might, of course, be made by methods originating in the United States).

E. *Make Sure That Each Species Is Separate and Distinct—That There Is No Overlapping.* The species of a classification must be mutually exclusive. This is clearly necessary, for the whole purpose of the classification is to list the *individual* members of a group or class; it would be misleading if each species listed were not separate and distinct from all the others. Usually, when the error of overlapping species is made, the writer lists the same thing under a different name or, without realizing it, he shifts his basis. Classification of reports as research, information, investigation, recommendation, and so on illustrates this error, for it is perfectly obvious that no one of these necessarily excludes the others; that is, a research report may most certainly be an investigation report or a recommendation report. To guard against this error, examine the listing of species you have made and ask yourself whether species A can substitute for species B or C or for any part of B or C. If so, you may be sure that you have overlapping species.

F. *Help Your Reader Understand the Distinction between Species.* When classification is being used as an expository technique, make your reader understand each individual species. Ensuring understanding may require that you discuss each species, giving a definition, description, or illustration of each—perhaps all three. In a discussion of steels a writer might, according to a basis of the number of alloying elements, list binary, ternary, and quaternary alloys. He would then want to inform his readers, unless he was certain they already understood, what each of these terms means, what alloying elements are used, and what special qualities each steel possesses. What we are talking about here is not peculiar to classification writing; it is the same old story of developing your facts and ideas sufficiently so that your reader can thoroughly understand you.

G. *Make Certain That in a Subclassification You Discuss Characteristics Peculiar to That One Subclassification Only.* Suppose you had classified grinding wheels according to the nature of the bonding agent used in them as vitrified, silicate, and elastic and then, in discussing elastic wheels, had pointed out that they are further distinguished by being made in several shapes, including saucer, ring, and so forth. Reflection will show that shape is not a distinguishing characteristic of elastic grinding wheels alone, but of all species, regardless of the bonding agent. It is clear, therefore, that while shape may be a suitable and useful basis for classifying grinding wheels, it is an unsuitable one for subclassifying elastic wheels. What is wanted for a thorough exploration of the subject of elastic wheels is a characteristic, or basis, peculiar to them and to none of the others. Thus you might subclassify elastic-bonded grinding wheels according to the specific elastic bonding material used, as shellac, Celluloid, and vulcanite. These would constitute subspecies which could not possibly appear under the heading of vitrified or silicate-bonded grinding wheels. In other words, you would have pointed out something significant about this particular kind of grinding wheel and not something characteristic of *any* grinding wheel.

Whenever you find yourself employing a basis for a subdivision which is applicable to the subject as a whole, you can either use it for the latter purpose, if you think it worthwhile, or incorporate the information into your prefatory discussion of the subject proper. An introduction to a discussion of grinding wheels, for instance, might contain the information that all kinds of grinding wheels, however made, come in various shapes.

When the process of subdividing a subject is followed to a logical end, a point comes when no further subdivision is possible. At this point, one is dealing with varieties of a species. We might classify safety razor blades according to the number of cutting edges as single-edge and double-edge blades. In further discussion of single-edge blades we could point out that they are of two specific kinds, depending upon whether they have reinforced backs or not. Then we could say that the Gem blade is a variety of single-edge blade with a reinforced back and that the Enders and Schick are varieties of unreinforced single-edge blades. And that's about as far as we could go with our subdivision. We would have reached the end of the possibilities along that particular line of inquiry. Note that in discussing single-edge blades we used a principle for subdivision peculiar to single-edge blades alone—obviously double-edge blades could not have reinforced backs.

## A Note on Partition

Earlier in this chapter we defined the term "partition"; now we would like to comment briefly on the use of partition in exposition. Classification, as we have seen, is a method of analysis (and exposition) which deals with plural subjects. You can classify houses, for instance, by considering them from the point of view of architectural style, principal material of construction, number of rooms, and so on. But you cannot classify *a house* except in the sense of putting it into its proper place in a classification which deals with *houses*. You can analyze a particular house, however, by naming and discussing its parts: foundation, floors, walls, and so on. This analytical treatment of a single thing (idea, mechanism, situation, substance, function) is called "partition" or, simply, analysis. As you know, it is a familiar and useful way of dealing with a subject.

The classification rules we have discussed also apply to partitioning. Let's review the especially pertinent ones:

1. Any breakdown of a subject for discussion should be made in accordance with a consistent point of view, or basis, and this basis must be adhered to throughout any single phase of the discussion. Furthermore, this point of view must be clear to the reader; if it is not unmistakably implicit in the listing of parts, it must be formally stated. You might, for instance, partition an engine from the point of view of functional parts—carburetor, cylinder block, pistons, and so on—or from the point of view of the metals used in making it, such as steel, copper, aluminum. The importance of

consistency in conducting such a breakdown is too obvious to need discussion.

2. Each part in the division must be distinctly a separate part: in other words, the parts must be mutually exclusive.
3. The partitioning must be complete, or its limitations must be clearly explained. It would be misleading to conduct an analytical breakdown of an engine and fail to name all of its functional parts. For special purposes, however, incompleteness can be justified by a limiting phrase, such as "the chief parts employed. . . ."
4. Ideally, a subpartitioning of a part should be conducted according to a principle or a point of view exclusively pertinent to the part. It would be inefficient, for instance, to conduct an initial breakdown of a subject according to functional parts and then turn to a subpartitioning of one part according to metallic composition if all parts had the same composition. Besides being inefficient, since a general statement about composition could be made about all parts at once, such a subpartitioning would be misleading if the reader were to get the idea that the metallic composition of the particular part under discussion *distinguished* it from other parts of the engine.

You do not need us to urge you to break down a subject for purposes of discussion. You would do it anyway, since it is a natural, almost inevitable, method of procedure. After all, a writer is forced into subdividing his subject matter for discussion because of the impossibility of discussing a number of things simultaneously. What we do want to emphasize is that you follow logical and effective principles in carrying out such divisions.

## Conclusion

Here is a restatement in practical terms of the fundamental ideas to keep in mind when you present information in the form of classification. These ideas will be stated as they would apply to the writing of an article of classification—such as the article on the subject of the abrasion of fire-clay refractories, reprinted at the end of this chapter. Remember that classification is a writing technique, not a type of report, and like all writing techniques it must always be adapted to the context in which it happens to appear.

1. Devote your introduction to general discussion, including definition when necessary, of the genus which is to be classified. Any-

thing you can say which will illuminate the subject as a whole is in order. It may be advisable to point out the particular value of classifying the subject, the limitations of the classification, a variety of possible bases for classifying the subject besides the one (or ones) which will be employed, and your own specific purpose. Be sure to state the basis unequivocally.

2. List the species, either informally or formally (as in the example on refractories), and then devote whatever amount of discussion you think is needed to clarify and differentiate the listed species. Subdivision of individual species, according to stated principles for division, may be carried out in the discussion.
3. Write a suitable conclusion (see Chapter 12).

## ILLUSTRATIVE MATERIAL

The following three items illustrate the technique of classification. All of them are clear and helpful; in each there is, nevertheless, an example of a common problem encountered in classifying.

The first item, which was written by a student, is clear and explicit in setting up the basis of classification. Toward the end, however, the author ran into the common problem of having to make a rather arbitrary decision about which genus a certain species belonged in. See what you think of his decision.

The second item, concerned with the grading of eggs, makes clear that a basis of classification is being presented, but it falls into the common difficulty of not defining the basis precisely.

The third item, on the subject of shrubs used in shelter belts, illustrates the problem of a shift in the basis of classification.

### Types of Sailboats

One way to classify modern sailboats is according to the number of hulls. On this basis, there are three general classes: single-hulled, twin-hulled, and triple-hulled. The single-hulled, the largest class and the most familiar, needs no comment. The popularity of both of the multihulled classes is relatively recent, however, having developed only since World War II. Of the two, the triple-hulled, or trimaran, is the more novel. A trimaran is comprised of a rela-

*This classification begins with an explicit statement of the basis: the number of hulls. The three species are then formally stated. Subsequently, each species is discussed in turn. However, the twin-hulled species is taken up last, instead of in second place where it was first listed. There seems little*

tively narrow central hull flanked on each side by a smaller hull to provide stability. This class ranges in size from small day-sailers to ocean-going cruisers. The twin-hulled class is made up of two groups: catamarans and outriggers. The distinction between them depends upon whether the two hulls are the same size or not. In catamarans, which make up much the larger of the two groups, the hulls are the same size and are either identical in shape or are mirror images of one another. That is, if you examined the individual hulls of some catamarans you would find that one side had more of a bulge just aft of the bow than the other side. This bulge might remind you of the camber in the upper surface of an airplane wing. When the two hulls are attached to one another by a "bridge" or deck to form the finished boat, the cambered sides face one another. The theory is that when the boat is sailing heeled over in a cross-wind, or on a "reach," with the windward hull partly out of the water, the cambered side of the other hull creates a force in the water that helps prevent the boat from drifting downwind. This theory is controversial. Like the trimaran, the catamaran ranges in size from small day-sailers to ocean cruisers. The outrigger, the second group in the twin-hulled class, is comprised of a relatively narrow main hull which is given stability by a slender pontoon fastened parallel to it on one side only. Since this pontoon could not very reasonably be called a hull, it is possible that the outrigger should be classified as a single-hull sailboat. On the other hand, boats in the single-hull class are not dependent for stability on anything not a part of the hull—a point on which the outrigger fails to qualify. The outrigger is a very ancient design. It is currently not widely used, and is probably to be found only in the form of day-sailers.

*reason for this change except possibly that the discussion of this class is the most complex of the three sections of the classification.*

*The twin-hulled class is subdivided. The basis of subdivision is explicitly stated in the words, "The distinction between them depends upon whether the two hulls are the same size or not." Then a subsubdivision follows: hulls that are of the same size are subdivided into those that are identical and those that are not identical but instead are mirror images. In this third-order division, the basis is not stated as explicitly as before. Is it clear enough?*

*An ambiguity in the classification is recognized in the sentence beginning, "Since this pontoon . . . ." In strict logic, the outrigger does seem to belong in the single-hull class because, as the author points out, the pontoon can scarcely be called a hull. Yet, as the author also points out, some kind of distinction should be made between outriggers and other single-hulled boats. The problem could be solved by creating one more species, for a total of four, as follows: single-hulled, single-hulled with pontoon support, twin-hulled, triple-hulled. Would this added complexity be worth the logical clarification? Apparently the author thought not, perhaps influenced by the fact that outriggers form a very small class indeed.*

## Eggs*

The four consumer grades for eggs, U.S. Grade AA, A, B and C, refer to specific interior qualities as defined by the "United States Standards for Quality of Individual Shell Eggs."

### Grades

Grade AA and Grade A eggs are of top quality. They have a large proportion of thick white which stands up well around a firm high yolk and they are delicate in flavor. These high-quality eggs are good for all uses, but you will find that their upstanding appearance and fine flavor make them especially appropriate for poaching, frying, and cooking in the shell.

Grade B and Grade C eggs are good eggs, though they differ from higher quality eggs in several ways. Most of the white is thin and spreads over a wide area when broken. The yolk is rather flat, and may break easily.

Eggs of the two lower qualities have dozens of uses in which appearance and delicate flavor are less important. They are good to use in baking, for scrambling, in thickening sauces and salad dressings, and combining with other goods such as tomatoes, cheese, or onions.

### Weights

Six U.S. Weight Classes cover the full range of egg sizes. Only 4 of these 6 classes are likely to be found on the retail market—Extra Large, Large, Medium, and Small. The other two are Jumbo and Peewee. Each of these size names refers to a specific weight class, based on the total weight of a dozen eggs. The weights per dozen eggs in each class, in ounces, are as follows: Jumbo—30, Extra Large—27, Large—24, Medium—21, Small—18, Peewee—15.

The grade letters (U.S. Grade A, etc.) indicate quality only. The weight class is stated separately and it indicates the weight of the dozen

*Eggs are here classified according to two different bases: grade, and weight. But what does "grade" mean? This article is not as explicit in regard to basis as the preceding article was. The meaning of "weight" as a basis is obvious, but all we know about grade, at least at first, is that it is related to quality. What, then, is quality? Comparison of the first two paragraphs reveals that quality has to do with flavor, and with the physical characteristics of the white and the yolk. What these characteristics are it is difficult to tell. The comments on the white might have something to do with viscosity. But what is a "high yolk"? In our experience, when a yolk begins to get "high" everybody immediately and positively loses interest in eating it. (Merriam-Webster's* Seventh New Collegiate Dictionary *says* "high *adj* 2b: beginning to taint."*) No doubt what the writer of the pamphlet actually had in mind was the fact that a fresh, good yolk will form a mound if deposited gently on a flat surface, rather than make a watery puddle. It's a perfectly good sort of test to apply, similar in principle to the familiar slump test for concrete, and probably the word "high" is commonly used in this way by people*

*From "Shoppers' Guide to U.S. Grades for Food," Home and Garden Bulletin No. 58, U.S. Department of Agriculture.

in ounces. Grade A eggs have the same quality whether they are small or large. The only difference is weight. Grade A eggs are not necessarily large; large eggs are not necessarily Grade A.

*who grade eggs. This pamphlet was not directed to experts, however, but to the average shopper. The pamphlet would have been clearer had the writer been more precise about the meaning of "quality," or had he simply declared that grade is determined by how fresh and good the egg is.*

## Insects and Diseases of Siberian Pea Shrub (Caragana) in North Dakota, and Their Control*

*Patrick C. Kennedy*

The Siberian pea shrub (*Caragana arborescens* Lamarck), commonly called caragana, is widely planted in farm and ranch windbreaks and shelterbelts in North Dakota. About 800,000 caragana shrubs have been planted annually in North Dakota in recent years. About 5 million additional caragana seedlings are planted each year in the adjacent Canadian provinces. Caragana is an important, long-lived component of shelterbelts, and its relatively short, dense growth makes it an excellent edge-row species.

The bases for this Note include an insect and disease survey of eastern North Dakota shelterbelts in 1960, an insect survey covering shelterbelts over the entire State during the summer of 1964, an intense insect survey of shelterbelts in five eastern North Dakota counties by Haynes and Stein, and observations by the author from May through October 1965.

Several insects and diseases were found on caragana. The combination of early summer defoliation by blister beetles, late summer defoliation by grasshoppers, premature leaf fall caused by a fungus, and the work of other pests caused considerable injury to caragana, and lessened the protective and esthetic values of many shelterbelts. These damages lessen the desirability of caragana for shelterbelt planting.

*The original of this report is an attractive four-page pamphlet. It is evidently designed to help people having some knowledge of botany who are expected to answer practical questions about keeping caragana shelter belts, or hedges, in good condition. One likely sort of reader would be a county agent who has been getting questions from farmers whose caragana shelter belts aren't doing well. To save space, we have eliminated a "Literature Cited" section from the end of the report, all citations of literature within the text, and photographs. The form of the scientific names of insects appears exactly as in the original.*

*The first paragraph indicates the importance or value of the subject by pointing out how widespread the use of caragana is. The second gives an account of the*

*U.S. Department of Agriculture Forest Service, Research Note RM-104.

Information that will help identify the most damaging insects and diseases found in the surveys, and the damage they cause, is presented below. In addition, currently recommended and practiced suppression measures are listed on page 4 [that is, at the end of the article].

### Grasshoppers

Of all the insects observed during this study, the spur-throated grasshoppers appear to cause the most damage to caragana in North Dakota. The three species of this group collected most prevalently were the two-striped grasshopper (*Melanoplus bivittatus* (Say)), the differential grasshopper (*M. differentialis* (Thomas)), and the red-legged grasshopper (*M. femur rubrum* (DeGeer)).

The major identifying characteristics of the adult two-striped species are two light tan stripes on the upper side which extend from the head to the wing tips. It is about 1-1/4 inches long. The differential grasshopper has the outer sides of the broad part of the hind legs marked with distinctive black bars arranged like chevrons; it is about 1-1/2 inches long. The red-legged grasshopper has bright red hind legs, and is only 3/4 inch long.

Plains grasshoppers usually feed on cereal grains and grasses, but they will feed on many species of trees and shrubs when their regular food is in short supply. L. F. Wilson noted that shelterbelt trees were fed upon and injured heavily only after field crops adjacent to the shelterbelts had been harvested. Since caragana commonly occupied the edge row, it was fed upon first and usually injured the most by defoliation and debarking.

### Blister Beetles

Three common and destructive blister beetles attack caragana in North Dakota: the ash-gray blister beetle (*Epicauta fabricii* (LeConte)), the caragana blister beetle (*E. subglabra* (Fall)), and the Nuttall blister beetle (*Lytta nuttallii* Say). The adult ash-gray blister beetle is 3/4 to 1 inch

*evidence on which this classification of organisms damaging to caragana was established. The third paragraph seems to imply that the genus to be considered will be something like "kinds of damage inflicted on caragana by organisms." Defoliation (two "kinds") and premature leaf fall are mentioned specifically. The fourth paragraph states what subjects will be taken up.*

*With minor exceptions, however, the information presented is actually classified according to two different bases: (1) kind of organism, and (2) kind of control. A shift from the first to the second basis appears in the last section of the report, entitled "Control Measures." Previously, on the basis of kind of organism, weevils formed a class, moths and butterflies another, and sucking insects a third. This third class was subdivided into aphids and plant bugs. In the last section, on the basis of kind of control, weevils, butterflies and moths, and plant bugs are all put into one class and aphids into another.*

*This report is actually fairly easy to use, but it does illustrate a common problem. Suppose, for example, that you are a county agent and you are asked to go out and examine some caragana. The leaves are sort of sickly looking and some have fallen off. The owner is standing right beside you waiting for your advice. You have this pamphlet on caragana in*

long and light gray; the caragana blister beetle is similar but black. The larger Nuttall blister beetle is 3/4 to 1 inch long and colored metallic purple blue with a green sheen.

The adults of all three species feed on caragana during late May and early June. Large numbers may cause severe defoliation, especially if the shrubs are small. The insects may either partially or completely consume the leaves and seed pods. The larvae of most blister beetles feed on grasshopper eggs, and for this reason are considered beneficial.

*your pocket and you get it out to check your conclusions. How quickly can you find the answer you need? (You might try timing yourself before reading our second comment below.)*

### Weevils

Three small weevils or snout beetles were collected from caragana. All three species are common pests of certain agricultural crops and trees, but are not very destructive to caragana.

The sweetclover weevil (*Sitona cylindricollis* Fåhraeus) is common throughout the State. The dark gray adult is about 3/16 inch long and usually feeds on the foliage of legumes, principally sweetclover (*Melilotus* spp.) during the spring and fall. Crescent-shaped notches in the leaves characterize the injury on caragana.

The lesser alfalfa weevil (*Sitona tibialis* Herbst =*S. scissifrons* Say) also occurs throughout North Dakota. It resembles the sweetclover weevil, but is 1/16 to 3/16 inch long, and more steel gray with paler stripes on the back. It is active from May to September. The adult normally feeds on several leguminous crop plants, but has been found on caragana miles away from the nearest cultivated legume crop. So far it has done little damage in shelterbelts, but has destroyed entire plots of caragana seedlings in Saskatchewan nurseries. The leaves are seldom completely eaten on larger caragana.

The red elm bark weevil (*Magdalis armicollis* (Say)) was collected from caragana, but I am not certain whether it will injure caragana. The reddish-brown adult, 1/8 to 3/16 inch long, normally feeds in the cambium of dying or recently dead trees, primarily elms.

### Moths and Butterflies

The larvae of the oblique-banded leaf roller (*Choristoneura rosaceana* (Harris)) and the variegated fritillary (*Euptoieta claudia* Cramer) were collected from caragana.

The leaf roller was the more common of the two species. It is about 3/4 inch long when full grown, and pale green to yellow green with a brown head. Most injury occurs in June, when the larva is found singly in a rolled leaf or in a leaf cluster tied with silk where it feeds on surrounding leaves.

The larva of the variegated fritillary is about 1-1/2 inches long when full grown. The body is orange-red with dark brown spines tipped with black. It feeds on the leaves of caragana, but its usual host plants are violets and pansies. Heaviest feeding occurs in August.

### Sucking Insects

The caragana aphid *Acyrthosiphon caraganae* (Cholodkovsky)) and the caragana plant bug (*Lopidea dakota* Knight) are commonly found on caragana. The aphid—small, green, and soft bodied—is usually found in clusters on the leaves, seed pods, and small twigs in mid-June. When populations are heavy the foliage wilts and leaves fall prematurely.

The plant bug is about 1/4 inch long, and has a black head and legs. The upper thorax and wings are dark brown or black; the abdomen is bright carmine. Populations so far have not been sufficiently high to cause damage.

### Septoria Leaf Spot

A leaf spot fungus, probably *Septoria caraganae* (Jacz.) Diedecke, attacks the shrubs during the summer. Infected leaves become chlorotic and fall 2 or 3 weeks prior to normal leaf fall. The fungus infects leaves of declining vigor. Shrubs infected 1 year are likely to be reinfected the next year because diseased leaves collect under them and become the source of reinfection. From my observations, however, certain individuals appear to be less susceptible to the disease. Such shrubs are always more vigorous and taller than susceptible ones.

### Other Organisms

The caragana seed chalcid (*Bruchoptragus caraganae* (Nikolskaja)), though unknown in North Dakota, is common in Manitoba and Saskatchewan. It infests the seed pods and has caused caragana seed losses up to 24 percent.

A damping-off organism (*Rhizoctonia solani* Kuhn) and a root rot (*Pullicularia filamentosa* (Pat.) Rogers) occasionally infect caragana. Little is known about their distribution and injury.

### Control Measures

Any one of the aforementioned organisms, with the possible exception of the seed chalcid, whether alone or together may weaken or kill caragana. Grasshoppers, blister beetles, and septoria leaf spot have already caused considerable injury in some localities. Small shrubs, newly planted shrubs, and shrubs weak from limited water or other causes need protection. However, caragana of all ages are susceptible to attack. Currently recommended suppression measures for each organism are listed below:

*Destructive organism* *Recommended control*

GRASSHOPPERS:
Apply malathion (57% emulsifiable concentrate), 1 to 1-1/2 pounds per acre, to a narrow strip of unharvested crop plants adjacent to a shelterbelt with power sprayer. Wait 7 days before harvesting grain. Apply before grasshoppers begin to feed on caragana (N. Dak. Ext. Serv. 1965, Wilson 1961).

BLISTER BEETLES:
Mix 2 tablespoonfuls of malathion (57% emulsifiable concentrate) per gallon of water. Fully cover foliage with mixture, using a knapsack or hydraulic sprayer when beetles become numerous late in May (N. Dak. Ext. Serv. 1965).

WEEVILS, BUTTERFLIES, MOTHS AND CARAGANA PLANT BUG:
Same as for blister beetles, except apply when

*As we suggested earlier, a basis of classification that might have been used in this pamphlet, but wasn't, is the kind of damage the shrub has received. Would it have provided a more easily used report? A fact to keep in mind in thinking about an answer is this: if inspection of the damage immediately indicated which organism was causing it, then the classification of the present concluding section is all right; if, however, the observer found himself uncertain as to what was causing the damage, then it would have helped to have everything*

particular insect is abundant and damaging caragana (see text).

CARAGANA APHID:

Mix 1-1/2 teaspoonfuls of malathion (57% emulsifiable concentrate) per gallon of water. Apply profusely to shrubs with a hydraulic sprayer in mid-June to obtain full coverage.

SEPTORIA LEAF SPOT:

Bury or destroy infected fallen leaves when possible. In August, spray foliage for complete coverage with Bordeaux mixture, using 1 pound of copper sulfate and 1 pound of burnt lime in 12 gallons of water. Follow with second and third applications at 10- to 14-day intervals (Walker 1957).

*classified on the basis of the kind of damage. In any event, it is always desirable to avoid shifting the basis of classification if possible. Perhaps avoidance of a shift was impractical here; what do you think?*

## SUGGESTIONS FOR WRITING

1. We can't claim that this first exercise is entirely serious, but it illustrates with great clarity how the selection of a basis of classification affects the way something is looked at. Under each genus below (identified by a capital letter) are two terms designating certain "roles" in our society. The two roles under "A," for example, are that of a member of a legislative committee on taxes and that of the owner of a car ferry. Of course the same man might have both roles, by chance, but that is a complication we don't need to attend to. All we need think about is whether the job of operating a car ferry service would lead a man to classify cars on a different basis from that most naturally chosen by a man with the duties of a member of a tax committee. For each of the two roles listed under each genus, indicate an appropriate basis of classification for the species making up this genus.

A. Automobiles

(a) a member of a legislative committee on taxes
(b) the owner of a car ferry

B. Musical instruments

(a) a composer
(b) an apartment house owner

C. Cattle

(a) the owner of a dairy farm
(b) the manager of a meat packing plant

D. Dogs

(a) an official of the American Kennel Association
(b) a postman

E. Public lectures

(a) a newspaper reporter
(b) a janitor

2. Write an article of classification of one of the following terms. Specify the imaginary reader. Be sure to pick a subject you know a good deal about.

Lawn mowers
Clothes washing machines
Adjustable wrenches
Glues
Paint applicators
Power saw blades for the home-craftsman type of saws
Lighted commercial signs
Fireworks
Detergents
Ink
Golf clubs
Automobile tires
Photographic films
Paper
Big game rifles
Tape decks
Bridges
Motor oil
Wind instruments (musical)
Road surfaces
Student organizations on campus
Intramural sports
Professional courses in ____________
(deal with courses in your field)

3. State a basis for partitioning each of the following terms, and make the partition.

A hacksaw
A straight chair
The electrical wiring in a small house
A tree

# 9
# Interpretation

## Introduction

Interpretation, as we'll use the word in this chapter, is the art of establishing a meaningful pattern of relationships among a group of facts. It differs from formal analysis (see Chapter 8) in that it does not attempt to be exhaustive and is freer of conventional form. It is nevertheless rigorously logical; and formal analysis naturally enters into interpretation rather frequently.

Interpretation, in the sense just indicated, is one of the most important elements of science and engineering. Practical decisions such as where to drill an oil well, or what lightning protection system to use on a stretch of electric power transmission line, are the result of interpretation of a body of facts. So are Newton's laws of motion. Interpretation is a creative activity, requiring both knowledge and imagination. Sometimes the results of interpretation can be at best only tentative, as in long-range weather forecasting. Sometimes the results are fairly certain, as in determining the cause of the failure of a particular gas engine. And there is an extreme in which the results are absolute, as in a mathematical equation where all the factors are exact quantities.

From one point of view the study of interpretation is simply the study of logic, with mathematics as its most stable reference point. From another point of view, however, the study of interpretation is a study of the art of communication, of communicating to other people what you have found out through the application of logic to a certain group of facts. It is the latter point of view with which we shall be concerned in this chapter.

Interpretative writing is often interesting to do because it is an integral part of the whole process of figuring something out. The subject matter itself is likely to be interesting to you. You've started with a challenge. Probably you're asking yourself, "Why did that happen?" Or, "What should I do?" The most abstract and, as we said, "rigorously logical" report often represents a personal experience of intellectual excitement and satisfaction. It is difficult to illustrate this fact with examples of technical writing because, as explained in the chapter on style, the personal element is usually subordinated in technical reports. In the writing done by a good journalist, on the other hand, there may be a strong emphasis upon the personal element. Logic suggests, therefore, that if we could find a skillfully written example of "technical journalism," we might be able to observe good technical interpretation in combination with the personal experience from which it was drawn. Printed below are parts of an article in which just such a combination is found. We are going to show it to you without further comment except to point out that, as will quickly become obvious, the style of the article is not designed for "professional" readers, but instead for the mass audience of readers who have just enough knowledge of flying to be interested in some of the finer points.

## So You'd Like to Fly a Fighter Plane*

*by Mike Dillon*

The F-51 skids a little as we turn from base to final. As the runway comes into line, level the wings and put down full flaps. Lower the nose a bit to maintain one hundred and thirty mph. Make one last cockpit check. Mixture "RUN." Prop set for twenty seven hundred rpm. Gear down and locked. Fuel selector to right tank.

All set. Take a deep breath. Try to relax. It's just an airplane.

Add a little power. Don't let that airspeed drop. The Mustang slides down final and crosses over the fence high. *Damn*, that runway looks short! Nothing to do now but ease the power back and hold her off. Stick back slow. Easy . . . easy . . . don't let her balloon. Add a little power to catch it. With half the runway gone the 51 touches, hops and then settles solidly on all three wheels. We swerve drunkenly as the long, wide nose blocks our vision. Runway is going fast. She's not going to stop! Go around, it's the only thing. Throttle forward. Easy now—remember this big machine has torque. Full throttle and the seven thousand pound

*From *Air Progress*, 23 (September, 1968), 19–20 *ff*. 

fighter begins to accelerate. Stick back hard to get off the ground quick. Can't be much runway left. Right rudder! More! Still more! The plane angles off to the left and into the grass. If it will just fly we can still make it. The trees are a thousand feet away. The big bird bumps and lurches as it hurdles across the turf.

Suddenly it's flying! But the left wing starts to drop. We already have full right rudder. Stick hard right to raise the wing. *Nothing!* It just keeps rolling over. Your head slams hard against the side of the canopy as that left wing plows into the ground. All is violence as the huge Hamilton Standard prop blades churn the earth sending dirt and aluminum flying. The battered F-51 thrashes, wildly, then cartwheels into the air. Back down it comes buckling the right wing and smashing the tail sideways.

Your sense of vision is lost, replaced by a blur. Only your hearing remains unimpaired. Just so you can listen to those horrible tearing, buckling sounds. Now it is the broken left wing that digs in, snapping the fighter around on its nose, right wing and tail. Mortally wounded the bird pauses vertically on its tail for a moment as if it were making one last reach for the sky. Finally, it topples over on its back smashing the Plexiglas canopy—flat.

As suddenly as it began, the violence stops. In the shattered cockpit you hang from the straps, painfully aware that you have just destroyed a great airplane and that you have allowed a great airplane to almost destroy its pilot. What happened? What went wrong?

The answers to these questions is what this report is all about. The accident described actually happened (not to your author).

For those who may think this sort of thing is rare, let us guide you for a few minutes as we nose around a little on this one, relatively small, southwest airport.

Here, inside the hangar, we find the carcass of another 51 that a doctor tore up on take-off in Albuquerque.

Off the runway is the almost buried remains of still another F-51 that a pilot spun in on his second Mustang flight.

Over thataway near the field boundary is the wreckage of an SBD whose owner spun in on his first flight. Note: All of these on one airport.

*Because of limitations of space, we have here deleted a portion of the article in which the author gives a case history of two accidents involving fighter planes. These histories are based on a study made by Dr. R. G. Snyder. The author of the article then asserts that, including the 25 accidents reported by Dr. Snyder, he is himself aware of a total of 55. You might be interested in his figures: one P-38, two Hawker Seafurys, three P-63's, seven P-40's, one SBD, five F8F Bearcats, thirty-four F-51 Mustangs and two Corsairs. In these accidents, 25 people were killed. The article continues as follows.*

There is a consistent pattern to most of these accidents. We believe the root cause can be traced to the pilots not accepting the fact that the surplus ex-WW-2 fighter is not just another airplane.

Four factors set such fighters apart. High wing loading, torque, asymmetrical thrust, and, in the case of the F-51, the laminar flow wing airfoil.

High wing loading is fairly easy to compensate for: You fly the plane faster. Thus, Common Sense Rule 1: Keep your speed up. Read the handbook; know the recommended speeds.

Torque, it seems, is much harder for first-time would-be fighter-type pilots to handle. We figure the reason for this is that many pilots just don't know what the word torque means. So what is torque? When you raise the nose of a Cessna 182 to climb, it will turn left if right rudder is not used to hold the plane straight. This left turning tendency, though often called such, is *not* torque. Torque is not a yawing force, but is, instead, a twisting, rolling force. It is the action-reaction between the propeller and airplane. The engine is twisting the prop clockwise (as viewed looking forward from the cockpit), therefore the prop is trying to twist the airplane counterclockwise (AVLFFTC). This tendency for the airplane to roll in opposite direction of prop rotation is directly proportionate to the amount of power the engine is producing, the rate with which the prop is being accelerated, and the ratio of prop weight to airplane weight. Bear in mind that an F-51 prop itself weighs more than four hundred pounds. When the pilot rams the throttle forward from idle the engine applies a tremendous force, 1000-plus-hp, to twist that heavy prop faster to the right. That same 1000-plus-hp, is also trying to roll your entire machine to your left. At low speeds this rolling force exceeds the amount of control available from the ailerons. To demonstrate this we'll slow the Mustang to 80-mph with full flaps and gear and just enough power to maintain level flight. Now, throttle all the way forward to the stop—As we push the throttle we also feed in full right stick and rudder. The F-51 rolls smoothly and smartly to the left onto its side and stops in a vertical slip, losing two hundred feet of altitude. Hm-m-m. Had we been at fifty feet instead of five thousand, we'd be like maybe dead.

Herein lies Common Sense Rule 2. Avoid abrupt power increases at low speeds in a Mustang. You have 1470-hp out in front. You don't need to use each one of those "horses."

O.K., you say. So that's torque. What *does* make my 182 turn left while climbing? Your author smiles and replies, "Glad you asked, ole buddy boy, 'cause that's our next point—P effect or, if you will, asymmetrical thrust. When any prop-driven airplane is at a positive angle of attack the prop blades at work on the right hand (starboard) side of the engine (assuming right-hand rotation) have a greater effective angle of pitch and therefore "pull" harder than when they get around on the left. . . .

This asymmetrical thrust (or pull) that is so easily controlled in your 182 gives to the fighter a characteristic normally thought of as applying only to multi-engined planes. It has a VMC. That is to say, single engine fighters have a speed below which the pilot cannot maintain control at high power due to asymmetrical thrust.

For instance, in the case of the P-40 the VMC at *cruise* power is about 75-mph. The author determined this empirically—not to mention, accidentally while trying to stall his own ex-fighter plane (a P-40) at cruise power. To shorten a long story I found that if the Warhawk's nose was raised smoothly an extreme angle of climb was necessary to keep the airspeed dropping. By the time the speed had dropped to 80-mph the plane was about seventy degrees nose up. As the speed fell to 75-mph the fighter, without buffeting, and against full right rudder and aileron, yawed and rolled to the left. The nose dropped to about the horizon and the plane sort of stopped inverted. This was a bit confusing at first but several more tries showed the same results. The plane had not stalled. There simply was not enough rudder or aileron control at that low speed to control the torque and that "uneven" pull of the prop blades. C.S. Rule #3, therefore, becomes a composite of Rules #1 & #2. Keep your speeds up to those recommended in the handbook, beware of high power settings at low speeds with high angles of attack. An F-51 in this configuration has much the same control problem as a twin Beech with its left engine feathered. Herein lies an interesting matter of judgment. Quite a few bold pilots think themselves capable of handling a Mustang, but few of these same chaps would be anxious to solo a twin Beech with no prior instruction—yet the Beech is no harder to fly.

The F-51 has one extra quirk—its laminar-flow wing. This low-drag, high-speed airfoil is the main reason for the Mustang's success as a long-range fighter. The prospective atomic-age F-51 pilot should realize that this airplane handles best at speeds above three hundred mph. For example, at 250-mph it takes only 4.5 G's to stall your Mustang. This, coupled with the fact that it usually requires as much as ten thousand feet to recover from a power-on spin, has claimed the life of more than one pilot.

Common Sense Rule #4: When your 51 starts to buffet in a tight turn, ease off a bit—that plane is trying to tell you something. Beware any lack of communication!

To return to our beginning. Why did the first pilot we introduced you to wrap his F-51 up in a ball? First, his decision to go-round was in error. The 51 decelerates very rapidly once on the ground. Said pilot could have stopped on the runway. Second, if the go-round was necessary, the pilot should have retracted his flaps immediately and retrimmed. With full-flaps and nose-up trim, the plane would break ground at a very low speed. Without the rudder being trimmed to compensate for torque and asymmetrical thrust, the pilot probably thought he was using full-right rudder when, in fact, he wasn't. The airplane was at or near its minimum control speed and full rudder travel would have been necessary even to have had a chance of retaining control.

We should explain that we have deleted the final paragraph of the article, but that in it the author says he did not mean to "knock" fighters; that they are a great pleasure to fly if the pilot has adequate training and a suitable temperament.

You probably found this article clear and easy to follow (perhaps with the exception of the comments on asymmetrical thrust, and on this point clarity suffered because we were unable to retain a graphic illustration of the meaning of this term that appeared in the original article). Why is this article so easy to follow, on the whole? One reason is that the author wrote clear, effective sentences. Another is that it is organized with professional skill. There are three specific points at which this skill in organization can be seen.

First, the author states exactly what the reader can expect to find out. And he states it not only once, but twice. The first time is at the end of the initial description of the accident. Remember that this is an article designed for a popular magazine, and that the writer is keenly aware that he must arouse the reader's interest quickly or lose him. So he begins with a few paragraphs of exciting action, and only at the conclusion of these paragraphs does he state what the reader can expect to learn from the article as a whole. The second time he indicates what the reader can expect to learn is when he turns from a description of accidents that have happened to a relatively abstract interpretation of why they happened. In this second statement, which appears directly after our italicized interpolation, he also becomes somewhat more specific. Here are the two statements again, one after the other. Both indicate that the author's purpose is to discuss the cause of the accidents.

(1) What happened? What went wrong? The answers to these questions is what this report is all about.
(2) There is a consistent pattern to most of these accidents. We believe the root cause can be traced to the pilot's not accepting the fact that the surplus ex-WW-2 fighter is *not* just another airplane.

As you will remember, this second statement is followed in the article by an extended interpretation in which an answer is given to the question of what happened, or what caused the accidents. Thus the reader has been told twice, in different terms, what he can expect to find out.

A second point in which the organization makes the article easy to follow is that before the relatively abstract interpretation is begun, the author carefully describes how he has acquired evidence about the seriousness of the problem he is examining. As noted earlier, this part of the article has been deleted and summarized in the italicized interpolation. The author also explains, later, how he has obtained another kind of evidence—that is, how he has learned, through personal experience, about such things as torque and asymmetrical thrust. From a technical point of view, however, this material is quite sketchy, a treatment forced upon the author by the fact that many of his intended readers are not technically trained.

The third point about the organization of the article is that the author indicates exactly how the interpretative part of it will be organized. He does so, as you will recall, in the following paragraph which appears directly after the second statement quoted above.

> Four factors set such fighters apart. High wing loading, torque, asymmetrical thrust, and, in the case of the F-51, the laminar flow wing airfoil.

Each of the four factors is then taken up in turn, in precisely the order stated. The reader is therefore never in doubt as to how the interpretation is organized.

The three points of organization just considered in relation to this skillfully written article are the three principal points requiring attention in virtually any interpretation. Leaving the article now, we turn to an examination of each of these three points in greater detail, with reference to their part in any interpretation you yourself may have to write. Following this examination, the discussion in this chapter will be concluded with a brief review of the place of the scientific attitude in interpretation. The three points of organization are, again, these three questions:

1. What is to be found out?
2. How was evidence obtained?
3. How will the interpretation be organized?

## What Is to Be Found Out?

The first job in writing an interpretation is to tell the reader what it was you wanted to find out when you began the work. The exact problem must be clearly stated. Probably no single part of an interpretation is more important.

What needs to be done in stating a problem clearly is simplified and made easy to grasp by dividing the whole job into its elements. There are, perhaps surprisingly, six of these elements. Each one is in itself a rather simple idea, however. And it isn't necessary to say something about every one of the six every time you state a problem. Just be careful to avoid an omission that would leave the problem unclear. We'll discuss each of the six elements briefly, one at a time.

A. *Acquiring a Thorough Grasp of All Available Information.* This step is, of course, a preliminary to the writing. It is only common sense to know all you can about a subject before writing about it. Sometimes, indeed, thorough knowledge will reveal that a supposed problem is only imaginary, as we discovered once when, after patiently trying to cure a certain plant of what appeared to be a disease attacking its leaves, we finally learned from a book that the leaves were just naturally supposed to look moth-eaten.

B. *Stating the Problem in Concise Form.* Although an expanded definition of a term may be several pages long, somewhere within that expanded definition there usually appears a single, formal sentence definition. Similarly, in the statement of a problem there should usually be a single sentence in which the problem is formally expressed in its most basic form, even though the full explanation of what the problem is may require considerable space. We'll start with some remarks on this concise statement of the problem, with the understanding that the concise statement is only a part of the whole job.

Boiling a complex problem down to one short, simple statement may prove a very keen test of your mastery of the subject. About the best insurance of success you can provide for yourself is to keep asking, "What am I really trying to do?" Try to avoid being dominated by conventional thinking and conventional phrases. Instead of saying, "The problem is to design a community center adequate for the needs of 3000 families," you might say, "The problem is to design a group of buildings in which 3000 families can conveniently secure food, clothing, furniture, hardware, drugs, automotive service, medical care, barber and beauty care, variety goods, and postal service." The second version is less concise than the first, but much plainer. The term "adequate for the needs," from the first version, is given concrete meaning in the second.

Here is another illustration. If you were explaining the problem of designing a tank type of vacuum cleaner, you would not begin with a discussion of nozzles, filters, exhaust areas, and so on. You might say that essentially the problem is to design a cylinder, open at both ends, in which dirt is filtered out of a stream of air drawn through at high velocity by a motor-driven propeller. In short, you would try to get to the very heart of the matter, eliminating for the moment all secondary considerations.

As always, however, the exact phrasing must be fitted to the reader. The statement of the problem of the vacuum cleaner given above would be more suitable for a layman than for an engineer who had been working on vacuum cleaner design for a long time.

C. *Defining Unfamiliar or Ambiguous Terms.* If your reader is to understand what you have to say about a problem, he will surely have to understand all the words you use. If you can't avoid using words he may not already know, tell him what they mean.

D. *Distinguishing between the Primary Problem and Subordinate Problems.* A given problem usually turns out to be made up of a number of subordinate problems. When this is true, the relationship between the primary problem and the individual subordinate problems must be shown. Each subordinate problem must be accurately stated, and its importance relative to the other subordinate problems indicated.

This principle of stating subordinate problems has already been observed in the listing of the four factors by the author of the article on the fighter plane. Moreover, an indication of relationships among these factors is found in the author's assertion that common sense rule 3 becomes a composite of rules 1 and 2. Another simple illustration of the idea of subordinate problems may be considered by imagining that you need to choose a new car. The primary problem here is, of course, the sum of numerous other problems: choice as to appearance, performance, economy, prestige, availability of maintenance services, and so forth. Many of us are good at persuading ourselves, with an assist from the advertisements, that the car we happen to want is the one that most nearly suits our needs. But this is not science. In a technical report, the subordinate problems would each be stated as precisely as possible, and an attempt would be made to evaluate the importance of each of them with respect to the purpose for which the car was intended. For a rural mail carrier, prestige would deserve attention only after such qualities as economy, body rigidity, and performance in mud had been considered. For a salesman, prestige might be more important.

Perhaps you've noticed a slight ambiguity in the terms used above. When is a subordinate problem a problem, and when is it a cause? Torque is a *cause* of airplane accidents, but we called it a problem. The *cause* of a buyer's choice of a particular car may be its economy, but we called this a problem also. The fact is, determination of which word is better (problem or cause) depends upon your point of view. For example: can fighter plane accidents be reduced by solution of the *problem* of torque? Yes. How? By training the pilot. On the other hand, from the point of view of a man investigating an accident it would be natural, as just suggested, to speak of torque as a cause. In the present discussion we will stick to the word "problem" because it is important in reference to the idea of standards of judgment, which will appear shortly.

One other fact should be noted here. If a problem is truly the sum of subordinate problems, a statement of the major problem itself is accomplished by listing the subordinate problems in somewhat the manner illustrated above in reference to the choice of a car. So discussion of the primary and subordinate problems really serves two purposes: it provides a thorough statement of the primary problem, and it clarifies the relationships among the subordinate problems.

So far we have discussed (in addition to the idea of the need for thorough knowledge) three elements: stating the problem in concise form, defining unfamiliar or ambiguous terms, and presenting the subordinate problems. The following short statement of a problem from a classic work on atomic energy will help to illustrate these elements.

> No one who lived through the period of design and construction of the Hanford plant is likely to forget the "canning" problem, i.e., the problem of sealing the uranium slugs in protective metal jackets. On periodic visits to Chicago the writer could roughly estimate the state of the canning problem by the atmosphere of gloom or joy to be found around the laboratory. It was definitely not a simple matter to find a sheath that would protect uranium from water corrosion, would keep fission products out of the water, and would not absorb too many neutrons. Yet the failure of a single can might conceivably require shutdown of an entire operating pile.*

In this paragraph the last portion of the first sentence is a concise statement of the problem, and at the same time a definition of a bit of technical jargon. In the next to the last sentence, the major problem is divided into subordinate problems. No comment is made on the relative importance of the various subordinate problems.

The two major elements that remain to be discussed are both concerned with the relation of the problem to background materials. Considering this relationship under the two different headings will, however, serve a practical purpose.

E. *Distinguishing between What Is Already Known and What Remains to Be Found Out or Decided.* The reader will want to know why the investigation being reported was undertaken at all. If the subject is one that has been investigated previously, one way of justifying the investigation, and of clearing the air in general, is to summarize the state of knowledge up to the point at which the investigation was begun, and to show what further information is needed. For instance, it has happened sometimes that the people of a certain community have engaged a firm of engineers to report on whether some civic improvement, like a power plant, should be undertaken,

*Henry D. Smyth, *Atomic Energy for Military Purposes*, rev. ed. (Princeton, N.J.: Princeton University Press, 1946), p. 146. Reprinted by permission of the author.

and then have fallen into dispute and hired a second firm to make another report. In the second report we should expect a very clear statement of what points have already been agreed on, and then of what new information is being considered or of what additional factors are entering into recommendations.

F. *Giving Background Information.* A professional astronomer would need no explanation of the importance of designing an adequately rigid mounting bracket for a small astronomical telescope, but for a layman a little background information, prior to a discussion of a particular mounting, would be very helpful. He would readily understand that even a slight gust of wind might set up a minute vibration in a poorly supported instrument. And when it was pointed out to him that the resulting vibration of the image would be magnified by as many times as the telescope would magnify the image, the strictly engineering problem of the mounting would take on significance. The amount of background information necessary in any statement of a problem depends upon the reader's familiarity with the subject, but in case of doubt there is little question that some information should be given.

Finally, then, we come to the question of organizing the six elements that have been discussed. What is their proper order in the statement of a problem as a whole? Discussion of the first, acquiring a thorough grasp of all the available information, can be eliminated at once, since it is not a part of the writing itself. We restate the others below before making some general remarks on organization.

Stating the problem in concise form
Defining unfamiliar or ambiguous terms
Distinguishing between the primary problem and subordinate problems
Distinguishing between what is already known and what remains to be found out or decided
Giving background information

A controlling principle in organization is the choice of explication versus synthesis. The problem may be stated near the beginning and then explained (explicated) bit by bit, or it may be stated only after considerable discussion, as the logical summation of the various subordinate factors entering into it (this would be synthesis). Usually explication is more practical, but if there is any reason to feel that your reader may be hostile to your way of phrasing the basic problem, synthesis may be the better. Imagine yourself telling a client that the problem is not whether his office building should be redecorated, but whether it should be torn down!

If explication is used, the order of the list above may be about what you would want. For synthesis, the order would be approximately reversed. Remember, however, that these remarks are no more than suggestions. Too many variables are involved in the process of stating a problem to permit any precise formula, or outline, to be written for all situations.

## How Was Evidence Obtained?

An interpretation can be no better than the data on which it is based. Consequently, a second major part of an interpretation is the provision of any necessary explanation about how the data were obtained, or of a statement of their probable reliability. In a large tank of crude naphthalene there may be considerable random separation of naphthalene and water, and a sample taken at a given point might prove to be 100 percent water. Any discussion of the contents of the tank would be useless unless carefully controlled sampling methods were used. And any reader should refuse to accept a statement about the contents of the tank which did not acquaint him with the method by which samples were taken, or at least with the probable accuracy of the results. Another example of the same principle can be seen in the botanist's complaint that carelessness in reporting the conditions under which an unusual plant has been found growing may rob the find of much of its value.

The point to remember is that the reader should have enough information about the data so that he can make his own inerpretation if he wishes.

## How Should the Main Part of the Interpretation Be Organized?

Having stated the problem, and possibly having commented on the source and validity of the data, the interpreter now must explain the significance of his evidence and state conclusions. He may have a great deal of evidence, and he may feel that, so far as getting his results down in writing is concerned, the situation is little short of chaotic. What he needs to do is to divide his material into units and deal with one unit or factor at a time. Our immediate purpose is to explain how to do this. We will examine the questions of how and where to take up the major factors and of how to present supporting data.

*Stating and Organizing the Major Factors in an Interpretation.* Although this subject as a whole could become extremely complex, there are only three major factors that we need to consider here:

1. The problem in concise form
2. The subordinate problems, or standards of judgment
3. Possible explanations, or possible choices

Before discussing the organization of these factors, we'll consider the meaning of the terms themselves.

We have already discussed the idea of stating the problem in concise form, and of stating the subordinate problems, so here we'll start with the term "standards of judgment." This term is new, but the idea is familiar. For instance, if you'll go back to the quotation from Smyth about the canning problem of uranium slugs on page 206, you'll see that what we called subordinate problems could just as well be regarded as standards of judgment. One of the subordinate problems was to design a sheath that would protect the uranium from water corrosion. To formulate the same idea as a standard instead of as a problem, one would merely write that the sheath must protect the uranium from water corrosion. The kind of sheath *chosen* from all the available kinds must meet this standard.

Let's take another simple example. Suppose your neighbor asks you to stop at the store on the way home to get a dozen eggs. You say you will; but because you have read the explanation of the grading of eggs in Chapter 8 you warily inquire, "What kind of eggs?" The only reply you receive is, "Nice big ones." On the way to the store you reason to yourself, "'Nice' means high in quality, so I'll get grade AA. And the biggest size is the Jumbo, so I'll get that." By this reasoning you have established two standards of judgment. To be acceptable, the eggs must meet the standards of being AA in grade and Jumbo in size. (Often the word "specifications" is used to mean the same thing as our term "standards of judgment.")

Likely enough, however, when you get to the store you discover they don't have any grade AA Jumbo. What they do have are grade A Large and grade AA Medium, plus some others too low in grade and too small to require consideration. So you undertake a quick mental interpretation of this data.

Before thinking about the results of your interpretation, let's ask once more—exactly what is the problem? If the problem is to obtain eggs for your neighbor's children to color on Easter morning, it looks at first glance as if the Large eggs would be an obvious choice; if the problem is to obtain eggs for your neighbor's breakfast, it looks as if the grade AA would be an obvious choice. But what about the price—another problem? If you are really serious about these eggs, what you will want to do is to transform your specific problems into standards of judgment and to make your choice with their help. If the eggs are for breakfast, you might transform your problems into the following standards of judgment. The eggs must (1) be as high as possible in grade, and not less than A; (2) be as large as possible; and (3) be as cheap as possible and in any case cost not more than $x$ cents per pound (eggs of different sizes would have to be compared by weight).

In summary, the difference between a problem and a standard of judgment can be put like this. If you say, "I have to obtain some eggs," you have expressed a problem. But if you say, "I have to obtain some eggs that are not lower than grade A in quality," you have expressed a problem in the form of a standard of judgment. If your object in interpreting a body of data is to make a choice, then you will naturally emphasize the standards of judgment by which your choice can be guided. On the other hand, if your object is merely to explain a body of data, as in the article on the fighter plane, then you may have little interest in standards of judgment.

Our principal purpose is to explain how to state and organize the three major factors in an interpretation—that is, the problem in concise form, the subordinate problems or standards of judgment, and possible explanations or possible choices. We have yet to comment on the meaning of possible explanations or possible choices before turning to the subject of the organization itself.

The term "possible explanations" simply calls attention to the fact that a problem, as well as its solution, may need a good deal of discussion. For example, the problem of torque is explained at some length in the article about the fighter plane. Similarly, the problem of asymmetrical thrust is explained, and some ideas about solving this problem by maintaining proper trim are suggested. The term "possible choices" refers to what might be called the options. In the example of the buying of eggs, two options were considered: grade A Large and grade AA Medium. We might take as another example the problem of buying a car, where many options or choices are available. In a report recommending purchase of a particular model of a car, there would need to be considerable attention given to a discussion of the choices available—perhaps all the choices in a given price range.

With these ideas in mind as to the meaning of the terms we are using, we can now summarize. (The comments that follow are numbered as they were on page 209, where we introduced this subject.) If a writer is taking an explanatory approach to his data—as in the article on the fighter plane—he will problably want to (1) define the problem concisely, (2) present and explain any subordinate problems, and (3) present and explain any possible solutions to these problems. If he is approaching his data with the purpose of recommending a choice or decision, he will then (1) define the problem, (2) establish standards of judgment, and (3) through his interpretation arrive at a choice or decision.

This brings us to the subject of the organization of these elements.

Let's begin with the problem of organizing an "explanatory approach." In simplified form, the organization would look like this:

I. Statement of problem
 A. In concise form
 B. In the form of subordinate problems
II. Presentation of subordinate problem number one, explanation, and (where appropriate) solution
III. Presentation of subordinate problem number two, explanation, and (where appropriate) solution

Additional subordinate problems would be treated as above.

Of course in practice some other matters usually have to be included. The most important of these additional matters are the presentation of data and the summary or conclusion. We will comment on these elements below. Still another possible complication is that sometimes the subordinate problems are interdependent, so that it is difficult to discuss them one at a time. No general solution can be found for this complication, but some of the principles considered below in relating standards of judgment to possible choices are helpful.

We will turn now to the problem of organizing an interpretation in which standards of judgment are employed. Consider the following.

I. Statement of problem
 A. In concise form
 B. In the form of standards of judgment
  1. Standard one
  2. Standard two
  3. Standard three
II. Statement of possible choices
 A. Choice one
 B. Choice two
 C. Choice three

Here, as you see, the writer is faced with the need for an organization that will permit the discussion of *every possible choice* in reference to *every standard.*

A specific practical illustration will help to clarify these principles. Let us consider the problem faced by a salesman in making a choice of a new car to use in his work. Let us assume that he does 90 percent of his driving within a large city, and that he has set up the following standards on which to base his choice:

1. There should be a large trunk, for samples.
2. Operating costs should be low.
3. The price should be low.
4. The performance should be good at low speeds.
5. The appearance should be neat and conservative.

In addition, let's assume that the salesman has narrowed his choice down to four cars: Brands A, B, C, and D.

Only a glance at the list of standards is needed to see that some of the standards will need considerable explanation. (Exactly how large a trunk does he need, and is the shape important? What does he mean by low operating costs? And so on.) With this need for explanation in mind, consider the following three ways of combining the major factors. The standards refer to the list above; the different possible choices of car are represented by capital letters.

*Version 1*

Statement of the problem
Explanation of all the standards
Explanation of why only four cars are to be considered
Judgment of each car in turn according to standard 1
Judgment of each car in turn according to standard 2
[Judgment according to standards 3, 4, and 5, same as above]
Summary of conclusions

*Version 2*

Statement of the problem
Listing of the standards (with very little explanation)
Explanation of why only four cars are to be considered
Explanation of standard 1, and judgment (according to this standard) of each car in turn
Explanation of standard 2, and judgment (according to this standard) of each car in turn
[Explanation of, and judgment according to, standards 3, 4, and 5, same as above]
Summary of conclusions

*Version 3*

Statement of the problem
Explanation of all the standards
Explanation of why only four cars are to be considered
Judgment of car A according to all five standards in turn
Judgment of car B according to all five standards in turn
[Judgment of cars C and D, same as above]
Summary of conclusions

All possible combinations of the major factors are not shown in the three versions above (for example, there might be a section or sections devoted to a general description of each of the cars, either near the beginning or later). However, the three versions do illustrate pretty clearly the kind of decision the interpreter has to make in organizing the major factors of a complex problem.

Which version is the best? For the car problem we would choose the second version. But for other applications it would be impossible to say which one is best without detailed knowledge of the whole situation. The chief point of these remarks, anyhow, is that if you are aware of the various possibilities, then you can select the most suitable organization for whatever subject and problem you have.

It should be remembered that the outlines above are by no means complete, being confined to illustrating relationships among the major factors. Later we shall illustrate a more nearly complete outline of an interpretation, but before doing so we shall make two additional comments about the major factors. These comments have to do with the elimination of possible choices and with the handling of conclusions.

If it should happen that one of the four cars being considered in the foregoing problem appeared to be a most likely choice according to every standard but one, it might nevertheless be necessary to rule out the car on that one point. Suppose, for instance, that car C was excellent in respect to four of the five standards, but that it had an extremely small trunk. If the salesman couldn't carry his samples in the car, it would be no buy for him. Once this fact had been shown, the car could be eliminated completely from any further consideration. This procedure speeds up and simplifies the whole interpretation. But a warning is needed here: Don't eliminate a possible choice on the basis of failure to meet a single standard if there is any chance that that possible choice (in our illustration, car C) would be the best one in spite of the one disadvantage.

A recurrent question about conclusions is whether they should be stated in the body of the interpretation and then restated at the end, or whether they should be stated only at the end. The answer is that almost invariably they should be stated at both points. If, when the four cars are judged by standard 1, car B is found to be superior to the others, a clear and rather formal statement of that conclusion helps prepare the reader to accept whatever final conclusion is offered at the end of the interpretation. Of course where anything like an introductory summary is used, the conclusions appear at the beginning as well.

Another important fact about conclusions in an interpretation is that considerable discussion of them may be needed at the end. This need can be seen in the simple example of the purchase of eggs. If grade A Large eggs are cheaper (by the pound) than grade AA Medium, which would you recommend? People might disagree about this choice. Where a choice becomes as subjective as it does here, the employment of standards of judgment may not be sufficient for a satisfactory decision. At the end of your interpretation, in such a case, you may have to point out the area in which the choice is essentially subjective. You can then either indicate your own choice or leave it to the reader, as circumstances suggest.

In bringing to a close these remarks on organization of the major factors, we shall add a somewhat more detailed outline. The outline below indicates one way of organizing a discussion about the choice of a car: it is by no means the only way, and it is more generalized than would be desirable in practice, but it has the virtue of filling out the introductory portion more than do the three short versions presented earlier, and thereby of removing some possibly misleading implications of the earlier versions. Reference to Chapter 16, "Forms of Report Organization," will illustrate differences in over-all organization that would be desired by certain companies.

I. Introduction
  A. Statement of the problem
    1. Discussion of the need for a recommendation
    2. Concise statement of the problem
    3. Concise statement of the standards of judgment
  B. Scope
    1. Statement of the cars to be considered
    2. Explanation of why the cars are restricted to the group named
  C. Comments on source and reliability of data
  D. Plan of development
    1. Comments on the presentation of data
    2. The over-all plan
II. Judgment according to the first standard
  A. Explanation of the standard
  B. Judgment of car A
    1. Presentation of data
    2. Interpretation of data
  C. Judgment of car B
    1. Presentation of data
    2. Interpretation of data
  D,E.—same as above
III,IV,V.—same as above for the remaining standards
VI. Summary of conclusions

As we said at the beginning of this discussion of interpretation, our objective has been to make suggestions to you about how to communicate to other people what you have learned from examination of a group of facts. In the ingenious essay by M. J. Gelpi entitled "Forcing a Good Decision," in Appendix F, however, you will find some supplementary ideas about the logic of the examination itself.

*Presenting Data.* In addition to organizing the major factors just discussed, some decisions have to be made about the presentation of data. You're likely to commence writing an interpretative report with a thick pile of data at your elbow and questions like the following going around in your head: How much of this data should I put into the report? Where should I put it? What form should it be in? How much should I try to tell the reader about what it means, and how much should I assume he will see for himself?

The question of how much data to include must be answered according to circumstances. A college instructor often asks his students for all the raw data they took, and sometimes all the raw data is included in industrial and research reports. Unless it is quite clear that the raw data should be included, however, it is better to leave most of this material out. If it is put in, it should usually go into an appendix. Don't clutter up the text with it.

Whether or not all the raw data is put in, it must, of course, be sufficiently represented in the body of the interpretation to convince the reader that he understands the situation as a whole. And so we come to the question, "In what form should the data be introduced into the body of the interpretation?" The answer is—in any form at all. But remember, as the architects like to put it, that form follows function. If your purpose is to communicate, then whatever form will best convey your idea is the one to choose. Graphic aids provide a tremendous range of possibilities for the illustration or presentation of factual material, often in very dramatic form (see Chapter 21). In addition, there are such possibilities as presenting small samples of data, providing short lists of key figures or facts from the data, working out a typical or illustrative problem or calculation, summarizing trends in terms of range and percentage of change, and many others. Actually, however, the only special knowledge you need about the form in which data can be presented is an acquaintance with the basic concepts of graphic aids. Other forms in which to present data will arise naturally out of the situation you are discussing. (We refer here only to the writing problem. Knowledge of statistical methods and of technical methods of interpretation in general is another matter entirely. There are some titles listed in the bibliography at the end of this book which are concerned with this subject.)

After the writer has decided upon the general organization of subordinate problems and (possibly) standards of judgment, his principal tasks are to decide how much data to put in, where to put this material, what form to present it in, and how to reveal significant relationships without, on the one hand, confusing the reader with a mass of detail or, on the other hand, failing to offer sufficient supporting evidence.

Success in this last task of revealing significant relationships is made most certain by a very clear decision, before the writing is begun, as to what relationships should be explained. A carefully worked-out outline is of inestimable value. Start with the assumption that your reader is intelligent but uninformed; then caution yourself that you cannot and should not discuss every detail.

There are three specific "don'ts" here that are of particular importance:

1. Don't put into writing the kind of information that is easier to grasp in the form of graphs or tables.
2. Don't restate all of the facts that have been put into tabular or graphic form. From our observation, this is a mistake students are especially likely to make.
3. Don't assume that, having made a table or a graph, *nothing* need be said about it. A little explanation of how to read the graph or table is often helpful. And almost invariably the significant relationships revealed by the table or graph should be pointed out.

These principles will be illustrated in the second report at the end of this chapter.

## Attitude

The attitude the interpreter brings to his writing should be the scientific attitude. This fact is perfectly self-evident, and yet it is not always easy to adhere to in practice. Detachment and objectivity are particularly difficult in the evaluation of evidence on an idea that one has intuitively felt at the outset to be true, but which has come to look less certain as investigation progressed. We are all in some measure the creatures of our emotions. A counterbalance to the natural human desire for infallibility even in intuitions, however, is the deep emotional satisfaction of feeling above and in command of a given set of facts, with no obligation beyond saying that a given idea is true, or false, or uncertain. It is this emotional "set" that should be brought to problems of interpretation.

An illustration of the kind of attitude that should *not* be taken turned up in some student papers we once read. We had given a class of engineers a sheet of data on the records of a number of football coaches and asked them to write an interpretation of the data. Personal loyalities evidently got mixed up in the analysis, for one student concluded solemnly that whatever the data might indicate, Coach X was definitely the best coach in the group because all the football players who had played under him said so!

Another problem of attitude that often arises in interpretation is that of adapting the manner of the interpretation to the individuality of a certain reader or readers. Human nature being what it is, novel or unexpected conclusions are almost certain to meet with opposition. A cool appraisal of probable opposition and an allowance for it in the manner of the presentation is not only wise and profitable, it is also kind. Kind, that is, so long as the conclusion being offered is an honest one.

## Summary

Interpretation, which is the art of informally establishing a meaningful pattern of relationships among a group of facts, has as its first important step the statement of the problem being investigated. Five elements that may enter into the statement of the problem are (1) presenting the basic problem in concise form, (2) defining unfamiliar terms, (3) distinguishing between the primary problem and subordinate problems, (4) distinguishing between what is known and what remains to be found out, and (5) providing background information. The probable accuracy of the data concerned in the interpretation should be discussed. In organizing the body of the interpretation, the major factors are the subordinate problems and (where present) the standards of judgment. Supporting data should be put into graphic or tabular form wherever possible; the writing should be devoted to pointing out significant relationships. Where a choice is to be made among a number of possibilities, early elimination of some possibilities speeds up the whole process. Conclusions should be stated as they are reached in the body of the interpretation, even if they are to be summarized elsewhere. The attitude throughout should be impartial and objective, although not without a little human consideration of the individuality of the intended reader.

## ILLUSTRATIVE MATERIAL

The first of the two illustrations of interpretation that follow, entitled "Sound Talk," is unusually interesting because of its concise handling of some of the principal elements in the art of interpretation. The second, a report from General Electric, is a fine example of a more extended treatment. Another very good example of interpretation may be seen in Appendix D.

### Sound Talk*

*by Dr. W. T. Fiala, Chief Physicist*

#### High Frequency Horns

The high frequency horn is an important part of any high fidelity speaker system. It must properly load the driver element, provide smooth distribution from its lower frequency limit to beyond the range of the human ear, offer no interference to the frequency response of the driver, and be free from resonances that introduce a "character" to the reproduced sound.

Horns available for high fidelity reproduction fall into four general types: diffraction horns, ring or circumference radiators, acoustic lenses and sectoral horns. Of these four, only one meets all the requirements for an acceptable high frequency horn. Diffraction horns provide no distribution control. At lower frequencies the distribution pattern is unusably wide. At higher frequencies it becomes progressively narrower, eventually becoming a narrow beam of sound. Good listening quality can only be found directly in front of the horn. Even there, since at lower frequencies the sound energy is wide-spread while it is concentrated as the beam becomes more directional, an un-natural accentuation of higher frequencies will be experienced.

The ring radiator, like the diffraction horn, makes no attempt to control high frequency distribution. It has the additional fault of phasing

*The first paragraph presents the standards of judgment. In a general way, the first sentence is a statement of the problem—i.e., how should a high frequency horn perform? The remainder of the paragraph is a statement of standards.*

*The second paragraph presents possible choices among horns.*

*The body of the discussion is organized in terms of the types of horn, in the order in which the horns were named in the second paragraph. The principle of elimination is used throughout. That is, each of the first three types is eliminated because of at least one alleged important weakness.*

*In conclusion, we'd like to point out that it would evidently be fairly easy to expand this highly condensed discussion into a long examination of the relative merits of these horns, and in this*

*From an advertisement of Altec Lansing Corporation appearing in *High Fidelity*, November, 1958, p. 131.

holes whenever the distance between the near and far sides of the radiator equal 1/2 the wave length of the frequency being reproduced.

The acoustic lens provides a smooth spherical distribution pattern at all frequencies. The lens elements used to achieve this distribution, however, act as an acoustic filter and seriously limit high frequency reproduction, tending to introduce a "character" to the reproduced sound.

Sectoral horns, when built to a size consistent with their intended lower frequency limit, provide even distribution control. The smooth exponential development of their shape assures natural sound propagation of the full capabilities of the driving element. They are the only horns that fully meet all of the requirements for high fidelity reproduction.

We believe that ALTEC LANSING sectoral horns, built of sturdy nonresonant materials, are the finest available. Listen to them critically. Compare them with any other horn. You will find their superior distribution and frequency characteristics readily distinguishable: their "character-free" reproduction noticeably truer.

*long examination to retain the organization found here.*

**Note:** *We are here taking the unusual step of reprinting an advertisement because it provides a remarkably concise illustration of some of the important principles considered in the preceding discussion. Our reprinting of it does not reflect any opinion of our own about, or even any interest in, the question of what kind of horn is really best.*

## What They Think of Their Higher Education*

*Educational Relations Service*

*General Electric Company*

### The Purpose of the Study

The General Electric Company has established three basic purposes for its policy of educational support:

1. The development of new and more effective manpower;
2. The development of new knowledge through better teaching and more adequate research;
3. The maintenance of the best possible social, economic, and political climate in

*For the second example of the writing of interpretation of data we have chosen a report which is interesting not only for its illustration of highly skillful writing and organization, but also for information and questions about relationships between college courses and subsequent careers. Parts of this*

*Educational Relations Information Bulletin, General Electric Company (New York, 1957).

which industry can grow and progress.

To promote an educational-relations program that will realize these desired objectives, it is necessary to discover, through study and research, the impact of various types of education upon the development of managerial and professional skills. The Company's own large number of college-trained personnel provided a ready field for such an investigation.

By careful analysis of the testimony of a large number of individuals as to the skills and strengths received or developed through their college experience, the Educational Research Section of Educational Relations Service endeavored to obtain a better understanding of the areas of college teaching most valuable to men and women in corporate enterprise.

Although it is not in the province of the American business corporation to plan or direct the establishment of college curricula, answers to the questions asked in this survey—thoughtful replies based upon actual personal experiences—may aid administrators and faculty members in the evaluation of academic programs. Thus it may be possible to focus attention on those areas of college study that tend, more than others, to promote the best joint interests of college campus and business office.

The development of this information within the General Electric organization does not signify that it is the complete or all-inclusive sampling necessary to pass judgment on all college programs of study. The nature of the corporate business is such that personnel selected from colleges must have specific educational qualifications and should not be considered, therefore, as a complete cross-section of all college graduates. Moreover, the highly developed interest of the General Electric Company in scientific research and engineering progress undoubtedly places more emphasis on the fields of science

*report have been omitted, because of lack of space.**

*The first section of the report, "The Purpose of the Study," is a very broad statement of the problem.*

*Omissions include, first, the foreword, the table of contents, and the last five paragraphs of the summary. Second, that part of the text concerned with discussion of the fifth, sixth, and seventh questions that are listed under the heading, "The Methods Used." And third, the list of tables in the Appendix, and the last 10 of the 22 tables in the Appendix. These omitted tables concern questions five, six, seven, and nine.

than might be found in industrial organizations of different character. Nonetheless, the college-graduate personnel of the General Electric Company does express a high degree of interest in things academic and has an abiding enthusiasm for the continued growth and improvement of the American system of higher education. Thus, the collective opinions and individual comments of this group may be of value to business management and educators alike.

## The Methods Used

The following questions were asked of approximately 24,000 college graduates employed by the General Electric Company:

1. What areas of college study have contributed most to your present position of responsibility with the General Electric Company?
2. What areas of college study have contributed least to your present position of responsibility with the General Electric Company?
3. Recognizing the value of a satisfying and rewarding use of leisure time, what areas of college study have contributed most to your leisure-time activity?
4. What specific areas of study or courses would you recommend most highly to a young high-school graduate entering college who aspires to a position of business responsibility?
5. Do college extra-curricular activities aid an individual in developing himself for a business career? If so, what type of activity lends itself best to such development?
6. When thinking of the benefits received from the college experience, are you most impressed by values arising from teaching personalities or from the subject matter of courses studied?
7. What types of financial support are recommended for students of today who desire college training, yet find themselves without sufficient family resources to finance such training?
8. Is college training worthwhile and necessary in the light of today's business operations?
9. If you were starting life again, would you attend the same college and take the same program of study?

*The second and third sections ("The Methods Used" and "The Group Studied") contain information about how evidence was obtained. They also have other functions, however. Together, they provide a statement of two major factors in the interpretation: (1) the subordinate problems (the list of questions under "The Methods Used,") and (2) the "solutions" (that is, the authors' interpretation of the employees' answers to the questions). Observe that there are several groups of employees answering each question.*

*Finally, the second section also reveals the plan of organization, since in the body of the report each of the questions in the second section is taken up in turn. (It would have been possible to organize the discussion according to groups of employees, rather than according to the questions.)*

Original plans for this survey called for interrogation only of the liberal-arts college graduates employed by the General Electric Company. After careful consideration, and to avoid a second mailing of questionnaires in the event it later seemed desirable, the same questionnaires were sent to all college graduates employed by the Company. As these were returned, they were arranged according to a preliminary classification, by college degrees, in engineering and non-engineering categories. Each group was tabulated separately and the replies were analyzed in depth.

Upon the return of as many completed forms as could reasonably be anticipated (approximately 60.4 per cent of the total college-graduate personnel), the non-engineering group was further divided. The first division was made according to the type of undergraduate college degree (i.e., Bachelor of Arts, Bachelor of Science, degrees in Business Administration, and degrees in Education). The second division was made according to the type of position with the Company (i.e., whether now engaged in technical or nontechnical work). Extreme difficulty was encountered in formulating this second classification. However, for the purposes of the study, each reported position was assessed as "technical" if the work was related to the direct application of science or engineering to the business process. Otherwise the person was listed as a non-technical employee.

A similar procedure was employed for the engineering group, which was divided into four major categories, according to degrees: Electrical Engineering, Chemical Engineering, Mechanical Engineering, and Other Engineering Degrees.

The questionnaires were originally designed for machine tabulation. However, a pilot sampling of the responses quickly showed that accurate tabulation required interpretation of a highly personal nature. Consequently, analysis of both groups has been predominantly a hand operation, with all comments and remarks noted and considered in the light of their application to specific questions. Thus, speed in computation was necessarily sacrificed for increased penetration into the interpretation of the meaning of the respondents. This seemed essential in the light of the interest in the study which has been expressed by educators who have known that it was in progress.

### The Group Studied

For this study of the college-graduate personnel of the General Electric Company, all employees holding degrees from accredited colleges and universities as of October 1, 1955 were queried. The representativeness of the group is shown by the fact that the responses came from managerial, professional, scientific, secretarial, and clerical employees. Less than five per cent of the total responses came from women. Position data provided by respondents clearly indicated that replies to questions were based generally upon the individual's own evaluation of his success with the Company.

Length of service with the Company is of interest. In the non-engineering group, the average length of service was 7.4 years, ranging from one year to more than 35. The engineering group averaged 11.49 years, with five per cent indicating service in excess of 30 years.

A grand total of 14,147 questionnaires was returned. 6429 (45.4 per cent) were from non-engineering graduates, and 7157 (50.6 per cent) were from engineering graduates. 561 (4.0 per cent) were incomplete or otherwise defective for analytic purposes, leaving 13,586 as a working total. On the basis of undergraduate college degree earned, the respondents can be divided as follows:

| Engineering Degrees | Per Cent | Non-Engineering Degrees | Per Cent |
|---|---|---|---|
| Electrical | 51.1 | Bachelor of Arts | 32.8 |
| Chemical | 7.8 | Bachelor of Science | 25.8 |
| Mechanical | 29.3 | Business Administration | 38.1 |
| Others | 11.8 | Education | 3.3 |

A further breakdown, based upon the type of job held at General Electric, provides the following data:

| | Engineers | Non-Engineers |
|---|---|---|
| On technical work | 69.3 per cent | 32.5 per cent |
| On non-technical work | 30.7 per cent | 67.5 per cent |

In order to avoid confusion, the above-described respondents will be referred to in succeeding pages by means of the following terminology:

Respondents holding engineering degrees will be called Engineering Graduates.

Those holding non-engineering degrees will be called Non-engineering Graduates.

Respondents employed in technical occupations will be called Technical Employees.

Those employed in non-technical occupations will be called Non-technical Employees.

## Summary of Findings

This study was undertaken in the hope that, from the testimony of college-graduate employees of General Electric, some relationship could be derived between the respondents' academic and extra-curricular college careers and their subsequent success and satisfaction in their jobs and leisure activities. The findings of this survey, with statistics included, are described in detail on the pages that follow. However, the unusual degree of unanimity of opinion from the group queried justifies a brief summary of the major results.

Four subject areas in the college curriculum were considered to be extremely valuable, regardless of the academic background or type of employment of the respondent, in contributing to career success. English communication—both written and oral—was reported high on the list. Non-engineering respondents placed this subject area first, while engineers rated it second only to Mathematics, which is also a communication tool. Other subject areas reported as important for career success by both groups of respondents included Physics, Economics, and Mathematics.

The least valuable subject areas, judged from a career standpoint alone, were felt to be History, Foreign Language, miscellaneous sciences (Biology, Botany, Geology, etc.) and certain social sciences (principally Government and Eco-

*The "Summary of Findings" is a good illustration of the common practice in technical writing of presenting a summary or abstract near the beginning of a report. See Chapter 4 for a discussion of this practice.*

*The presentation of the evidence and the discussion of the evidence are handled with great skill. One of the most important points to note is that the bulk of the evidence is presented in the form of tables. Those tables that are of most immediate interest are put in the body of the text, and the rest are put in the Appendix. The discussion is used chiefly to emphasize the principal findings, to explain relationships, and to give information not found in the text.*

*One criticism of the report arises at this point, however. In the discussion of how the data were gathered, and of how the interpretation of the data was carried out, the report fails to give enough information for a thorough*

nomics). Some indication was offered that techniques of teaching certain courses left much to be desired, particularly in the social science area, where, it was felt, attention was often directed to theory at the expense of practical applications.

It is interesting to note as well that engineers often reported certain engineering courses as "least valuable," particularly if such courses were not in line with interests and occupations. It should be borne in mind that these subject areas were reported as least valuable from the career standpoint only, there being no indication of their over-all value to the educated man.

This last qualification is brought home even more dramatically by the fact that some of these same courses ranked among the most important from the standpoint of value in leisure time. Both engineers and non-engineers reported English Literature as the most valuable course from the leisure-time point of view. Other courses noted by both groups as valuable in this respect included History, Science and Engineering, Economics, Physics, Mathematics, and Philosophy. As might be expected, the liberal-arts graduates tended to indicate a greater breadth of "value courses" in the non-science areas.

When asked to name the program of studies most recommended for success in a business career, respondents gave almost equal emphasis to the four major study areas (Sciences, Social Sciences, Humanities, and Business). Except for differences in ranking as to importance, survey respondents indicated that a good collegiate program for business management training should include basic work in English, Science and Engineering, Mathematics, Economics, and General Business. Liberal-arts graduates also stressed the fundamental value of work in Psychology and the Humanities. All of this can be interpreted as a strong vote of confidence for a broad liberal education.

### Detailed Analysis of Results

*I. "What Areas of College Study Have Contributed Most to Your Present Position of Responsibility with the General Electric Company?"*

In this question, the basic assumption is that the respondent has achieved some degree of

*understanding of some important conclusions. The principle that we want to call to your attention here is that great care is needed to ensure a reader's understanding and acceptance of the method used in gathering and interpreting data. If you'll look at Table 3A, for example, you'll see that the nontechnical groups are apparently given a weight equal to the technical groups in the determination of the consensus, or final ranking of subjects. But, as stated earlier in the report, the ratio of nontechnical to technical employees is only approximately three to seven. At first glance, these facts, and other related problems, seem to invalidate some of the rankings in the "Consensus" – and perhaps they really do. On the other hand, it is conceivable that had the method of interpreting the data been clarified, the rankings would appear justified. For instance, if Table 3A reflects the number of times the various courses were cited in the responses, as opposed to a scalar ranking of the courses, the weighting problem could possibly be affected. But we can only conjecture, because of the lack of information. For our purposes, as we indicated earlier, the principle that is illustrated here is that when you gather and interpret data, you must be very careful to make your methods clear.*

*In summary, we might say that this report is a fine example of the technique of organizing and writing an interpretation of data, but*

permanence within the framework of the Company. Since he has had at least a year of service, his task is now sufficiently familiar for him to review the contributions of his collegiate background to his business efforts.

Even employees with 30 or more years of service found it possible, upon reflection, to pinpoint college courses that had marked value. A very few respondents, often secretarial, declined to answer this question, saying that the word "responsibility" did not apply to their positions. Also a few respondents indicated that all of their college courses provided a certain degree of assistance in making their careers possible.

The great majority of the Non-engineering Graduates reported the most helpful and valuable subject area was English communication. Both written and spoken English were cited as of extreme value in business success. This showed up in replies both from Technical and Non-technical Employees. Many went to some length to comment on the importance of an individual's ability to communicate easily and clearly. Engineering Graduates, on the other hand, put English second to Mathematics in importance. The inference is that both study areas tend to provide the communication skills so essential to modern business success.

The other most helpful and important subject-matter areas listed by the Non-engineering Graduates were not quite so clear-cut and well-defined. Technical employees signified that some form of business understanding was particularly helpful in adjusting to their present positions. Non-technical Employees indicated that both Economics and Mathematics were of equal importance. It is interesting to note that both Economics and Mathematics were listed as of great importance to a business career by many who chose to comment on specific course values. Economics appeared to be far and away the most valuable social science among the repondents. Economics achieved third place in the replies of Technical Employees, while Mathematics tied with Accounting and Psychology for fourth place. Physics held fifth position.

Among Non-technical Employees, business courses and Accounting tied for third position,

*that the content of the discussion of method needs to be improved if all the conclusions are to be accepted with assurance.*

while Psychology and Physics were tied for fourth place.

Engineering Graduates, after classifying Mathematics and English as the two most important subjects, gave third position to the courses that produced their specific engineering skills. There was ample indication that this group feels that some knowledge of basic engineering is helpful as a part of the core of any collegiate academic program. Following these areas in importance, the Engineering Graduates listed Physics, Economics, and Chemistry, in that order.

The following table indicates the relative ranking of courses by both Engineering and Non-engineering Graduates.

**Ranking of Courses**
**Reported Most Valuable to Career**

| Engineering Graduates | Non-Engineering Graduates |
|---|---|
| 1. Mathematics | 1. English |
| 2. English | 2. Economics |
| 3. Engineering | 3. General Business |
| 4. Physics | 4. Mathematics |
| 5. Economics | 5. Psychology |
| 6. Chemistry | 6. Physics |

Another indication of relative importance is the percentage of replies testifying that a subject area is of value:

| Course Area | Per Cent of Replies: Engineering Graduates | Per Cent of Replies: Non-Engineering Graduates |
|---|---|---|
| 1. English Communication | 58.40 | 73.68 |
| 2. Economics | 21.60 | 55.59 |
| 3. Mathematics | 72.21 | 53.24 |
| 4. Business | – | 43.67 |
| 5. Accounting | – | 33.80 |
| 6. Psychology | – | 25.55 |
| 7. Physics | 55.21 | 25.00 |
| 8. Engineering | 53.84 | – |

*II. "What Areas of College Study Have Contributed Least to Your Present Position of Responsibility with the General Electric Company?"*

Extreme care must be taken in the interpretation of the replies to this question. It would be easy to assume that the results indicated a lack of value for many college study areas. This is not necessarily the case, since respondents often pointed out that their replies signified only a lack of direct contribution to their personal business career. In fact, many courses mentioned as lacking career value were reported by the same people as valuable for leisure-time pursuits. Some failed to answer this particular question because they felt that all courses taken had provided a certain benefit in their personal development.

There was, however, a great uniformity in the replies of Non-engineering Graduates to this question. Almost a standard pattern showed History, Foreign Language, and Miscellaneous Sciences contributing least to a business career. The sciences most frequently mentioned as lacking in business value included Biology, Botany, Zoology, and Geology. Although History was indicated as of little value in business career development, other social sciences such as Government, Sociology, and Economics were often mentioned in that order. As might be anticipated, however, except in cases where foreign contact work is involved, the value of foreign languages was seriously questioned, although there was indication that languages do offer certain mental disciplinary benefits.

The same Non-engineering Graduates suggested that various business courses of a specialized nature had little career value, since the information contained in them could have been achieved in a much more practical fashion in the business world. Many indicated, too, that some specialized business courses taken in college had little connection with the type of business task encountered later in industry. This view was most often expressed by those who had technical responsibilities within the business organization.

There was little variation in the replies to this question by the Engineering Graduates. With the exception that their listing included Engineering and Chemistry and excluded Business and Accounting, the same general thinking prevailed. Engineering Graduates appeared to be critical of engineering offerings in areas other than their personal interest and specialization. Some who were not using their engineering training in their immediate jobs were inclined to belittle the whole province of engineering as a career asset. Such a reaction might reasonably be expected and should develop no concern.

The comparative ranking of "least useful to a career" courses follows:

**Ranking of Courses**
**Reported Least Valuable to Career**

| Engineering Graduates | Non-Engineering Graduates |
|---|---|
| 1. Foreign Language | 1. Foreign Language |
| 2. History | 2. Miscellaneous Sciences |
| 3. Engineering | 3. History |
| 4. Economics | 4. General Business |
| 5. Government | 5. Accounting |
| 6. Chemistry | 6. Economics |
| 7. Literature } Tie | 7. Mathematics } Tie |
| 8. Mathematics } Tie | 8. Physics } Tie |
| 9. Miscellaneous Sciences | 9. Government |

The least valuable courses for a career in business, listed as percentages of total replies mentioning a course, were as follows:

| Course Area | Per Cent of Replies | |
|---|---|---|
| | Engineering Graduates | Non-Engineering Graduates |
| 1. Miscellaneous Business | 14.74 | 55.30 |
| 2. History | 46.81 | 52.45 |
| 3. Language | 59.84 | 52.24 |
| 4. Miscellaneous Science | 12.11 | 48.15 |
| 5. Miscellaneous Humanities | 1.99 | 22.09 |
| 6. Government | 20.42 | 18.78 |
| 7. Chemistry | 25.10 | 18.04 |
| 8. Physics | – | 17.70 |
| 9. Engineering | 45.16 | – |
| 10. Economics | 23.32 | – |

*III. "What Areas of College Study Have Contributed Most to Your Use of Leisure Time?"*

In corporate personnel practice, there is increasing emphasis upon the importance of employees' leisure-time activities. Such activities, when satisfying and rewarding, can frequently contribute to the development of better and more valuable employees. So it appeared desirable to evaluate the type of courses that have best served the college graduates in the development of their leisure-time pursuits.

Among Non-engineering Graduates, both Technical and Non-technical Employees indicated strongly that a variety of business courses, too numerous to specifically mention by name, contributed greatly to their non-vocational activities. Apparently college graduates within the Company have a lively interest in business activities not directly associated with their own work. In some instances, local government activity was cited as a leisure-time application of knowledge gained through business courses.

Both Technical and Non-technical Employees reported English Literature as a strong second contender for leisure time, contributing to the ability to relax and to develop non-business thinking. (The unsympathetic might equate these answers with a rationalization of "escape" reading!) Indications were also present that the type and quality of personal leisure reading were directly related to literature courses at college. It will be remembered that this area of study was mentioned with some frequency as one of the least valuable career courses.

Miscellaneous Sciences (Biology, Botany, and Geology) and History (specifically American and European) divided honors for third and fourth position. Technical Employees indicated History as third in value while Non-technical Employees offered Miscellaneous Science. Fourth position was the reverse for both groups. Miscellaneous Humanities (including Arts, Music, and Religion) was in fifth place among Technical Employees, while Mathematics placed fifth with Non-technical Employees. Although not numerous enough for tabulation, frequent mention was made of courses in Religion, especially by Technical Employees. In general, then, these respondents tended to choose, for leisure and enjoyment, those subjects they did not get in college.

Again the replies from Engineering Graduates followed a strikingly similar pattern with largely identical courses listed, but with a somewhat different ranking. It was the expressed regret of many that academic time had not been available to develop possible leisure-time interests. Some suggested a longer academic program to provide such time.

**Ranking of Courses**
**Reported Most Valuable for Leisure Time**

| Engineering Graduates | Non-Engineering Graduates |
|---|---|
| 1. English Literature | 1. General Business |
| 2. Engineering | 2. English Literature |
| 3. History | 3. History } Tie |
| 4. Economics | 4. Science and Engineering } Tie |
| 5. Physics | 5. Mathematics } Tie |
| 6. Mathematics } Tie | 6. General Humanities } Tie |
| 7. Philosophy } Tie | 7. Economics |
| | 8. Physics } Tie |
| | 9. Philosophy } Tie |

There was great enthusiasm in many of the replies for the liberal-arts curriculum for personal satisfaction outside the business office. Many respondents indicated that their enjoyment of life was directly traceable to a broad background of interest generated by the program of study in the liberal-arts college.

Upon the basis of percentage of total replies mentioning course areas as most rewarding for leisure-time pursuits, the following listing was obtained:

| | Per Cent of Replies | |
|---|---|---|
| Course Area | Engineering Graduates | Non-Engineering Graduates |
| 1. Miscellaneous Business | – | 58.83 |
| 2. English Literature | 66.43 | 49.43 |
| 3. History | 37.05 | 38.82 |
| 4. Miscellaneous Science | – | 37.18 |
| 5. Mathematics | 22.16 | 30.17 |
| 6. Miscellaneous Humanities | – | 27.08 |
| 7. Engineering | 39.11 | – |
| 8. Philosophy | 24.54 | – |
| 9. Physics | 24.78 | – |
| 10. Economics | 35.87 | – |

*IV. "What Specific Areas of Study or Courses Would You Recommend Most Highly to a Young High-school Graduate Entering College Who Aspires to a Position of Business Responsibility?"*

One of the remarkable features of the results of this tabulation was the relative equality of importance attached to each of the four broad areas of study.

| | Per Cent of Replies Mentioning Area | |
|---|---|---|
| **Most Valuable Area of Study** | **Engineering Graduates** | **Non-Engineering Graduates** |
| Science and Technical | 34.52 | 26.00 |
| Humanities | 23.88 | 26.00 |
| Social Sciences | 21.17 | 25.00 |
| Business | 20.43 | 23.00 |

Thus, the obvious inference is that the composite General Electric college graduate is specifically in favor of a broad program of study encompassing some work in each of the above categories. The average pattern of response for Non-engineering Graduates (allowing each person three choices of course or subject-matter areas) was (1) English communication and expression, (2) Economics, and (3) Mathematics, Engineering, or Business.

Engineers, as might be expected, placed heavy emphasis upon their specialty but ranked English as the number one course, with Engineering, Economics, Business, and Mathematics respectively following in line. Several expressed caution, however, and said that engineering was their first choice *only* in the event that the student planned to enter a business similar to General Electric. Otherwise they would have placed English in first position.

Many comments from both groups pointed to the need for a balanced program of studies without undue specialization. Specialization, they held, should be reserved for graduate training or for special educational courses offered by the industry in which the student accepts employment.

Typical comments emphasized that the program should be broad, should teach mental discipline and the ability to think, and should allow the student to take "all available engineering courses that might be scheduled." This, of course, reflects the influence of engineering in the General Electric organizational structure.

The following table indicates the over-all percentage of respondents indicating the named course of study:

| Course Area | Per Cent of Respondents Indicating Value: Engineering Graduates | Per Cent of Respondents Indicating Value: Non-Engineering Graduates |
|---|---|---|
| English Communication and Expression | 75.81 | 62.96 |
| Economics | 68.84 | 55.27 |
| Business Courses | 61.52 | 41.64 |
| Mathematics | 25.72 | 36.23 |
| Engineering | 74.06 | 29.29 |
| Psychology | 17.46 | 25.66 |

The comparative ranking chart follows:

**Ranking of Courses Most Recommended for Management Responsibility**

| Engineering Graduates | Non-Engineering Graduates |
|---|---|
| 1. English | 1. English |
| 2. Engineering | 2. Economics |
| 3. Economics | 3. General Business |
| 4. General Business | 4. Mathematics |
| 5. Mathematics | 5. Engineering and Science |
| | 6. Psychology |
| | 7. Humanities |

There was little difference in the specific "most valuable" course recommendations of both Technical and Non-technical Employees in either of the groups surveyed. In fact, engineers of all classifications reported favorably on the same five course areas listed above with relatively small differences in ranking. Such unanimity of opinion among more than 7000 widely scattered engineers appeared remarkable.

*VIII. "If You Were Starting Life Again, Would You Attend the Same College and Take the Same Program of Study?"*

One of the best indications of personal satisfaction with the college attended or the course pursued is the answer to whether, upon reflection, the same patterns would be followed if one could live the period over again. This question was asked not only because it provides a check against other statements about the college program, but because it also provides a yardstick for the analysis of the General Electric Corporate Alumnus Program. An unsatisfied graduate is less likely to contribute financially, even on a matching basis, to the college he attended.

Three out of four Non-engineering Graduates would return to the same college if they had the choice again. Of the dissatisfied quarter, the chief reasons seem to be:

1. The college attended was chosen because of low costs during days of financial hardship;
2. College was attended at night, and respondent would rather now attend during the day;
3. Respondent attended a large university and would now choose a smaller school with more personal environment;
4. Respondent attended a small school but now believes that better facilities might be found at a large institution;
5. If a different program (now desirable) had been taken, another college would have been selected.

The most numerous reasons given were numbers one, three, and five.

The greatest degree of dissatisfaction was found among those who had attended teachers colleges and received a degree in Education. Generally, they had entered these schools with a professional desire to teach and now, employed in industry, they believe a broader education would have been of more general and lasting value.

Engineering Graduates as a group were even more satisfied with the educational institutions of their choosing. If the opportunity of choice were again available, 85 per cent would go to the same institution. Since a career in engineering requires considerable planning, it would appear that more care had gone into the selection of the college, thus creating an atmosphere of satisfaction.

Non-engineering Graduates were far less satisfied with their course selections. Forty-six per cent of the replies indicated a different course would be pursued if the respondent might choose again. Business Administration degree recipients were most satisfied with the program they pursued; 67 per cent so indicated. Education graduates as a group were the least satisfied; 31 per cent indicated that they would now change their program of study if the opportunity were offered. Undoubtedly the same reasoning applies to both cases. Business Administration graduates are apparently in work with the Company closely related to their program of study, whereas the Education majors find that industrial tasks call for things quite different, for the most part, from teacher preparation.

Almost 56 per cent of the Bachelor of Arts personnel would choose the same course again, and approximately 57 per cent of the Bachelor of Science majors would repeat their same course work.

Three out of four Engineering Graduates would choose the same course of study again. The most satisfied group was the Electrical Engineers, with over 80 per cent reporting a desire to repeat the same program. Mechanical Engineers, Miscellaneous Engineers, and Chemical Engineers followed in that order. Some replied that course variations would be small, consisting in some cases of greater breadth in non-engineering courses.

The group as a whole indicated strong preference for a broad general educational background, amply fortified with English, Economics, and Mathematics. However, it was suggested many times that any good program in preparation for a business position should include as much training in science and engineering as possible. This reflects the Company complexion to a marked degree, but it is one major complaint of those who avoided such course work in their personal academic programs.

*IX. Comments concerning the Liberal Arts.*

The prime purpose of undergraduate college education, according to many respondents, is not the acquisition of specialized information and operational techniques. Rather, it was volunteered time and time again that the power to think and to analyze a wide range of problems successfully is the true goal of college education. Even if some consideration is given to the technical aspects of education, failure to produce an individual with these abilities is in essence a failure of the college program itself. The "whole man" concept seems quite strong in the thoughts of this group of employees.

The ability to get along well with others is also a factor that respondents feel should be stressed in the college curriculum. Those courses that aid the individual in the better understanding of his or her associates come in for high praise, because of the complicated interconnection in the lives of all of us. In this same vein, there was some emphasis upon the theory that college should develop within the individual a burning desire to associate himself with religious, social, community, and service drives to aid in the improvement of living conditions for his fellow man.

The importance of concentrated study in the areas of English and Mathematics is also deserving of a final note. The fact that both fields have become indispensable to human expression and understanding is accepted by this survey group. Heavy concentration upon both areas of study was deemed essential in the shaping of tool courses for successful living.

In general, the broad background offered by the liberal-arts curriculum can be tailored, in the light of many comments, to the fashioning of a highly successful career in industry, particularly on the managerial or professional level.

Appendix

## TABLE 1A

### Types of Degrees and Work Assignments Engineering Graduates

| Type of Degree | Number Employed in Technical Work | Number Employed in Non-technical Work | Total Respondents | Per Cent of Grand Total |
|---|---|---|---|---|
| Electrical Engineers | 2446 | 1212 | 3658 | 51.1 |
| Chemical Engineers | 442 | 113 | 555 | 7.8 |
| Mechanical Engineers | 1449 | 654 | 2103 | 29.3 |
| Other Engineers | 626 | 215 | 841 | 11.8 |
| Totals | 4963 – 69.3% | 2194 – 30.7% | 7157 | 100 |

## TABLE 1B

### Types of Degrees and Work Assignments Non-Engineering Graduates

| Type of Degree | Per Cent of Respondents |
|---|---|
| Bachelor of Arts – Liberal Arts Course | 22.6 |
| Bachelor of Arts – Science Course | 10.2 |
| Bachelor of Science – Liberal Arts Course | 3.6 |
| Bachelor of Science – Science Course | 22.2 |
| Business Administration | 38.1 |
| Education | 3.3 |

| Nature of Work | Per Cent of Total |
|---|---|
| Technical | 32.5 |
| Non-technical | 67.5 |

## TABLE 3A

### Ranking of College Courses Most Valuable for a Career Engineering Graduates

| Type of Degree | Nature of Work | Order of Importance: Mathematics | Physics | Chemistry | Engineering | English | Economics |
|---|---|---|---|---|---|---|---|
| Electrical Engineers | Technical | 1 | 2 | – | 3 | 4 | 5 |
| | Non-technical | 1 | 3 | – | 4 | 2 | 5 |
| Chemical Engineers | Technical | 1 | 4 | 3 | 2 | 5 | – |
| | Non-technical | 2 | 5 | 4 | 3 | 1 | – |
| Mechanical Engineers | Technical | 1 | 2 | – | 3 | 4 | 5 |
| | Non-technical | 1 | 3 | – | 4 | 2 | 5 |
| Other Engineers | Technical | 1 | 3 | – | 4 | 2 | 5 |
| | Non-technical | 1 | 5 | – | 3 | 2 | 4 |
| Consensus–All Engineers | | 1 | 4 | 6 | 3 | 2 | 5 |

## TABLE 3B

### Ranking of College Courses Most Valuable for a Career Non-Engineering Graduates

| Type of Degree | Nature of Work | Order of Importance: English | Economics | Business | Math. | Psychology | Accounting | Physics | Misc. Sciences |
|---|---|---|---|---|---|---|---|---|---|
| B.A. in Science Courses | Technical | 1 | 2 | 3 | 4 | – | – | 5 | – |
| | Non-technical | 1 | 2 | 4 | 3 | 5(6) | – | 5(6) | – |
| B.A. in Liberal Arts Courses | Technical | 2 | 3 | 1 | 4 | – | – | 5 | – |
| | Non-technical | 2 | 3 | 4 | 1 | – | – | 5 | – |
| B.S. in Science Courses | Technical | 1 | 3 | 2 | 5 | 4 | – | – | – |
| | Non-technical | 1 | 2 | 3 | 4 | – | – | 5 | – |
| B.S. in Liberal Arts Courses | Technical | 1 | 3 | 2 | 4 | – | – | 5 | – |
| | Non-technical | 2 | 3 | 1 | 4 | – | – | – | 5 |
| Business Administration | Technical | 1 | 2 | 5 | 3 | – | 4 | – | – |
| | Non-technical | 1 | 3 | 5 | 2 | – | 4 | – | – |
| Education | Technical | 3 | – | 1 | 4 | 2 | – | 5 | – |
| | Non-technical | 1 | 3* | – | 2 | 5 | 3* | – | – |

*Tie.

## TABLE 4A

### Ranking of College Courses Least Valuable for a Career Engineering Graduates

| Type of Degree | Nature of Work | History | Language | Engineering | Economics | Government | Chemistry | Literature | Misc. Sciences | Math. |
|---|---|---|---|---|---|---|---|---|---|---|
| Electrical Engineers | Technical | 1 | 2 | 3 | 4 | 5 | – | – | – | – |
| | Non-technical | 2 | 1 | 3 | – | 4 | 5 | – | – | – |
| Chemical Engineers | Technical | 2 | 1 | 3* | 3* | – | – | 5 | – | – |
| | Non-technical | 2 | 1 | 3 | – | 5 | 4 | – | – | – |
| Mechanical Engineers | Technical | 1 | 3 | 2 | 4 | 5 | – | – | – | – |
| | Non-technical | 2 | 1 | 3 | – | 4 | 5 | – | – | – |
| Other Engineers | Technical | 1 | 2 | 4 | 3 | 5 | – | – | – | – |
| | Non-technical | 2 | 1 | 3 | – | – | – | – | 4 | 5 |
| Consensus – All Engineers | | 2 | 1 | 3 | 4 | 5 | 6 | 7* | 9 | 7* |

*Tie.

## TABLE 4B

### Ranking of College Courses Least Valuable for a Career Non-Engineering Graduates

| Type of Degree | Nature of Work | Misc. Business | Language | Misc. Science | History | Economics | Accounting | Math. | Humanities | Government | Philosophy | Physics | Chemistry |
|---|---|---|---|---|---|---|---|---|---|---|---|---|---|
| B.A. in Science Courses | Technical | – | 2 | 1 | 3 | – | – | – | 4 | – | 5 | – | – |
| | Non-technical | – | 1 | 4 | 3 | 2 | – | – | – | – | – | 5 | – |
| B.A. in Liberal Arts Courses | Technical | – | 2 | 1 | 3 | – | – | 4 | – | 5 | – | – | – |
| | Non-technical | – | 1 | 3 | 2 | – | – | 4 | – | – | – | 5 | – |
| B.S. in Science Courses | Technical | 3 | 2 | 1 | 4 | – | 5 | – | – | – | – | – | – |
| | Non-technical | – | 3 | 4 | 2 | – | 1 | – | – | – | – | – | 5 |
| B.S. in Liberal Arts Courses | Technical | – | 2 | 1 | 3 | 4* | – | – | – | 4* | – | – | – |
| | Non-technical | 1 | 2 | 3 | 4 | 5 | – | – | – | – | – | – | – |
| Business Administration | Technical | 3 | 1 | 5 | 2 | – | 4 | – | – | – | – | – | – |
| | Non-technical | 1 | 2 | 4 | 3 | – | – | – | – | – | – | 5* | 5* |
| Education | Technical | 3 | 5 | 1 | 2 | – | – | – | 4 | – | – | – | – |
| | Non-technical | – | 3 | 2 | 1 | – | – | – | 4 | – | – | 5 | – |

*Tie.

**TABLE 5A**

**Ranking of College Courses Most Valuable for Leisure Activities Engineering Graduates**

| Type of Degree | Nature of Work | Literature | Engineering | History | Economics | Physics | Math. | Philosophy |
|---|---|---|---|---|---|---|---|---|
| Electrical Engineers | Technical | 1 | 2 | 3 | 4* | 4* | – | – |
| | Non-technical | 1 | 4 | 3 | 2 | 5 | – | – |
| Chemical Engineers | Technical | 1 | 3 | 2 | 5 | – | 4 | – |
| | Non-technical | 1 | 2* | 4 | 2* | – | – | 5 |
| Mechanical Engineers | Technical | 1 | 2 | 3 | 4 | 5 | – | – |
| | Non-technical | 1 | 2 | 4 | 3 | 5 | – | – |
| Other Engineers | Technical | 1 | 2 | 3 | 4 | – | 5* | 5* |
| | Non-technical | 1 | 4 | 3 | 2 | – | – | 5 |
| Consensus All Engineers | | 1 | 2 | 3 | 4 | 5 | 6* | 6* |

*Tie.

**TABLE 5B**

**Ranking of College Courses Most Valuable for Leisure Activities Non-Engineering Graduates**

| Type of Degree | Nature of Work | Misc. Business | English | History | Misc. Science | Economics | Physics | Math. | Misc. Humanities | Philosophy |
|---|---|---|---|---|---|---|---|---|---|---|
| B.A. in Science Courses | Technical | 1 | 3 | 2 | 4 | – | – | – | 5 | – |
| | Non-technical | 1 | 2 | 4 | 3 | – | – | 5 | – | – |
| B.A. in Liberal Arts Courses | Technical | – | 2 | 1 | 3 | – | 4* | 4* | – | – |
| | Non-technical | 1 | 2 | 4 | 3 | – | – | 5 | – | – |
| B.S. in Science Courses | Technical | 1 | 4 | 2 | 3 | – | – | – | 5* | 5* |
| | Non-technical | 1 | 2 | 3 | 4 | 5 | – | – | – | – |
| B.S. in Liberal Arts Courses | Technical | 1* | – | 3 | 1* | – | – | – | 4 | 5 |
| | Non-technical | – | 2 | 3 | 1 | – | – | 5 | 4 | – |
| Business Administration | Technical | 2 | 1 | 4 | – | 3 | – | 5 | – | – |
| | Non-technical | 1 | 2 | – | 5 | 4 | – | 3 | – | – |
| Education | Technical | – | 1 | 2 | 3 | – | 5 | – | 4 | – |
| | Non-technical | – | 2 | 3 | 1 | – | – | 5 | 4 | – |

*Tie.

**TABLE 6A**
**Ranking of Courses Most Recommended for Management Responsibility Engineering Graduates**

| Type of Degree | Nature of Work | English | Engineering | Economics | General Business | Math. |
|---|---|---|---|---|---|---|
| Electrical Engineers | Technical | 1 | 2 | 3 | 4 | 5 |
| | Non-technical | 1 | 2 | 3 | 4 | 5 |
| Chemical Engineers | Technical | 1 | 2 | 4 | 3 | 5 |
| | Non-technical | 2 | 1 | 4 | 3 | 5 |
| Mechanical Engineers | Technical | 2 | 1 | 3 | 4 | 5 |
| | Non-technical | 2 | 1 | 3 | 4 | 5 |
| Other Engineers | Technical | 1 | 2 | 3 | 4 | 5 |
| | Non-technical | 3 | 1 | 2 | 4 | 5 |
| Consensus—All Engineers | | 2 | 1 | 3 | 4 | 5 |

**TABLE 6B**
**Ranking of Courses Most Recommended for Management Responsibility Non-Engineering Graduates**

| Type of Degree | Nature of Work | English | Economics | Misc. Business | Math. | Engineering | Misc. Science | Psychology | Misc. Humanities |
|---|---|---|---|---|---|---|---|---|---|
| B.A. in Science Courses | Technical | 1 | 2 | 3 | 5 | — | 4 | — | — |
| | Non-technical | 2 | 1 | 3 | 4 | — | — | — | 5 |
| B.A. in Liberal Arts Courses | Technical | 1 | 2 | 4 | 3 | — | — | 5 | — |
| | Non-technical | 1 | 2 | 3 | 4 | 5 | — | — | — |
| B.S. in Science Courses | Technical | 1 | 2 | 3 | 5 | — | 4 | — | — |
| | Non-technical | 1 | 2 | 3 | 4 | — | 5 | — | — |
| B.S. in Liberal Arts Courses | Technical | 1 | 2 | 3 | 5* | 4 | 5* | — | — |
| | Non-technical | 2 | 3 | 4 | 1 | 5 | — | — | — |
| Business Administration | Technical | 1 | 2 | 3 | 5 | 4 | — | — | — |
| | Non-technical | 1 | 2 | 3 | 4 | 5 | — | — | — |
| Education | Technical | 1 | 2 | 4 | — | — | 5 | 3 | — |
| | Non-technical | 1 | 3 | 2 | 4 | — | — | 5 | — |
| Consensus—All Respondents | | 1 | 2 | 3 | 4 | 5* | 5* | 7 | 8 |

*Tie.

## SUGGESTIONS FOR WRITING

The following list represents a wide range of possible subjects for practicing interpretative writing, and will probably suggest many additional subjects. In selecting a subject, you will need to make several decisions. Let's consider the first subject in the list as an example of what will need to be thought about. In doing so, let's assume that a decision has already been made as to the approximate length of the completed interpretation.

Your first step, then, would be to decide whether to evaluate only one lawn mower, or to compare several. If you evaluate only one, your second step should probably be to select your imaginary reader. With these two decisions made, you are ready to start obtaining any necessary information you don't have and organizing your interpretation.

Should you decide to compare several lawn mowers, you would presumably compare different brands of the same type, or different types. You might, for example, compare all the brands you could find of self-propelling reel mowers. Or you could compare any reasonable combination of types of mowers. Self-propelling reel mowers and small riding mowers might make a reasonable combination. With this combination, you would not be concerned with particular brands, except perhaps incidentally, but rather with the characteristics of the two types.

As always, however, what is reasonable depends on your decision about your reader. The combination just mentioned would surely not be reasonable for a reader who was an active young man with only 500 square feet of lawn to mow, but it would be entirely reasonable for an elderly man with 5000 square feet to mow.

The choice of reader also, as you will remember, has much to do with the length of the finished interpretation. If you picked a reader who just wanted to know about the advantages and disadvantages of all available types and brands of lawn mowers, and who wanted technical details, you would be confronted with an impossibly long project.

With such principles as the foregoing in mind, then, choose one of the following subjects for an interpretation. In your paper, specify the reader.

Lawn mowers
Wrist watches
Snowmobiles
Small astronomical telescopes
Slide rules
Shrubs for a hedge or windbreak
Portable typewriters

Automobile brakes
Trees for lumber
Electrical test sets for the amateur
Small sailboats for the combined purpose of family use and racing
Feed for a particular type of stock animal in a specified locality
Outboard motors
Reels for fishing rods
Preservation and storage of hay
Trail bikes
Gas engines for model airplanes
Garden tractors
Finishes for home-made furniture, excluding paint and enamel
35-mm cameras

# SECTION THREE

# Transitions, Introductions, and Conclusions

*The duties of a writer are somewhat like those of a highway builder. A highway builder must know how to construct a good road, a road that will carry weight. A writer, for his part, must know how to make a sentence that will carry meaning. Again, a highway builder must know how to lay out a system of roads so that the traveler can go from one place to another easily and quickly; he doesn't want to find that he must go through Kansas City to reach Chicago from New York. For the writer this is organization. Finally, the highway builder must know how to devise and locate signs that will keep the traveler informed as to what lies behind him and what lies ahead. The writer's comparable duty is to write introductions, conclusions, and transitions. The purpose of this section is to discuss these elements.*

*In the following three chapters we shall offer some examples of writing in which there are clear transitions, introductions, and conclusions, and we shall make suggestions about how to do such writing. We believe you will see that a route marker in a report, as on the highway, is a good thing.*

# 10
# Transitions

A transition is an indication of what is going to be said, a reference to what has already been said, or both. It may be a single word, a phrase, a sentence, a paragraph, or an even longer passage. In form, transitions may be quite mechanical and obvious, or unobtrusively woven into sentences with other purposes. We shall discuss what a transition is, how to write a transition, and where to place a transition.

## What a Transition Is

We said a moment ago that a transition may be a word, a phrase, a sentence, a paragraph, or an even longer passage. Let's begin with words and phrases. Below are two passages which differ only in the presence or absence of transitional words.

> 1. Evidently the creation of a plutonium production plant of the required size was to be a major enterprise even without attempting to utilize the thermal energy liberated. By November 1942 most of the problems had been well defined and tentative solutions had been proposed. These problems will be discussed in some detail in the next chapter; we will mention them here.

> 2. Evidently the creation of a plutonium production plant of the required size was to be a major enterprise even without attempting to utilize the thermal energy liberated. Nevertheless, by November 1942 most of the problems had been well defined and tentative solutions had

> been proposed. Although these problems will be discussed in some detail in the next chapter, we will mention them here.*

The second version differs from the first only by the addition of two words ("nevertheless," and "although"), and yet it is noticeably smoother than the first. Careful reading of the two passages reveals very clearly the marked effect that two such apparently minor changes can create. Moreover, a definite change in meaning occurs, especially after "nevertheless." This word adds force to the idea that it is remarkable that the accomplishments mentioned were achieved in so short a time.

Numerous words and phrases are frequently used as transitions in this manner. The following is a partial list:

however
on the other hand
in spite of
moreover
furthermore
consequently
also
now
so
as a result of
therefore
of course
for example
besides
in addition
indeed
in fact
as previously noted
in comparison
in the first place
secondly
finally
next
then
in other words
and
but

Perhaps you will feel that such words and phrases as these do not exactly "indicate what is going to be said," as we claimed they do. Yet they do indicate, very often, the logic of the relationship between two units of thought. Such terms as "moreover," and "furthermore" indicate that "more of the same" is coming; "however" suggests that a different point of view is to be introduced or a refutation or qualification offered; "consequently" establishes a cause-and-effect relationship; and so on.

There is another way in which words and phrases serve a transitional purpose besides the one just described: that is through the repetition of key terms. Consider the italicized terms in the following: "This experiment can be carried out successfully only under certain *conditions*. These *conditions* are . . . ." The second statement might have begun, "One of these . . . ," or "The first of these . . . ." It is a good idea to remember that repetition of the main subject of discus-

*From Henry D. Smyth, *Atomic Energy for Military Purposes*, rev. ed. (Princeton, N.J.: Princeton University Press, 1946), p. 104. Reprinted by permission of the author.

sion itself helps keep the reader's eye on the ball and leads him from one thought to another. Suppose the above experiment were Millikan's oil-drop experiment: every now and then in a discussion of it, it would help to substitute "Millikan's experiment," or "the oil-drop experiment" for the term "experiment" or whatever other term might be used as the subject of sentences concerning it.

Sentence transitions and paragraph (or longer) transitions are associated less with stylistic qualities and more with organization than are the shorter ones just discussed. Usually these longer forms consist of a statement of what has been or will be said. The last sentence in the passage quoted a moment ago is an example of a transitional sentence which is obvious in form. The first sentence of the present paragraph is an example of one less obvious in form. Both, however, serve the same purpose: to provide information about the content of a coming passage. This function may be seen again in the illustrations below, which represent both the obvious and the less obvious forms.

1. Having considered the economic feasibility of this alloy as a transformer core, we turn now to the problem of hysteresis loss.

2. Even if this alloy is economically feasible as a transformer core, however, there still remains the problem of hysteresis loss.

This sentence, in either form, would serve as a good transition. The first of the two forms calls attention forcibly to the change of topic; the second performs the same function but less obtrusively. The first might be the easier to remember; the second makes smoother reading. A choice between the two forms will follow consideration of this difference.

There is no important difference in principle between a sentence transition and a paragraph transition. In fact, our chief reason for using both terms was to make it perfectly clear that an entire paragraph, as well as a sentence, may be used to make a transition. Naturally, the same differences in form that may be found among sentence transitions hold true for paragraphs, as the following illustrations show:

1. In Chapter 1 and other early chapters we have given brief accounts of the fission process, pile operation, and chemical separation. We shall now review these topics from a somewhat different point of view before describing the plutonium production plants themselves.*

2. In previous chapters there have been references to the advantages of heavy water as a moderator. It is more effective than graphite in slowing down neutrons and it has a smaller neutron absorption than graphite. It is therefore possible to build a chain-reacting unit with uranium and

*Smyth, *ibid.*, p. 130.

> heavy water and thereby to attain a considerably higher multiplication factor, *k*, and a smaller size than is possible with graphite. But one must have the heavy water.*

The second example is the less mechanical. In the original text it is followed by discussion of work done on heavy water—and thus it serves as a transition.

So far, we have considered some typical examples of transitions with special attention to their forms, ranging from single words to paragraphs, and from obvious, rather mechanical types to those which are less obtrusive. Finally, we will note the various purposes or functions that a transition may have. There are six important ones:

1. Smoothing out style, principally by clarifying logical relationships through the use of single transitional words or short phrases. Example: "however," "on the other hand."
2. Indicating what topics are to be discussed. Example: "This section is devoted to an analysis of the effect of temperature on bearing noise with a given lubricant."
3. Reminding the reader of topics discussed. Example: "It is evident, then, that neither increasing the number of workmen nor increasing the speed of the line can, in the present circumstances, increase the output of Final Assembly."
4. Announcing a change of subject. Example: "In addition to the major advantages just described, the proposed changes in design would offer several secondary advantages."
5. Making reference to an earlier or later statement of a similar, related, or pertinent idea. Example: "As was said in the preceding section . . . ." "As will be shown in Chapter X . . . ."
6. Keeping attention focused by the repetition of key terms. Example: "The first step in aligning the circuit . . . ." "The next thing to do in aligning the circuit . . . ."

In conclusion we must say that the foregoing discussion has not by any means been an exhaustive statement of what a transition is. To go further, however, would be primarily to enlarge upon the theoretical rather than the practical aspects of the subject.

## How to Write Transitions

On the subject of learning how to write transitions, we have two practical suggestions to make, and that is all.

1. Don't hesitate to be quite mechanical about it at first. If you can't think of any better way, just say, "This concludes the discussion

*_Ibid._, p. 147.

of so-and-so. Next thus-and-so will be discussed." As you continue to practice writing, you will find that you acquire a habit of using transitions, and develop an ability to make them as obvious or unobtrusive as you wish. But don't expect the process to become completely automatic.

2. When you have completed a rough draft of a report, read through it once with the sole purpose of spotting points at which transitions should be added. After locating and marking all such points, go back and write the transitions. Possibly, also, you will want to delete some of the transitions that you originally wrote. It is possible to have too many, and it is possible to get them in the wrong place. This raises the question of how to decide where transitions should appear.

## Where to Put Transitions

There is no formula according to which transitions can be located. Every report is unique and presents its own problems. But if there is no formula, there is a principle—and the principle is simply this: don't give your reader a chance to get lost. Again, it's like putting up highway signs. In looking over your report for short transitions, it is wise to keep trying out the effect of adding a word or phrase to a sentence. This experimenting can be done quite deliberately. In checking the location of longer transitions, it sometimes helps to start with the outline. Put an asterisk where the shift in thought is great enough to require a strong transition; then examine the text itself at the points noted.

A special problem that comes up here is the effect of the use of subheads, like the one above, on the handling of transitions. Clearly, the subhead itself is a transitional device, and can be expected to inform the reader fairly accurately of changes in subject. On the other hand, it is easy to overestimate the amount of attention a reader gives to subheads (see the comments on titles in Chapter 11, "Introductions"). Our advice is to write the transition pretty much as if the subhead weren't there. And almost without exception our advice is to avoid using the subhead as an antecedent for a pronoun in the sentence that follows it. This point will be immediately clear if you compare the following undesirable sentence with the one that actually appears under the subhead above ("Where to Put Transitions"): "There is no formula for this."

Illustration is probably more valuable than advice with respect to almost every aspect of the art of writing transitions, and so for the rest of our exposition we shall turn to an extended illustration.

## ILLUSTRATIVE MATERIAL

The material that follows consists of two paragraphs taken from Henry D. Smyth's famous work, *Atomic Energy for Military Purposes.* These paragraphs are fairly representative of good technical writing. They are not loaded down with transitions, but they do make clear, easy reading. (Observe, however, the abuse of the word "such" in the second paragraph.)

We have italicized transitional elements.

Numerous pronouns (not in italics) also serve a transitional purpose. This is one of the problems we had in mind in our previous assertion that our discussion of transitions was by no means exhaustive. Even though pronouns are often clearly transitional in function, it seems best to exclude them from a consideration of transitions because their inclusion would add difficult theoretical problems without adding much of practical value.

### The Equivalence of Mass and Energy

1.4. One conclusion that appeared rather early in the development of the theory of relativity was that the inertial mass of a moving body increased as its speed increased. This implied an *equivalence* between an increase in energy of motion of a body, that is, its kinetic energy, and an increase in its mass. To most practical physicists and engineers this appeared a mathematical fiction of no practical importance. Even Einstein could hardly have foreseen the present applications, *but* as early as 1905 he did clearly state that mass and energy were *equivalent* and suggested that proof of this *equivalence* might be found by the study of radioactive substances. He concluded that the amount of energy, E, *equivalent* to a mass, m, was given by the equation

$$E = mc^2$$

where c is the velocity of light. If this is stated in actual numbers, its startling character is apparent. It shows that one kilogram (2.2 pounds) of matter, if *converted* entirely into energy, would give 25 billion kilowatt hours of energy. This is *equal* to the energy that would be generated by the total electric power industry in the United States (as of 1939) running for approximately two months. Compare this fantastic figure with the 8.5 kilowatt hours of heat energy which may be produced by burning an *equal* amount of coal.

1.5. The extreme size of this *conversion* figure was interesting *in several respects. In the first place,* it explained why the *equivalence* of mass and energy was never observed in ordinary chemical combustion. We *now* believe that the heat given off in *such* a combustion has mass associated with it, *but* this mass is so small that it cannot be detected by the most sensitive balances available. (It is of the order of a few billionths of a gram per mole.) *In the second place,* it was made clear that no appreciable quantities of matter were being *converted* into energy in any familiar terrestrial processes, *since* no *such* large sources of energy were known. *Further,* the possibility of initiating or controlling such a *conversion* in any practical way seemed very remote. *Finally,* the very size of the *conversion* factor opened a magnificent field of speculation to philosophers, physicists, engineers, and comic-strip artists. For twenty-five years *such* speculation was unsupported by direct experimental evidence, but beginning about 1930 *such* evidence began to appear in rapidly increasing quantity. *Before discussing* SUCH *evidence and the practical partial conversion of matter into energy that is our main theme, we shall review the foundations of atomic and nuclear physics. General familiarity with the atomic nature of matter and with the existence of electrons is assumed. Our treatment will be little more more than an outline which may be elaborated by reference to books such as Pollard and Davidson's* APPLIED NUCLEAR PHYSICS *and Stranathan's* THE "PARTICLES" OF MODERN PHYSICS.*

*Smyth, *op. cit.*, pp. 2-3.

# 11

# Introductions

The introduction of a technical report has several very definite functions. In fact, it is scarcely an exaggeration to say that the word "introduction" has a special meaning in technical writing, and that you might find it helpful to forget whatever meanings you have associated with the term. The introduction is, of course, the first portion of the text. It may or may not be preceded by a title page, letter of transmittal, preface, table of contents, list of illustrations, and abstract. Whether any or all of these elements are present, however, the introduction should be a complete and self-sufficient unit.

The primary purposes of an introduction to a technical report are to state the subject, the purpose, the scope, and the plan of development of the report. In addition, it is sometimes necessary to explain the value or importance of the subject. Often it is desirable to summarize principal findings or conclusions.

The organization of the introduction and the degree to which any of its parts is developed depend upon circumstances. It should not be supposed that a good introduction is necessarily a long one; sometimes only a sentence or two is sufficient. The organization is affected particularly by the need of stating a key idea in the opening sentence.

We shall discuss the four primary functions of an introduction in the order stated above, concluding with some comments upon the problems of initial emphasis and of the statement of the importance of the subject.

As noted above, introductions to technical reports often include a summary of the major conclusions or results that are presented in the body of the report. Since the presence of such an introductory summary does not affect the fundamental character or functions of an introduction, no consideration is given to it in this chapter. A discussion of the introductory summary may be found in Chapter 4.

## Statement of the Subject

At the very beginning, the reader should be given a clear understanding of what the exact subject of a report is. How this information is best presented depends, as usual, upon what the reader already knows. To some extent the title of a report is, or can be, a statement of the subject; but almost without exception the subject should be stated again in the introduction, and the title should never be used as an antecedent for a pronoun in the introduction. That is, if the title were "The Arc Welding Process," you should not begin the introduction with the words, "This process . . . ."

The effectiveness of the title of a technical report depends upon making it as informative as possible while still keeping it reasonably short. Titles which are merely ornamental, or even misleading, are a source of constant annoyance, as you have no doubt discovered when using periodical guides. Try to think of the title as it will appear to the reader. We recall a university commencement in which one of the doctoral dissertations listed on the program was entitled, "The Life of an Excited Atom." From the titters provoked in the audience it was clear there were numbers of people present who, if not persuaded that this subject was risqué, considered it at least rather unnecessary.

The statement of the subject in the introduction itself may involve one or more of the following three problems: definition of the subject, theory associated with the subject, and history of the subject.

It may prove necessary to define the subject and the terms used in stating the subject. For example, in a discussion intended for an uninformed reader, entitled "Hydroponics—Gardens without Soil," we would want to explain both what the word "hydroponics" means and what gardens without soil are. We might write, somewhere in the introduction:

> The word "hydroponics" is simply a name given to the process of growing plants in a liquid solution instead of in soil. Our chief interest will lie in the commercial application of this principle—that is, in "gardens" in which tanks filled with gravel and a mineral solution replace soil as the source of food for the growing plants.

In this illustration it is primarily the concept of soilless gardening which must be conveyed to the reader; but since that concept is expressed by the unfamiliar term "hydroponics," it is necessary to be certain that the relationship between the term and the subject is clear. In short, the writer should give special attention to making clear both the subject itself and any unfamiliar terms associated with it. If the subject is already familiar in some degree to the reader, the writer should adapt his statement of the subject accordingly.

Sometimes, however, even a well-written formal definition is insufficient, and it may become necessary to give the reader some background information. This background information is usually either theoretical or historical, or both.

For large land surveys, it is occasionally desirable to stop the survey and re-establish the true north direction, in order to localize instrument errors. The new determination of true north may be accomplished by sighting on the pole star. For a student of surveying with no knowledge of taking astronomical "sights," a report on the procedures of establishing true north might well begin with a section on the theory of such an operation. Only through a comprehension of the theory could the operation itself (the principal subject) be fully understood. If such a section on theory were short, it could be included in the introduction; if it were long, it might better go into a section by itself, immediately following the introduction. In either case, the writer should remember that what he is trying to do is to make clear what the subject of his report is.

The purpose of discussing the history of a subject in a technical report is much the same as that of discussing the theory. It gives the reader an understanding of the total situation of which the particular subject is a part. For instance, in a report on the methods of manufacture of the "buckets" on a jet-engine turbine, a brief history of the development of the jet-engine turbine might help a good deal in showing why the buckets are now made as they are. A warning is in order here, however. Don't allow yourself to start discussing history simply because you can't think of any other way to get started. Ask yourself if the history is clearly contributing toward the basic purpose of the report. If it is, good; if not, out with it.

Like the theory, the history can go either into the introduction or into a section by itself. In fact, theory and history are often combined —which is perfectly all right and natural.

## Statement of Purpose

It is imperative that the reader understand the purpose of a report. And remember that we are concerned here with the purpose of the report, not of the subject. The purpose of a drill press is to drill holes; but the purpose of a report about a drill press might be to discuss the most efficient rate of penetration of the drill.

There can seldom be any objection to saying simply, "The purpose of this report is . . . ." Frequently the statement of both the subject and the purpose of a report can be accomplished in the same sentence, often the first sentence in the report. If the statement of the scope of the report and of the plan of development of the report can be included in this same sentence without awkwardness or lack of clarity, there is no reason for not putting them there. The fundamental requirement of a good introduction is that it perform the four basic functions; there can be no rules about how they are accomplished, nor can there be rules for a fixed order of these functions.

## Statement of Scope

The term "scope" refers to the limits of a subject. The problem in the introduction is to explain what the limits are so that the reader will expect neither more nor less than he finds.

Limits may be stated in several ways. One way is concerned with the amount of detail: a report may be described as a general survey of a subject or as a detailed study. Another way has to do with how great a range of subject matter is included. For example, a report on standardizing the location of the pilot's controls in aircraft might include all types of aircraft or only one type, like multi-engine aircraft. The reader must be told what the range is. A third way is to note the point of view from which the report will be written. There is a good deal of difference, for instance, between announcing that a report is on the subject of the plumbing in a certain hotel, and announcing that the subject is the plumbing in this hotel from the point of view of a sanitation engineer.

These ideas may be of some value in helping you think how to say what the scope of a given report is, but the basic idea to remember is simply that you must keep defining and qualifying your subject until it is certain that the reader will know what to expect.

## Statement of Plan of Development

The statement of the plan of development of a report is simply a detailed application of the slogan, "First you tell the reader what you're going to tell him. . . ." It is a simple idea, easy to carry out, and unquestionably one of the most important elements in the introduction. The phrasing may be straightforward and formal: "This report will be divided into five major parts: (1) ______________, (2) ______________,

. . . ." Or it may be more "literary": "The most important aspects of this subject are ______________, ______________, ______________, . . . ." The manner should suit the situation. Usually the statement of the plan of development comes at or near the end of the introduction.

## Other Problems

Two other problems that should be mentioned are the need for a proper initial emphasis and the occasional desirability of an explanation of the importance of a subject.

The first few statements made in an introduction are especially critical because on this very limited evidence the reader is forming an impression of the report as a whole. His impression as to the content and purpose of the report should be accurate. If he later finds that his first impression was wrong, confusion and irritation will be the probable result.

Ask yourself how much the reader already knows about the subject. Has he requested this particular report, or will it reach him unannounced? Is it about a subject he is interested in, or a project which he approves? Or is he likely to be, at the outset, indifferent or even hostile? Usually there can be no objection to some variation of the "The purpose of this report . . ." beginning, but it would be a mistake to suppose that this is always true. Consider the following opening sentence:

> When it became apparent, in the fall of 19___, that the water supply of ______________ City would soon be inadequate to support the industry now located in the city, the City Council requested the firm of Smith and Rowe to prepare a preliminary report on the outlook for the immediate future, together with tentative recommendations of measures to be taken.

The initial emphasis here falls upon the urgent need for action. In comparison, an opening consisting of a statement of purpose would be less effective; and an opening consisting of the first sentences of a history of the water supply problem might be quite misleading.

The importance of a failing water supply needs no explanation. But suppose the water supply was adequate and a report was being written to show that steps should be taken to prevent a probable shortage at the end of another ten years. The writer would face a quite different problem. He would have to devote considerable space to proving that a merely probable event of ten years in the future was of immediate practical interest. The fundamental principle is to analyze your reader and estimate his needs and attitudes. The fourth introduction that is quoted at the end of this chapter illustrates an extended comment on the importance or value of a subject.

## Summary

The major functions of an introduction are to state the exact subject of the report, its exact purpose, its scope, and its plan of development. The statement of the subject is primarily a problem in definition but may require extended discussion of background material, particularly of history or theory, or both. On the other hand, for an informed reader, the subject need only be named. The statement of purpose is often combined with the statement of subject. The statement of the scope of the report may be conveniently considered in three aspects: the "range" of the subject matter, the detail in which the subject is to be discussed, and the point of view from which the subject is to be discussed. The statement of the plan of development presents no difficulties, but is extremely important; it normally appears at or near the end of the introduction. The organization of the whole introduction is affected by the selection of the proper initial emphasis. Sometimes it is desirable to explain the importance of a subject.

## ILLUSTRATIVE MATERIAL

The following pages contain six examples of introductions. The third and fourth are examples of students' work; the others were written on the job. The last one is accompanied by the entire report of which it is a part.

### A Method of Calculating Internal Stresses in Drying Wood*

*(Forest Products Laboratory Report No. 2133)*

#### Introduction

As wood dries, it is strained by a complex pattern of internal stresses that develop as a result of restraints characteristic of normal shrinkage. Such stresses are found in all lumber during normal drying and are responsible for most of the defects associated with the drying process.

Although such stresses have been known and recognized for many years, no suitable method of calculating their magnitude and distribution has been available. As a result, the development of schedules for drying wood without excessive losses due to drying defects and without unduly

*This introduction comes first in this group of examples because of the question it immediately raises about the initial emphasis. The question is this: Can you imagine a reader who would need the information in this report but who would be ignorant of the facts in the first paragraph? Before trying to answer this question, let's ask another. If the au-*

*Courtesy Forest Products Laboratory, Madison, Wisconsin.

prolonging the drying process has been almost entirely by empirical procedures.

In recent years, investigations of the stress behavior and perpendicular-to-grain mechanical properties of drying wood have laid the groundwork for a more fundamental approach to the problem of improved wood drying. However, effective use of such data requires a method of evaluating drying stresses at any point on the cross section of a drying board. Such a method has not been available up to this time.

This report describes a method for calculating the perpendicular-to-grain stresses associated with the drying process and illustrates the application of the method to one condition of wood drying.

*thor wants to prevent* excessive *losses, as he says in paragraph two, why emphasize* normal *conditions at the beginning? Shouldn't he emphasize the contribution his work has made to the elimination of excessive losses?*

*One possible revision is to delete the first paragraph and to reword the first sentence in the second paragraph as follows: "Although it has been known for many years that defects appear in drying lumber because internal stresses are developed, no suitable method of calculating the magnitude and distribution of these stresses has been available."*

## Stability Study of 220-KV. Interconnection between Philadelphia Electric Company, Public Service Electric & Gas Co. of N. J., Pennsylvania Power & Light Co.*

The effects of line to ground short circuits on the stability of the interconnection have been investigated by careful mathematical calculation based on the best available data as to line and system characteristics. While, on account of unavoidable differences in actual and assumed conditions, and on account of the methods by which the problem has been simplified for purposes of calculation, extreme accuracy cannot be hoped for, nevertheless most of the essential factors have been considered and evaluated, and it is therefore felt that the final results obtained are substantially correct.

The report has been divided into three main sections as follows:

*This introduction gives the impression that it was written in haste, with no pleasure. It is certainly no pleasure to read. The second sentence would have a hard time getting by a freshman English teacher. And yet the introduction performs, at least to a limited degree, all of the four major functions of an introduction. The subject of the report is stated in the first sentence, and—for the technical reader for whom the report is obviously*

*From a General Electric Company report, Engineering General Department (Schenectady, N. Y.), p. 1.

I Results
II Basis of Study
III Method of Calculation Employed
IV Representative Curves and Diagrams

*intended—the purpose is made fairly clear as well. The second sentence is concerned chiefly with scope. And the plan of the report is perfectly plain. (The little mix-up about how many main sections there are will be found in the original.)*

## Problems of Control of Wheat Rust and White Pine Blister Rust

This report describes problems encountered in efforts to control wheat rust and white pine blister rust by eradication of one of the two hosts required by each disease. As its two hosts, wheat rust requires wheat and barberry. White pine blister rust requires white pine and either currants or gooseberries.

Both diseases have been combatted by efforts to eradicate, in a given area, one of the hosts. These efforts, however, have not always been successful because certain spores of these diseases may survive for a time independently of a host. In addition, it is not always easy to locate and get permission to destroy all the objectionable plants. This latter problem, however, lies beyond the scope of this report.

The life cycle of each disease will be described, and then the circumstances will be explained in which eradication of one of the hosts may fail to eliminate the disease.

*This introduction, an example of a student's writing, illustrates several of the elements of a good performance. The initial emphasis is appropriate, the subject and purpose are clear. There is enough background information to indicate the significance of the subject; the scope is limited; and there is a good statement of the plan of development. One criticism that might be made, however, is that a little more does need to be said about scope. Will the description of the life cycle of the diseases be presented in a way suited to the interests and knowledge of the average gardener, or will it be directed to—say—an advanced student of botany? One or two more sentences on this point would be helpful.*

## Report on The Direct Hydrogenation and Liquefaction of Coal*

### 1. Introduction

In recent years much time, money, and energy have been spent on the problem of obtaining synthetic liquid fuels. In European countries, where domestic supplies of crude oil are rela-

*The initial emphasis of this introduction is upon the importance of the subject. The general subject is stated*

*From a report written by Mr. Don R. Moore while a student at The University of Texas, and reprinted here with his permission.

tively very low, the production of synthetic liquid fuels has become imperative to their self-sufficiency. Today, even in the United States, where reserves of crude petroleum are seemingly very great, scientists are devoting great emphasis to the production of liquid fuels from other sources.

*in the first sentence. The first three paragraphs of the report are devoted to the historical background of the subject.*

Because of its great abundance and accessibility, one of the principal organic raw materials which has received consideration in recent years as an important source of synthetic liquid fuels has been coal. The known supply of coal in the world today is tremendously great compared to the known reserves of crude petroleum. Although new discoveries of petroleum have boosted supplies, there is little doubt that the supply of petroleum in the United States will run short many, many years—even centuries—before coal supplies are exhausted. Scientific estimates have placed the life of petroleum reserves in the United States at between ten and fifty years while estimates have placed the life of coal reserves well in excess of one thousand years.[1]

*This paragraph and part of the next are concerned with theoretical background.*

It is because of this possibility of an impending shortage of crude petroleum that the conversion of coal to oil by hydrogenation processes has become so important. As yet, the production of fuel oils from coal is not economically feasible in the United States. Gasoline produced from the direct hydrogenation of coal would cost 22.6 cents per gallon if produced by a plant which had a daily production of 3,000 barrels or between 15 and 16 cents per gallon if produced by a plant which had a daily production of 30,000 barrels; the same fuel produced from crude petroleum by the common thermal cracking refinery ·process would cost 8.5 cents per gallon.[2] However, in the future it is believed that engineering achievements in the field of coal-hydrogenation coupled with a rise in the price of fuel oils produced from crude petroleum (which will surely occur should a shortage of crude petroleum arise) will possibly make the production of gasoline and other motor fuels from coal-hydrogenation economically feasible.

[1]*Synthetic Liquid Fuels*, Hearings before a Subcommittee of the Committee on Public Lands and Surveys, United States Senate, Seventy-Eighth Congress, p. 137.

[2]*Ibid.*, p. 53.

Although this conversion of coal to oil appears to be a mysterious and complicated process, it may be discovered from the discussion appearing in the second section of this report that the composition of certain bituminous coals which have been freed from ash resembles the composition of crude petroleum to a great extent.

The actual chemical conversion of coal to oil can be accomplished by either of two hydrogenation processes—the direct, or Bergius,[3] process or the indirect Fischer-Tropsch[4] process. The material presented in this report, however, will concern only the primary reaction involved in the conversion of coal to oil by the direct hydrogenation process. This report will discuss the conversion from a chemical aspect and will not cover engineering details and difficulties involved in such a conversion by commercial-scale continuous-phase[5] processes.

It is the purpose of this report to discuss the mechanism and yields of the primary reaction involved in the synthesis of coal to oil by the direct hydrogenation process, the operating variables involved in the reaction, and the effect of the rank and type of different samples of coal upon the total liquefaction yields from the reaction. These topics will be discussed in the order stated.

*Here the scope of the report is limited. The sentence beginning, "The material presented . . ." limits what we earlier called the range of the subject. In the next sentence, point of view and detail are mentioned.*

*The subject is clarified by definition of terms in the footnotes.*

*The introduction concludes with a formal statement of purpose and plan.*

## Design, Construction, and Field Testing of the BCF Electric Shrimp-Trawl System*

*by Wilber R. Seidel*

### Introduction

In 1961, we—that is, the staff at Gear Research Unit of the Exploratory Fishing and Gear Re-

*Here is an excellent introduction—and yet in our opin-*

[3]The Bergius Process (named after a German who was a pioneer in the field of coal-hydrogenation) is a process in which hydrogen is forced into the reactive intermediates formed by a thermal decomposition of the complex molecular structure of the coal.

[4]The Fischer-Tropsch Process, devised by the two German scientists, is a process in which the coal is burned to form "water-gas" which is then hydrogenated to form oils.

[5]A continuous production process in which coal is constantly fed to a liquefaction converter and in which the liquefaction yields are constantly removed for further hydrogenation.

*From *Fishery Industrial Research 4,* No. 6, U.S. Department of Interior Fish and Wildlife Service, Bureau of Commercial Fisheries.

search Base, Pascagoula, Mississippi—evaluated electric systems that could be used on shrimp trawls to increase the harvest of brown and pink shrimps. Because these shrimp burrow in the ocean floor during daylight, commercial fishing has been restricted to night trawling—that is, to the period when they are not in their burrows. The aim of these preliminary evaluations was to develop a system that would force the shrimp out of their burrows during daylight. If successful, it would allow around-the-clock fishing and more efficient use of harvesting gear.

Initial trials with electric shrimp-trawl systems, though encouraging, were not satisfactory because the rates of catch during the daytime were much smaller than were those at night. However, they did show that, before a successful electric trawl could be developed, the exact electrical requirements that would cause optimum shrimp response had to be determined. In 1965, Klima studied the response of shrimp to an electric field. His data provided the background information for the needed design of an adequate electric shrimp-trawl system.

The goal of the work reported here was to design a full-scale prototype shrimp trawl that would permit a test of the commercial feasibility of electric trawling during daylight. The aim was not to build a fully engineered production-model trawl, but simply to develop one that would show whether daylight electric trawling is practical. Accordingly, the work was divided into two main parts. The first was concerned with designing and developing a full-scale electric shrimp trawl; the second with testing it under actual fishing conditions.

*ion it violates an important principle. That is, we don't think it has a good initial emphasis. We think this deficiency could be taken care of very simply. All that is needed is to start the introduction as a whole with the sentence that now stands as the first sentence of the last paragraph: "The goal of the work . . . ."*

*This introduction is so well written that there can be little doubt that what we have called its "deficiency" was a matter of deliberate choice and not of carelessness or lack of skill. We feel confident that the author of this report has the skill to handle the initial emphasis however he wants to.*

*Why did he do it this way, then? We don't know. We would say that his way gives a slower, easier pace to the introduction, and that is pleasant. But we still consider it less effective than it would be with the revision we suggest. Of course people's preferences may differ about such things. What do you think?*

*Here's a suggestion. If you like the rather slow, indirect beginning in this introduction—fine; but before you try writing introductions in this way, practice getting the initial emphasis exactly on the most important aspect of your subject. Practice until you have proved to your instructor you can do it every time. Then you can experiment, confident that you know what you are doing.*

# Rapid Method for the Estimation of EDTA (Ethylenediaminetetraacetic Acid) in Fish Flesh and Crab Meat*

*by Herman S. Groninger and Kenneth R. Brandt*

## Abstract

EDTA, a quality stabilizing additive, is usually applied to seafoods by spraying or dipping, and the amount of EDTA retained by the treated product must be determined by an analytical method. A titration method based on the chelation of EDTA with thorium ion was modified for use in the determination of EDTA in fish flesh and crab meat. The modified method is both simple and rapid and gave about 90-percent recovery of added EDTA from samples of fish flesh and crab meat.

## Introduction

EDTA has been reported to be useful or potentially useful as an additive to seafoods to stabilize color and retard the formation of struvite (National Academy of Sciences, 1965), inhibit enzyme-catalyzed changes in flavor (Groninger and Spinelli, 1968), and inhibit the growth of bacteria (Levin, 1967).

Often, EDTA is applied to seafoods by spraying or dipping the product. The amount of EDTA actually added must then be determined by a suitable quantitative method.

A number of methods have been developed for the determination of EDTA in various materials (Belot, 1964; Brady and Gwilt, 1962; Cherney, Crafts, Hagermoser, Boule, Harbin, and Zak, 1954; Darbey, 1952; Haas and Lewis, 1967; Kratochvil and White, 1967; Lavender, Pullman, and Goldman, 1964; Malat, 1962; Vogel and Deshusses, 1962). In all of these methods, the principle of measurement is based on the chelating capacity of EDTA. In general, each method was developed for use on a specific type of product.

*This example of an introduction is presented in the context of the entire report of which it is a part. Since it is directed to research chemists in the specialized area of the subject matter of the report, you may find the content not very inviting or informative—unless you happen to share their interests. If, however, you will look at the way the authors have gone about their job of writing, rather than at the technical content, we believe you will discover that this is actually a highly interesting and useful illustration of how to write a report. It is, in fact, a report in miniature, with all the parts of a long report compressed into a small space—except for a table of contents.*

*From *Fishery Industrial Research 4,* No. 6, U.S. Department of Interior Fish and Wildlife Service, Bureau of Commercial Fisheries.

Efforts were made to adapt several of these methods (Brady and Gwilt, 1962; Haas and Lewis, 1967; Vogel and Deshusses, 1962) for the analysis of EDTA in fish muscle and cooked crab meat. None were satisfactory. During the testing of the EDTA methods, we found, however, that the method of Pribil and Vesely (1967), which was developed for the determination of EDTA during the commercial synthesis of EDTA, could be modified satisfactorily for use in the determination of EDTA in fish flesh and crab meat.

The purpose of this paper therefore is to report on the modified method. The paper gives the details of the method, the recoveries of added EDTA from fish flesh and crab meat, the precision of the method, and the precautions to be observed when EDTA is used in the presence of interfering substances.

## I. Details of the Method

1. Prepare extracts of fish flesh or crab meat by disintegrating 20 grams of material with 40 milliliters of 5-percent trichloroacetic acid for 1 minute in a blender.
2. Filter the mixture through Whatman No. 2v filter paper.[1]
3. Collect 2 milliliters of filtrate and adjust the pH to 11.0 with 10-percent sodium hydroxide.
4. To the filtrate, add 5 milliliters of 2-percent calcium acetate and readjust the pH to 11.0.
5. Remove the precipitated calcium phosphate by centrifugation and follow by filtration through Whatman No. 1 filter paper.
6. Wash the precipitate with a dilute solution of alkaline calcium acetate and combine the washings with the phosphate-free filtrate.
7. Adjust this filtrate to pH 3.5 with 0.5 *N* hydrochloric acid.
8. Add 5 milliliters of 0.2 *M* acetate buffer, pH 3.5, and from 1 to 2 drops of a 0.16-percent aqueous solution of xylenol orange to an aliquot.

*This report, as you may have noticed, is taken from the same source as the preceding example of an introduction. Perhaps not surprisingly, it contains, in our opinion, exactly the same deficiency with respect to the initial emphasis. What do you think about this one?*

*The many references in this introduction to relevant publications make for slow reading. We ourselves would prefer to use only footnote numbers in the text and to put the names at the bottom of the page or at the end of the article. On the other hand, there are certainly some advantages in putting the names into the text as these writers have done. One of these advantages is that for the chemists reading the report some of the names would probably be familiar and would have the immediate effect of adding information to the background that this part of the introduction is supplying. That is, the reader might think, "Oh, yes, I know that man's work."*

*As we remarked back in Chapter 7 (Description of a Process), this report contains a fine illustration of an appropriate use of itemized directions intended for a reader who is thoroughly familiar with the subject matter.*

*Finally, we suggest that, with the next chapter on conclusions and summaries in mind, you take a close look at the similarities and differences among the ab-*

[1]Use of trade names is merely to facilitate description; no endorsement is implied.

9. Titrate the aliquot with 0.001 *M* thorium nitrate to a red-violet endpoint.
10. Calculate the content of EDTA from a standard curve prepared by the titration of solutions containing 0 to 5 $\mu$M EDTA (Figure 1).

*stract, introduction, and summary in this report. You may feel that in so short a report the value of a formal summary is reduced almost to the vanishing point. Yet we must admit we enjoy having it there. In any event, the differences in the content of the three elements just named are easy to see in this highly compressed form. Study of this skillfully written miniature report provides many suggestions of much practical value for the writing of longer reports.*

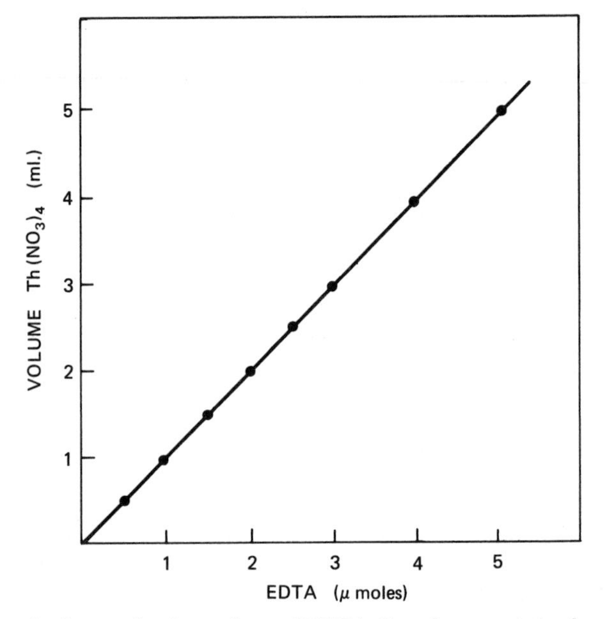


Figure 1. Standard curve for the analysis of EDTA. Samples containing known amounts of EDTA were titrated with 0.001 M $Th(NO_3)_4$.

II. Recoveries of Added EDTA

Table 1 shows the efficiency of recovery of EDTA added to fish flesh and crab meat. At all the concentrations of added EDTA tested, the recovery appeared to be adequate for the purposes of this determination.

TABLE 1.—*Recovery of EDTA added to fish flesh and crab meat*

| EDTA added | EDTA recovered from: | |
|---|---|---|
| | Fish flesh | Crab meat |
| $\gamma$/g. | *Percent* | *Percent* |
| 37.5 | 95 | – |
| 75 | 96 | 95 |
| 150 | 93 | 90 |
| 300 | 90 | 91 |
| 600 | 92 | 92 |

## III. Precision of the Method

Table 2 gives a statistical evaluation of the recovery from samples containing 300 gammas of EDTA per gram of fish flesh and crab meat. These recovery results show that the method is suitable for determining the amount of EDTA additive in a fishery product. Also these results compare favorably with those obtained with similar methods.

TABLE 2. *Precision of recovery of EDTA from fish flesh*

| Trials | EDTA added | EDTA recovered | | | Standard deviation |
|---|---|---|---|---|---|
| | | Range | Mean | | |
| | $\gamma$/g. | $\gamma$/g. | $\gamma$/g. | *Percent* | $\gamma$/g. |
| 13 | 300 | 246–276 | 266 | 89 | 11 |

## IV. Precautions to be Observed

Phosphates interfere with this method. The presence of phosphates gives high results in terms of EDTA, because these compounds combine with the thorium ion. This interference is completely eliminated by removal of the phosphates with calcium.

Citrate also interferes with the method. Usually, however, not enough citrate is present in fish and crab meat to cause a problem. In samples of crab meat in which citrate has been added, this interference can be eliminated by the removal of the citrate from EDTA by chromatography on cation resin. This separation is accomplished as follows:

1. Adjust the filtrate from the phosphate-removal step to pH 8.8 with 0.5 milliliter of 0.2 *M* tris buffer.
2. Pass the filtrate over a 5- to 6-centimeter AG 50 W-X8 (+H) column (200–400 mesh).
3. Wash the column with from 10 to 20 milliliters of water containing 0.5 milliliter of 0.2 *M* tris buffer, pH 9.0.
4. Elute the EDTA with from 10 to 20 milliliters of *N* hydrochloric acid in 0.5 *M* potassium chloride.
5. Adjust the eluate to pH 3.5 with 0.5 *N* hydrochloric acid.
6. Add 5 milliliters 0.5 *M* acetate buffer (pH 3.5).
7. Titrate the eluate just as you would a regular sample.

The overall recovery obtained when citrate is present is about 80 percent. This loss of 20 percent can be attributed to a portion of the EDTA passing through the Dowex 50 column along with the citrate.

Apparently, metal ions such as $Cu^{2+}$, $Mg^{2+}$, and $Ca^{2+}$ do not interfere.

## Summary

The Pribil and Vesely (1967) titration method based on the chelation of EDTA with thorium ion was modified for use in the determination of EDTA in fish flesh and crab meat. The modified method, which is simple and rapid, gave about 90-percent recovery of added EDTA from samples of fish flesh and crab meat.

Phosphate and citrate interfere. Techniques are given for their removal from samples containing EDTA.

## Literature Cited

Belot, Y.
1964. Determination of EDTA in the presence of complexing metals. Analytical Abstracts 11: Abstract No. 3739.

Brady, G. W. F., and J. R. Gwilt.
1962. Colorimetric determination of EDTA. Journal of Applied Chemistry 12: 79–82.

Cherney, P. J., Barbara Crafts, H. H. Hagermoser, A. J. Boule, Ray Harbin, and Bennie Zak.
1954. Determination of ethylenediaminetetraacetic acid as the chromium complex. Analytical Chemistry 26: 1806–1809.

Darby, Albert.
1952. Colorimetric determination of the sodium salts of ethylenediaminetetraacetic acid. Analytical Chemistry 24: 373–378.

Haas, Erwin, and La Vera Lewis.
1967. Microdetermination of cadmium and EDTA in protein solutions. Archives of Biochemistry and Biophysics 118: 190–199.

Groninger, H. S., and J. Spinelli.
1968. EDTA inhibition of inosine monophosphate dephosphorylation in refrigerated fishery products. Journal of Agricultural and Food Chemistry 16: 97–99.

Kratochvil, Byron, and Martha C. White.
1965. Spectrophotometric determination of microgram quantities of (ethylenedinitrilo) tetraacetic acid with bis (2, 4, 6-tripyridyl-*s*-triazine) iron (II). Analytical Chemistry 37: 111–113.

Lavender, A. R., Theodore N. Pullman, and David Goldman.
1964. Spectrophotometric determination of ethylenediaminetetraacetic acid in plasma and urine. The Journal of Laboratory and Clinical Medicine 63: 299–305.

Levin, Robert E.
1967. The effectiveness of EDTA as a fish preservative. Journal of Milk and Food Technology 30: 277–283.

Malat, M.
1962. Photometric determination of microgram amounts of EDTA with bismuth-pyrocatechol violet complex. Chemist-Analyst 51: 74–75.

National Academy of Sciences—National Research Council.
1965. Chemicals used in food processing.

National Academy of Sciences—National Research Council, Publication No. 1274, 294 pages.

Pribil, R., and V. Vesely.
1967. Determination of EDTA, DPTA, and TTHA in their mixtures. Chemist-Analyst 56: 83.

Vogel, J., and J. Deshusses.
1962. Determination of ethylenediaminetetraacetic acid (EDTA) in foodstuffs, particularly in wines. Analytical Abstracts 9: Abstract No. 5447.

# 12 Conclusions and Summaries

In this chapter we shall discuss the chief considerations in bringing a report, or a section of a long report, to an end.

One of these considerations is an aesthetic one: how to give a sense of finality and completeness to the discussion. We shall make some comments on this problem, but for the most part we shall be concerned with the content rather than with the possible aesthetic function of the conclusion or summary. First we shall discuss conclusions and then summaries.

We've been using these two words, "conclusion" and "summary," together so far, and it may have seemed that one or the other alone would do as well. The reason for retaining both is that we want to use them in different and rather specialized ways. That is, a conclusion, in this chapter, is not going to mean the same thing as a summary.

A conclusion is, of course, an end. In technical writing, however, there are three different kinds of conclusions or ways of bringing a report to an end. As we said above, the idea of a summary is not included in any of these three kinds of conclusions—with the exception that sometimes conclusions are summarized, as we'll explain.

The first of these three kinds of conclusions is what we will call the "aesthetic." Its function is merely to bring the discussion smoothly to a stop. This need is felt particularly when there seems to be little point in reviewing what has been said, and yet it seems awkward just to stop. *Often it is wise just to stop;* but not always. For instance, at the end of a description of how to develop film at home, you might want to close: "Although, as has been shown, developing and printing film is not a difficult process, it is one which affords a great opportunity for experimenting with effects, and thus provides a continuing novelty and challenge. Reasonable caution in carrying out the steps just described will start you on the way to a most pleasant and interesting hobby." Nothing significant in the way of review or of decisions has been said here, but a reasonably graceful conclusion has been made.

The second of the three kinds of conclusions is one in which the results of an investigation or study are stated. For example: "The conclusion reached as a result of this study is that the toe of the dam is being undermined by water flowing through fissures in the limestone bed of the stream." This kind of conclusion is often called "findings," or "results."

The third kind is the decision reached at the end of a discussion concerning a choice of action, or concerning a practical problem for which a solution must be presented in the form of a forthright recommendation as to what action should be undertaken. A typical example: "The inescapable conclusion is that the wisest course of action is to fabricate the panels of asbestos rather than to provide heat shielding for the present wooden panels."

Obviously, the difference between the second and third kinds of conclusions is sometimes very slight. And the fact is that the terms "conclusion," "findings," and "results" are used rather indiscriminately for both kinds. The difference between the two kinds does become important, however, when it is necessary that the reader understand whether he is merely being given some information (the second kind of conclusion) or told that a decision about a course of action has been made (the third kind).

Conclusions of either the second or third kind should be presented clearly and forcibly. Needless to say, a long report devoted to a complex train of reasoning that terminates in an obscure conclusion is exasperating. Often it is desirable to present the conclusion in a paragraph of its own, with the subhead "Conclusion." In a short report the conclusion may appear only at the end; and in a long report devoted to reaching only one important conclusion, that conclusion may also appear only at the end (except for a possible appearance in the abstract or introductory summary). On the other hand, there may be a series of "conclusions" (findings, decisions) in the body of the report, which are restated at the end of the report and sometimes formally entitled "Summary of Conclusions." The following paragraph is an illustration of this type of conclusion:

### Conclusions

Based on the work conducted on the hydraulic unit at this laboratory and reported herein, it is concluded that:

1. The oil in the system under normal operating conditions contained small amounts of air.
2. The amount of air in the oil varied considerably with different operating conditions.
3. There appeared to be no significant difference in the operation of the unit from a force vs. speed standpoint when operated with varying amounts of air in the oil.
4. Excess air introduced into the oil through the pump intake readily dissolved in the oil when it was subjected to high pressures so that there appeared to be no air mechanically entrained in the oil during normal operation of the unit.
5. The greater efficiency reported for this hydraulic machine is probably due to the fact that it is operated with forces nearer the actual cutting force so that the system is always fairly near equilibrium, thus always placing the maximum load on the cutter, but never overloading it.
6. There appeared to be no significant difference in the operation of this unit when using either Oil A or Oil A (R & O).

It is not necessary to number the conclusions, although in this instance it was surely a good idea.

Finally, there are the summaries. A summary is, as we have already implied, a review or concise restatement of the principal points made in the discussion. It is more useful at the end of a report engaged chiefly in presenting a body of information than in an analytical or argumentative report, or in a descriptive report which would not justify anything more than an aesthetic conclusion.

The writing of a good summary requires a very clear grasp of each one of the fundamental ideas of the report. In fact, writing a summary may serve as a test of whether you have actually seen and formulated clearly the fundamental ideas in the report. And a summary should contain no new ideas or information. It is a restatement of information —not an addendum.

The following illustration contains not only a summary but also the paragraph which immediately precedes the summary. The purpose of including this extra paragraph is to show how it is summarized in the last sentence of the summary.

### Cooperation between the Metallurgical Laboratory and du Pont

Since du Pont was the design and construction organization and the Metallurgical Laboratory was the research organization, it was obvious that close cooperation was essential. Not only did du Pont need answers to specific questions, but they could benefit by criticism and suggestions on the many points where the Metallurgical group was especially well informed. Similarly, the Metallurgical group could profit by the knowledge of du Pont on many technical questions of design, construction, and operation. To promote this kind of cooperation du Pont stationed one of their physicists, J.B. Miles, at Chicago, and had many other du Pont men, particularly C.H. Greenewalt, spend much of their time at Chicago. Miles and Greenewalt regularly attended meetings of the Laboratory Council. There was no similar reciprocal arrangement, although many members of the laboratory visited Wilmington informally. In addition, J.A. Wheeler was transferred from Chicago to Wilmington and became a member of the du Pont staff. There was, of course, constant exchange of reports and letters, and conferences were held frequently between Compton and R. Williams of du Pont. Whitaker spent much of his time at Wilmington during the period when the Clinton plant was being designed and constructed.

Summary

By January 1943, the decision had been made to build a plutonium production plant with a large capacity. This meant a pile developing thousands of kilowatts and a chemical separation plant to extract the product. The du Pont Company was to design, construct, and operate the plant; the Metallurgical Laboratory was to do the necessary research. A site was chosen on the Columbia River at Hanford, Washington. A tentative decision to build a helium-cooled plant was reversed in favor of water-cooling. The principal problems were those involving lattice design, loading and unloading, choice of materials particularly with reference to corrosion and radiation, water supply, controls and instrumentation, health hazards, chemical separation process, and design of the separation plant. Plans were made for the necessary fundamental and technical research and for the training of operators. Arrangements were made for liaison between du Pont and the Metallurgical Laboratory.* [NOTE: this last sentence summarizes the preceding paragraph.]

As you see, the style of this summary is distinctly "choppy." One bald statement follows another. This is probably a good idea. The summary can be regarded almost as a list of the major ideas, and there is little reason to try to escape very far from the form of a list. Indeed, summaries are sometimes broken down into numbered statements, as was the "conclusion" quoted earlier.

Finally, we must say plainly that conclusions and summaries cannot be written by formula. The principles we have discussed are of

*From Henry D. Smyth, *Atomic Energy for Military Purposes,* rev. ed. (Princeton, N.J.: Princeton University Press, 1946), pp. 128–129.

considerable value, but they are only principles, not prescriptions. As always, the important thing is the successful accomplishment of the function itself, not the particular method adopted.

## Summary

For convenience, the final section of a report can be classified as either of two types: conclusions or summaries. Conclusions can be subdivided into those that are primarily aesthetic, those that announce the results of an investigation or study, and those that present a decision concerning a course of action. An aesthetic conclusion merely brings the report to a graceful close. The other two kinds of conclusions may appear only at the end of the report (with the exception of a possible appearance in the abstract or introductory summary), or both in the body of the report and at the end. A summary is a restatement of important information. It should contain no new ideas; and, in comparison with the length of the report, it should be short.

# SECTION FOUR

# Types of Reports

*So far we have been concerned with various fundamental skills and techniques needed in technical writing. Now we turn our attention to the forms of writing in which these skills and techniques are used.*

*In some organizations there is little formality attached to report writing. Each writer decides what form is best suited to what he has to say. Elsewhere, particularly in large organizations, numerous and sometimes elaborate forms are devised and given names. Thereafter, within the organization, these forms are spoken of as types of reports, and young men are given instructions on how to write them. This is exactly as it should be, if the forms devised satisfy the needs of the organization. But it does result in the creation of a tremendous lot of "types." In a casual search that took no more than thirty minutes, we once turned up the following examples of so-called types of reports:*

| | | |
|---|---|---|
| *preliminary* | *service* | *examination* |
| *partial* | *operation* | *examination-trip* |
| *interim* | *construction* | *inspection* |
| *final* | *design* | *investigation* |
| *completion* | *failure* | *memorandum* |

| | | |
|---|---|---|
| *status* | *student-laboratory* | *notebook* |
| *experimental* | *industrial-research* | *short-form* |
| *special* | *industrial shop* | *periodic* |
| *trade* | *evaluative* | *information* |
| *formal* | *test* | *work* |

*It is possible that the foregoing list could be boiled down to a few fundamental types. No one, however, has ever succeeded in winning general acceptance of a working system of classification of reports, and it seems unlikely that any attempt will ever succeed. Your best preparation for writing whatever sort of report you may be asked for is (1) a mastery of the fundamental techniques and skills of technical writing, and (2) an acquaintance with some widely used types of reports, so that the word "report" will have concrete meaning for you. Of course a fairly wide variety of reports, or excerpts from reports, has already been presented as illustrative material in earlier portions of this book.*

*The purpose of Chapters 13–16 is to introduce, in some detail, a few generally accepted types of reports. In addition, we shall take up some composite forms which, although not precisely reports, can conveniently be considered at this point. All forms discussed have in common the fact that the special techniques described earlier appear in them in combination. The reports discussed are the progress report, the recommendation report, the proposal, and what we shall here identify loosely as the form report. These are followed by Chapters 17–19, which discuss the composite forms —oral reports, business letters, and writing for professional journals.*

# 13

# The Progress Report

## Introduction

One easily distinguishable type of report is the progress report —distinguishable because of its purpose and general pattern of organization. This chapter explains how to prepare a progress report.

The progress report's main objective is to present information about work done on a particular project during a particular period of time. It is never a report on a completed project; in some ways it is like an installment of a continued story. Progress reports are written for those who need to keep in touch with what is going on. For instance, executives or administrative officials must keep informed about various projects under their supervision to decide intelligently whether the work should be continued, given new direction or emphasis, or discontinued. The report may serve only to assure those in charge of the work that satisfactory progress is being made—that the workers are earning their keep. Not the least important function of the progress report is its value as a record for future reference.

It is neither possible nor worthwhile to list here the extent of the activities on which progress reports are made; the extent is tremendous. Any continuing, supervised activity may have progress reports made on it—anything from research projects in pure science to routine construction jobs. Nor is it possible to be dogmatic about the frequency with which such reports appear: often progress reports are made on a monthly basis, but sometimes the week may be the time unit, or the quarter-year. Anyhow, the time covered in the report has little to do with the way the report is organized and presented.

## Organization and Development

About the best way of getting at the problem of what should go into a progress report, and how, is for the writer to ask himself what the reader will want to find in the report.

Common sense tells us that the reader will want to know at least three things: (1) what the report is about, (2) what precisely has been done in the period covered, and (3) what the plans are for the immediate future. Quite naturally, he will want this information given in terms he can readily understand, and he will expect it to be accurate, complete, and brief. Great emphasis is often placed on brevity.

The foregoing suggests a pattern of organization as well as some clues regarding development of the report. From the standpoint of organization, there should be three main sections: a "transitional" introduction, a section giving complete details of progress made during the current period, and a "prophetic" conclusion.

*The Transitional Introduction.* In the first of these sections, the transitional introduction, the reporter must identify the nature and scope of the subject matter of his report, and he must relate it to the previous report or reports. He may be expected to summarize earlier progress as a background for the present account. Finally, if circumstances warrant—or it is expected of him—he may present a brief statement of the conclusions reached in the present unit of work and, possibly, some recommendations. This latter function is especially applicable in progress reports on research projects. It is not as pertinent in an account of the progress on a construction or installation project.

In serving as a transition between the current report and the preceding one, this part of the report need not be lengthy, for it is essentially a reminder to the reader—a jog to his memory. Reading it gives him an opportunity to recall the substance of the previous reports so he can read the present one intelligently. The title may partially bridge the gap between reports, for it may name the project and number the report. Something like "Boiler Installation in Plant No. 1, Progress Report No. 5" is characteristic. But even such a descriptive title is not enough, and many reports do not bear such titles (see the example reprinted at the end of this chapter). The discussion—or the briefing—is needed to hook the current report securely onto the preceding one.

*The Body of the Report.* With the introduction out of the way, the reporter must next tackle the body of his report—the detailed account of current progress. The first point that needs to be stressed here is the importance of making this part of the report complete, accurate, and clear. This is much easier said than done, mainly because it is easy to forget the reader. Remember that the report is not a personal record for the writer, but information for some particular reader or readers about the work done. If you keep this in mind, you should have very little trouble.

The second thing that needs to be said concerns organization. Although some progress reports are organized chronologically with subsections covering parts of the over-all period (a monthly report might have four subdivisions, each being a running narrative account of the work done during a week's time), most of them are organized topically. For instance, a report of progress made on a dam construction job contained the following subdivisions: (1) General [interpretative comments], (2) Excavation, (3) Drilling and Grouting, (4) Mass Concrete, and (5) Oil Piping. A report of progress made on the production of an aircraft model contained these topical subdivisions: (1) Design Progress, (2) Tooling, (3) Manufacture, (4) Tests, and (5) Airplane Description. The sample progress report included in this chapter provides another example. But these illustrations should not be regarded as prescriptions. The important thing is that the development of the main section of the report should grow logically out of the subject matter itself and the requirements of those who want the report.

Giving a careful, detailed account of work done may require the presentation of quite a mass of data. Usually such data, particularly numerical data, cannot be presented in the conventional sentence-paragraph pattern; it would be unreadable. Tables, of course, are the answer. But since you will want to make your reports as readable as possible, you will do well not to interrupt your discussion with too many tables. It is better to put them in an appendix at the end of the report and confine yourself to evaluative or interpretative remarks about the data in the body of the report itself. Don't forget to tell the reader that the tables are in the appendix. For instance, the report on a dam construction mentioned in the paragraph above contained a table giving an estimate of quantities of material used, one on unit and concrete costs, and another giving the type and number of employees along with the amount of money paid out for each. Here is the first of these tables:

*Estimate of Quantities—Week Ending April 25, 19____*

| Bid Item | Description | Unit | Previous Total | This Period | To Date |
|---|---|---|---|---|---|
| 1 | Mass concrete | cu yd | 787,686 | 18,792 | 806,478 |
| 2 | Steel reinforcing | lb | 2,369,350 | 29,883 | 2,399,233 |
| 3 | Black steel pipe | lb | 213,107 | 666 | 213,773 |
| 4 | Cooling pipe | lin ft | 317,417 | 188 | 317,605 |
| 5 | Electric conduit | lb | 367,480 | 309 | 367,789 |
| 6 | Copper water stop | lb | 35,424 | 856 | 36,280 |

The presentation of data such as this in connected reading matter would be difficult, to say the least; moreover, this is a short table —each of the others contained four or five times as much data. Although tables are a great convenience and sometimes a necessity, remember that they should not be allowed to stand alone without comment.

*The Conclusion.* With one exception, the requirements of the conclusion to a progress report will depend entirely on the nature of the work reported on. If progress on research is being reported, for instance, it may be necessary to present a careful, detailed statement of conclusions reached—even though these conclusions have been briefly stated in the introduction. It may also be desirable to make recommendations about action to be taken as a result of present findings or about future work on the project. On the other hand, it is not likely that a report on the progress made on a simple machine installation would require formal conclusions or recommendations. But you are not likely to have trouble with this problem, for the nature of the subject matter will suggest naturally what should go into the last section.

There is one thing, however, which you will do in almost all progress reports, regardless of subject matter, and it is suggested by the term "prophetic" used earlier. You must tell the reader approximately what he may expect the next report to be about and what its coverage, or scope, will be. Along with this forecast it may be advisable to estimate the time necessary for completion of the entire project. Here is an important caution: don't promise too much. It is very easy for the inexperienced worker to overestimate the amount of work that can be covered in a forthcoming period. You will naturally want the forecast to look promising, but you will not want it to look so promising that the reader will be disappointed if the progress actually made does not measure up to your prediction.

A final word of advice: be brief but complete and use the simplest terminology you can.

## Form

We have discussed the content and presentation of the three main parts of the progress report—introduction, body, and conclusion. There remains the problem of form. Two forms are used for progress reports, the choice depending on the length and complexity of the report. They are the letter form and the conventional or formal report form. The first is used for short reports submitted to one individual or to a small number of persons. The second is used for longer reports, submitted perhaps to an individual but more often for circulation to a number of company officials and perhaps to stockholders and directors as well.

The letter report has a conventional heading, inside address, and salutation. Many are in military form (especially those written on government contract for research and development projects; see the illustrative form on page 284). The opening paragraph makes reference to the preceding report and identifies the nature and scope of the present one. The parts of the rest of the report are usually labeled by means of marginal headings, these corresponding to the subject matter divisions. The conventional ending is the complimentary closing, "Respectfully submitted," followed by the signature. This form is especially suitable in those organizations where the report serves primarily as a means of "accounting for" the reporter's activity. Besides, it has the advantage of the personal touch.

The letter, however, is not suitable for long reports of progress on elaborate projects submitted for wide circulation to sponsors or directors. For one thing, the letter loses its identity as a letter if it extends over a large number of pages, especially since marginal subheadings are usually employed. There may be, of course, a letter of transmittal. But the report proper will follow the pattern described in the chapter on report format (Chapter 20).

## ILLUSTRATIVE MATERIAL

The material on the following pages includes a portion of a progress report and an illustration of typical format for a letter-form progress report. The first illustration, you will note, is not a complete progress report; it does, however, illustrate the letter format and suggests the kinds of headings which appear. We would have presented the actual report, but security consideratons made it impossible. We should point out that in the actual report, there were seven subdivisions under the "Summary of Activity" division, each dealing with progress on a specific research area. Two brief reports of trips made for conferences with individuals in cooperating research organizations were attached to this progress report. The progress report itself was five pages long.

The second illustration should be examined carefully, not necessarily as a model, but as a fairly typical example of the progress reports commonly written today. Note the extent to which it conforms to the pattern just discussed. Do you think it was written for a reader thoroughly acquainted with the technical subject matter or for one without such a background?

THE UNIVERSITY OF TEXAS AT AUSTIN
APPLIED RESEARCH LABORATORIES
POST OFFICE BOX 8029 10000 FM ROAD 1325 AUSTIN, TEXAS 78712

AC 512, 836-1351

Serial No. – – –
(Date)

Sponsoring Agency
Attn: (Code or Office Symbol and name of Project Engineer as appropriate)
(Contract Number)
Contractual Line-Item Reference (e.g., as identified on DD Form 1423)
(Address of Sponsoring Agency)

Subject: Research and Development Technical Performance Summary, (Applicable Monthly Period)

This is the – – – research and development summary submitted under Contract – – –, Project – – – . ...

Summary of Activity

(Normally this section is limited to two or three very concise paragraphs summarizing activities for the month. Reference is made to attachments such as Trip Reports, Visit Reports, Technical Opinions, Technical Memoranda, and Professional Resumes of newly-assigned personnel.)

Summaries of Activity for Specific Tasks. (Reasonably detailed, but brief, nontechnical summaries of individual tasks, identified by title, which received attention during the reporting period are presented. Depending upon the level of effort, from two to twelve or fourteen such summaries may be treated.)

Summary of Man-Hours

(Man-hours expended during the reporting period are tabulated by classification of the investigators or by name. A cumulative total of man-hours through the reporting date may be required.)

Items Requiring Coordination

(Specific items requiring action and/or coordination on the part of the Sponsor are identified.)

Work Schedule for Next Reporting Period

(A projection is made of specific action items to be considered during the ensuing period. Also, planned trips and expected visits are identified.)

(Signature of Responsible Official)
Name and Title

– – –/– – –
Enclosures (n)

*Figure 1. Typical Format for a Letter-Form Progress Report*

## Flight Dynamics of Ballistic Missile Interception*

*Study Coordinator: Dr. James Ash*

### Statement of the Problem

This study concerns the problem of active missile defense against a ballistic missile. The study is confined to the elementary case of interception of a single ballistic missile by a single defensive missile. The primary objectives of the study are to:

a) Determine what information is required for the specifications of an interceptor system for an ICBM.

b) Specify the performance requirements for an AICBM system.

c) Formulate the mathematical description of the system.

d) Compare and evaluate various systems under various flight situations.

The results of this study should furnish information necessary for the planning of more complex regional defense systems against actual multiple ICBM attacks.

### Current Progress

Study 14 terminated 31 August 19__. The results are contained in a forthcoming final report entitled, WADC TR 59-516, *Flight Dynamics of Ballistic Missile Interception*. A résumé of that report follows.

The problem of an active unitary interceptor system operating against a ballistic missile is studied to determine the most suitable functional forms of the system. Particular attention is given to the intercontinental ballistic missile, and it is assumed that detection and tracking are accomplished by radiation means from friendly territory. Analytical methods and procedures are presented for the investigation of the ballistic missile vs. countermissile duel with consideration of the ballistic missile approach speed and angle, detection range and tracking range capabilities of the interceptor system, preparation

*The first two paragraphs of this progress report are a statement of what the report is about. An unusual feature here is the fact that the clearest statement of the relationship between this report and previous work is contained in the first sentence of the second major section, rather than at the beginning. Moving that sentence ("Study 14 terminated 31 August 19__") to the beginning of the report would be an improvement—perhaps in the following form: "This report is concerned with work done under Study 14, which was terminated 31 August 19__." Then the first sentence under "Current Progress" might begin, "The results of Study 14 are contained . . . ."*

*The general organization of this progress report is characteristic of the type. At the end, however, the author is evidently suggesting what work could be done if the contract were renewed, rather than promising what will be done under a contract still in force.*

*University of Chicago, Laboratories for Applied Sciences. Work sponsored by United States Air Force under Contract No. AF 33(616)-5689.

time of the interceptor missile, and lethal radius of the interceptor warhead.

Methods of computation for reaction time and range relationships are developed for both minimum-energy and nonoptimum ICBM elliptic trajectories. Refinements of the Keplerian elliptic trajectory for the effects of air drag and nonsphericity of the earth are considered for accuracy computation. Expressions have been developed for the effects of observational errors on the predicted orbital elements. The re-entry phase has been considered and equations are provided for the estimation of path deflections and energy emission due to air drag. Frequency distributions of probable United States targets have been compiled to provide estimates of expected ranges and azimuth angles in the event of an ICBM attack. The geometrical limitations of detection and tracking, and visibility zones for observation stations have been graphed in relation to the parameters of the ICBM trajectory and reaction time.

The AICBM radar factors, including the short-range requirements and dependence on attenuation, the propagation factor, antennae, power sources, radar cross-sections, noise effects, and the present and projected capabilities of AICBM radar systems are discussed and analyzed. Expressions and graphs are formulated for the computation of expected elevation angle error, range error, and Doppler velocity errors for radio propagation through successive layers consisting of the troposphere, the ionosphere, and free space. The possibility of improved detection by means of low-frequency radio waves is examined, and a theoretical analysis of plasma shock waves has been made in an effort to understand the physical mechanism underlying recent low-frequency reflection observations. In addition to radar, the feasibility of detection and tracking by means of passive infrared techniques, including consideration of the radiation from the rocket exhaust, optical limitations, and atmospheric effects have been examined.

The general dynamic equations of the interceptor rocket have been formulated with consideration of the effects of variable mass and changing of inertia as the fuel is expended and of the secondary effects of Coriolis and centrifugal accelerations due to the motion of the earth. Generic interceptor equations of motion suitable for flight simulation studies are given. The theoretical background for the estimation of high altitude drag by means of free-molecule flow considerations is reviewed. The interceptor performance parameters and their interrelationships involving weight ratios, specific impulse, slant range, and reaction times have been developed for both the ballistic and powered phases, as well as a method for computing dispersions by means of perturbations about the basic trajectory.

### Future Work

The restrictions under which Study 14 was pursued include (1) free flight phase of a ballistic missile trajectory, (2) single interceptor-ballistic missile duel, and (3) land-based interceptors and land-based ballistic missiles. Future work would call for a relaxation of these hypotheses to include more realistic situations involving a more detached analysis of some specific interceptor systems operating against specific ballistic missiles in certain selected flight situations.

The publication of WADC TR 59-516 obviates the necessity for a classified supplement to this Quarterly Report.*

## SUGGESTIONS FOR WRITING

1. Write a progress report in letter form (see Chapter 20 on report forms) giving an account of the progress you have made to date on a long report assignment. If you are writing a research report, include in your progress report an account of library research (indexes consulted, books available on the topic, general reference works, and so on), note taking, making of illustrations, rough draft, and the like —anything pertinent. Include a statement about what remains to be done and a prediction of the anticipated date of completion. Additional reports may be made later on this same project.

2. If you are engaged in any sort of extended laboratory experiment

*An appendix lists trips and visits made during the quarter and another lists technical notes and reports issued under the research contract.

in one of your technical courses, make a progress report on the work accomplished to date. Assume that it is being made to someone unfamiliar with the technical nature of the subject matter. Use conventional report form—title page, table of contents, and so on—but omit the letter of transmittal, since the introduction can perform its function.

3. Assuming that your technical writing course is the project, write a report of the progress made during the preceding month. Do not forget to include in the beginning section a brief statement or synopsis of earlier progress which (you will assume) has already been reported. Put this in letter-report form and address it to a hypothetical educational adviser.

# 14
# Recommendation Reports

## Introduction

Because the term "recommendation report" is so frequently used in technical writing, both in textbooks and in the field, one would naturally suppose that this is a type of report with an easily identifiable kind of content and organization. In fact, however, any report that contains recommendations is a recommendation report —and almost any report may contain recommendations.

Examination of numerous recommendation reports will show that their basic characteristics differ widely. The bulk of the content of a recommendation report is most often interpretative, but it is not uncommon to find more description than interpretation. There is no standardized organization for recommendation reports, with the exception that the recommendations are usually stated near the beginning, or near the end, or both. The format may be any one of the many varieties in use. The function of a recommendation report, on the other hand, would seem at first to be fixed and stable— that is, to persuade the reader to take a certain course of action; and usually this function is indeed evident in practice. But a consultant might conceivably be indifferent as to whether his recommendations were acted upon and reflect his indifference in the tone of his report. We must conclude, in brief, that we are dealing with an ambiguous concept when we discuss recommendation reports: a considerably less exact concept than that of the progress report.

Nevertheless the vitality of the idea of a recommendation report, as shown by the wide currency of the term, is a warning that we should not treat the concept too casually. Of course what we are dealing with fundamentally is the situation (in report form) in which the abstract thinking of the laboratory and the study passes over into the realm of practical action. The importance of this action probably accounts for the vitality of the idea of a recommendation report. Furthermore, it would be a mistake to suppose that people don't know what they mean when they use the term "recommendation report." Probably they don't always know exactly; but if your boss or your college instructor tells you to write a recommendation report about something, we strongly urge that you do not stop to itemize the ambiguities we have just been pointing out. Get busy and analyze the problem you have been given, decide upon the proper course of action, and make a forthright recommendation. That kind of procedure is what your boss or your college instructor will mean when he speaks of a recommendation report.

The only writing problems that will be new to you in a recommendation report are how to phrase recommendations and where to put them. Before we take up these problems, however, we will give some special attention to reader analysis and style.

## Reader Analysis and Style

Ideally, the art of persuasion should never enter into the professions of science and engineering. The scientist or engineer would investigate physical laws, or apply them to a specific problem, report his findings, and be through. The reader would need no convincing or persuading; he would be governed solely by logic. Practically, of course, things are seldom so simple.

There are two somewhat different situations to be considered: first, that in which you are given definite instructions to prepare a recommendation report, and second, that in which you volunteer a recommendation. A volunteered recommendation may be inserted into a report written primarily for a different purpose (like a progress report) or it may be made the chief subject and purpose of a special report.

When you have been instructed to make a recommendation, you may find that it is fairly obvious what action should be recommended, and also that everybody agrees it is the proper action. This is wonderful—and not uncommon at all. But you may on other occasions find that after studying the subject (1) you don't think any action at all should be taken, whereas you either know or suspect that your superior or associates feel that some action is desirable; (2) you think action should be taken, but foresee unwillingness to act; (3) you think a certain course of action should be taken, but expect that a different course of action will be favored; (4) you think action should be taken, but cannot see a clear advantage between two or more possible courses of action.

If you find that the evidence does not indicate a clear-cut decision, the best policy is simply to say so, with an especially thorough analysis of advantages and disadvantages. This does not mean you should not make a concrete recommendation; you should—but not without making clear what the uncertainties are. This situation is found in some measure in the report printed at the end of this chapter.

When you expect opposition to recommendations you are convinced should be made, you should give a good deal of thought to the tone of your report and to methods of emphasizing the points which clarify the logic of the situation.

Don't let yourself fall into an argumentative tone. We have in our files a report from a research organization that begins with the statement, "This is a very important report." Our own immediate reaction, when first reading this statement, was a suspicion that maybe the report wasn't really very important or the writer wouldn't have thought it necessary to try so unblushingly to persuade the reader it was. It would have been a more effective report had the writer prepared a good, clear introduction, stated his major conclusions in the proper place, and added this sort of statement: "The great importance of this discovery arises from the fact that . . . ."

In general, be forthright in tone or manner, but not blunt. Instead of saying, "The present method is wasteful and inefficient," remind yourself that whoever designed the present method was probably doing the best that could be done at the time, and was not unlikely proud of his work. You may prefer to write something like, "The proposed new method offers a considerable increase in efficiency over the present method." Or, "Certain changes in the present method will result in an increase in efficiency."

Putting emphasis on the proper points demands first of all an analysis of the probable attitudes of your reader. If you do not expect opposition, there is little problem here. If you decide that opposition is probable, a good general policy is to discuss, first, the advantages and disadvantages of the recommendation you think might be preferred to the one you intend to make—being careful to state fairly *all* its advantages; second, to present the advantages and disadvantages of the course of action you prefer; third, to give a summary and recommendation. This approach provides emphasis through relative position—the value of the preferred action being shown after the weaknesses of the alternative have been explained. Emphasis may also be achieved through paragraphing and sentence structure. For example, use of a series of short paragraphs written in short declarative sentences when you sum up the advantages of the preferred course of action will result in an especially forceful impression.

The problem of whether to volunteer a recommendation is most likely to arise when you have a positive suggestion about work in which your own official part is only routine, or about work which is not a part of your official duties. In the long run there is no doubt that the more ideas you have and the more suggestions you make, the faster you will be promoted and the more fun you will have with your work. But when it comes to volunteering recommendations in writing, two cautions ought to be observed.

First, be very sure that your recommendation is sound and that you have shown clearly that it is sound. (In this connection, a review of the discussion of interpretation in Chapter 9 will be helpful.) Your superiors aren't going to be pleased with mere opinions.

The other caution is to be careful not to give the impression that you are trying to "muscle in" on something. This is likely to be a delicate point, and you'll be wise to think about it very deliberately. To a certain extent this difficulty can be met by avoiding the kind of phrasing used in a formal recommendation (see below) and by presenting your recommendation in the form of a conclusion. Instead of saying, "It is recommended that the temperature of the kiln be lowered 15 degrees and the drying time prolonged to 84 hours," you could say, "Better results would evidently be obtained by lowering the temperature of the kiln 15 degrees and prolonging the drying time to 84 hours."

Altogether, then, when you have recommendations to make, your first problem is to determine precisely what course of action or what decision is best justified by the evidence. Your second problem is to estimate your reader's probable attitude toward your recommendations. Your third problem is to prepare a report that will be effectively organized to make clear the logic of your recommendations to the specific reader or readers you expect to have.

## How to Phrase Recommendations

To a certain extent it is possible to classify recommendations as formal and informal. An informal recommendation may consist merely of a statement like, "It is recommended that a detergent be added to the lubricant." Or, "Therefore a detergent should be added to the lubricant." In a sense, any suggestion or advice constitutes a recommendation. The formality with which it should be presented is determined by its relation to the major problem being discussed and by the tone of the whole report, as stated above. Usually, the more important a problem is, and the longer the discussion of it, the more need there is for a formally phrased recommendation.

A highly formal recommendation is illustrated by the following:

> After consideration of all the information available concerning the problems just described, it is recommended:
> That the present sewage disposal plant be expanded, rather than that a new one be constructed;
> That the present filter be changed to a high-rate filter;
> That a skilled operator be employed.

Sometimes each main clause in a recommendation like this is numbered. Sometimes the recommendations are presented as a numbered list of complete sentences preceded by the subhead "Recommendations" and without any other introduction. And, of course, sometimes they are simply written out in sentence form as shown in the "informal" example in the preceding paragraph, without unusual indentation and without numbers.

It is occasionally advisable to accompany each recommendation with some explanation, in contrast to limiting each recommendation to a single statement as illustrated above. An example will be found at the end of this chapter. This method is often particularly useful when recommendations are stated twice in the same report, at the beginning and again at the end. The explanations usually accompany the second statement.

## Where to Put Recommendations

Recommendations almost invariably appear at the end of a recommendation report. If the report is long, and especially if an introductory summary is used, they are likely to appear near the beginning as well, immediately after the statement of the problem. When they appear both at the beginning and the end, however, those at the end are likely to be stated informally, whereas those at the beginning are more formal, usually with the heading "Recommendations."

If the report is long, it is desirable to put the recommendations at the beginning so that a reader may at once find the major results. And it is always wise to state the recommendations at the end so that they will be the last ideas impressed on the reader's mind.

## ILLUSTRATIVE MATERIAL

The report on the following pages is a recommendation report in which the recommendations appear only at the end, accompanied by considerable explanation. On the whole the report is done well, but the introduction is weak (the initial emphasis particularly), and part of what is presented under "Discussion of the Results" seems to belong more logically to the preceding section on "Procedure." The English is only fair. The "Conclusions" and "Recommendations" sections are the strongest in the report.

For an example of a much more elaborate recommendation report, see Appendix D.

### Lindberg Engineering Company

*Research Laboratory**

*Determination of the Deposit That Collects on the Element Coils and the Cause of Failure of the Element Coils in Type 2872-EH Furnace, Serial 2162 at Olds Motor Works, Lansing, Michigan.*

#### Introduction

The Olds Motor Company of Lansing, Michigan have been experiencing abnormal element failures in their Type 2872-EH furnace. Mr. W. Bechtle of our Detroit Office sent in a peculiar, light brown, fluffy deposit he found on the burnt-out wire coils. He found that this substance was very retentive and difficult to shake off the coil.

His report pointed out that the work being drawn in this furnace was heat treated in an Ajax salt bath and quenched into oil. He stated that although the work was washed and appeared to be clean, small quantities of salt might still be present to cause the deposit on the elements and the subsequent failure. A request was made to analyze the material to determine whether it came from the Ajax salt bath and also to determine if it were responsible for the element failure.

#### Procedure

Conductivity tests and a complete chemical analysis were made of the coating taken off the heating element that failed.

#### Discussion of the Results

The resistance of the material as received was tested with a sensitive meter, and it was found to be very high and thus not a conductor when cold. Good contact may not have been made when making the above test because of the powdery condition of the material. To verify this and also to check on the solubility, some of the substance was placed in distilled water. The resistance of the distilled water was found to be 80,000 ohms before the substance was placed in it. After the substance was placed in the water and stirred, the resistance dropped to 15,000 ohms, showing very slight solubility and conductivity. Tests at high temperatures were not made because the conditions existing in the furnace with a high voltage of about 450 to 480 could not be reproduced.

The chemical analysis of the fluffy substance removed from the elements is as follows:

| | |
|---|---|
| Silicon ($SiO_2$) | 28.78% |
| Iron Oxide ($Fe_2O_3$) | 46.70 |
| Alumina ($Al_2O_3$) | 13.06 |
| Nickel Oxide ($NiO$) | 1.60 |
| Chromium Oxide ($Cr_2O_3$) | 0.30 |
| Barium Chloride ($BaCl_2$) | 1.88 |
| Barium Oxide ($BaO$) | 0.57 |
| Total | 92.89% |

(Balance may be Potassium and Sodium compounds that were not determined.)

The presence of barium definitely indicates that the Ajax heat treating salt is getting into the draw furnace. Barium Chloride is one of the common constituents of heat treating salts, and there are no traces of this substance in materials used in the construction of the Lindberg Furnace.

The presence of the high percentage of silica, alumina, iron oxide, and other oxides can be easily explained. The silica, alumina, nickel, and chromium oxides come from the dust produced by the wear of the furnace refractory and alloy parts rubbing together under constant vibration. The iron oxide comes chiefly from the scale of the heat treated work or the scale produced in tempering the work. All of this dust is being constantly carried in the recirculating air stream over the work and through the element chamber. Under normal conditions this dust will not stick to the elements because the melting point is very high and the velocity of the air is too great to allow the dust to settle out.

With the introduction of only a slight amount of heat treating salt, however, this condition is entirely changed. The heat treating salt spalls off the work due to the difference in its thermal expansion over steel when heated in the tempering furnace and is carried with the other dust in the air stream. As work is continuously being tempered, the concentration of the heat treating salt particles becomes greater. When the heat treating salt particles are carried over the heating elements by the recirculating air stream, the salt particles strike the element and melt to its surface. (The heating element temperature may go as high as 2000°F., depending on the furnace temperature and the load.) The remaining dust from the furnace refractory and scale from the heat treated work then stick to the element.

The failure of the element then results from the material covering the insulators and making contact to the frame and the element coil. The high voltage of 440 to 480 arcs across, burning out sections of the wire from the element. The material covering the element also acts as an insulator, causing the element to run abnormally high in temperature which also reduces its life.

### Conclusions

1. The material coating the heating element in the furnace is a combination of heat treating salt from the hardening operation and refractory and oxide dust from the tempering furnace. The heat treating salt dust melts when it strikes the elements and the remainder furnace dust sticks to it.

2. The element failure is due to the conductivity of the melted heat treating salt covering the element and insulators and thus shorting to the frame. The high voltage of 440 to 480 arcs across and cuts sections out of the element coil. The substance covering the coils also acts as an insulator, causing an abnormally high element temperature which also greatly reduces its life.

### Recommendations

1. The most logical thing to do is to prevent the heat treating salt from entering the draw furnace. This, however, is not quite as easy as one may first think because heat treating salts containing barium are not very soluble in hot water or caustic solutions. The solubility depends upon the amount of barium salts present in the mixture. Barium chloride itself is only slightly soluble in hot or cold water or in caustic solutions. Heat treating salts containing barium are very difficult to dissolve out of blind holes, recesses, and etc.

The only sure way that all traces of barium heat treating salts can be cleaned from the work is by the following procedure: 1st—Clean salt and oil from quenched work in hot Oakite solution; 2nd—Rinse in hot water; 3rd—Clean in an acid solution (one part of hydrochloric to four parts of water, or a standard pickling bath can be used if care is observed); 4th—Rinse in hot water to remove acid.

A better method would be to use a heat treating salt that is readily soluble in water. Such a salt is a compound of a mixture of sodium and potassium chlorides and is the best type of neutral hardening salt that can be used. One such commercial salt on the market is known as Lavite #130, made by the Bellis Heat Treating Company. This salt has a useful temperature range from 1330°F. minimum to 1650°F. maximum. In as much as we are not familiar with the hardening requirements, only the above general suggestion can be given in reference to cleaning all traces of salt from the work.

2. Reduce the voltage that can exist between the coils and the frame by redesigning the heating element and thus have less chance of burning out sections of the element by arcing due to the conductivity of the heat treating salt.

We are now designing an element for this furnace in which the element hook-up will be changed to a series star circuit instead of the delta circuit now employed. Both will operate from the 460 volt line, but the voltage between the coil and the frame over any one insulator will be reduced to about 130 volts instead of the 460 volts on the present arrangement.

This will, of course, reduce the tendency for arcing when the heat treating salt and furnace dust collect on the elements and insulators. This is not, however, a substitute for cleaning the heat treating salt from the work. The subject company must do everything possible to remove the traces of heat treating salt if low maintenance is to be achieved. The redesign of the element will act as a safety factor to prevent excessive burn outs in case some heat treating salt gets into the furnace occasionally.

3. If it is found impractical to follow out recommendation 1, and the new redesigned coil we are supplying for recommendation 2 does not reduce the maintenance, then consideration should be given to rebuilding the subject furnace to gas fired instead of electric.

## SUGGESTIONS FOR WRITING

One very good subject for a recommendation report is the suggestion of a topic for a long paper you are going to write, either in this course or some other. Discuss the purpose of the paper, and the standards that the subject of the paper should meet. That is, the subject should be one that interests you, one on which there is sufficient information available in the library or elsewhere, and one that is suitable in scope for a paper of the length required. You will think of additional standards appropriate to your own project. If it is to be a paper requiring research in the library, consult bibliographical guides that will help you locate information, and take a preliminary look at some of the materials to which these guides refer you. Include in your recommendation report a discussion of these bibliographical guides and of the materials you examined, remembering that your objective is to persuade your reader that there is enough material available to justify undertaking a report on the subject you've chosen, and that it is material you will actually be able to use.

Other suggestions for recommendation reports may be found by consulting the list at the end of Chapter 9, on interpretation. The topics there may easily be reworded to suggest recommendations.

# 15
# Proposals

## Introduction

The following statement by a vice-president of a major electronics company testifies to the importance of the presentational form we wish to discuss with you in this chapter—the proposal.

> The technical proposal is our division's primary sales tool. Our proposals can have a major effect on our performance in the highly competitive . . . market, and it is imperative that they be as good as we can possibly make them. They must not only be technically sound, but they must also be logically and clearly presented.*

In this research and development oriented age we live in, the proposal is unquestionably one of the most important types of technical writing you are likely to be called upon to do. Just exactly what is this important presentational form?

Before answering this question, we want to explain one unusual aspect of this chapter on proposals. Often a proposal is a very long report, and instead of being prepared by a single author a considerable number of people may collaborate in the writing. Still other people contribute to the proposal through oral consultation with those who do the actual writing. In one respect, then, this chapter is not a discussion of a kind of report you will write, but rather an explana-

*H. J. Wisseman, Asst. Vice-President, Texas Instruments Inc., in John A. Walter, *Some Notes on Proposal Writing*, Texas Instruments, Inc.

tion of a large-scale project in which you might be involved. On the other hand, some proposals are quite brief, and are prepared by a single author. This chapter will also make suggestions relevant to such brief proposals.

A proposal is a written offer to solve a technical problem in a particular way, under a specified plan of management, for a specified sum of money. Put a slightly different way, a proposal is a document designed to convince a "customer" that the company (or organization) presenting it is better qualified to supply a desired product or service than are all the other companies submitting proposals. Let's take a closer look at these statements.

The *written offer to solve a technical problem* describes, often in minute detail, the design or plan proposed, sometimes along with some discussion of alternate plans and designs. Strictly speaking, it is this written offer to solve the technical problem that is known as the "technical proposal."

The *specified plan of management* mentioned in the definition above is commonly called the "management proposal." In general, the management proposal explains to the prospective client precisely how the entire project will be managed, tells who (often by name) will manage it, and suggests a time schedule for completion of the phases of the project. One of its important purposes is to assure the customer that his problem will be worked on by competent personnel during every stage, from prototype design through manufacturing if the proposal is for "hardware," or from initial exploratory study to final solutions if the proposal is for "software." ("Software" is a report providing answers to a basic problem or problems.) Moreover, the management component of the proposal makes clear that the lines of responsibility for quality and reliability—and for efficient communication between customer and supplier—will be firm and clear.

The phrase *for a certain sum of money* refers to what is called the "cost proposal." This part of the proposal gives a detailed breakdown of costs in terms of labor and materials.

Often all of these three basic elements of a proposal are contained between one set of covers. Indeed, the entire proposal may be quite brief, as noted earlier. Sometimes, however, each of the three elements is so long that they are bound as separate documents. And sometimes the customer requires separate volumes because evaluation of the different elements will be conducted simultaneously by different groups of evaluators. More will be said about evaluation later. In form and appearance, except for the fact that sometimes it becomes very long, the proposal usually adheres to the general characteristics of a formal technical report. An example will be found at the end of this chapter.

Proposals are commonly referred to as *solicited* or *unsolicited.* The former, by far the most common variety, are submitted in response to an *invitation to bid* (sometimes called a "bid request," a "purchase request," or a "request for proposal"). Invitations to bid may be published or they may be mailed directly to companies from a "source file" maintained by the company or organization issuing the invitation. Unsolicited proposals, on the other hand, are prepared by a company or organization in the hope that the excellence of the idea or plan proposed will attract the interest of the potential client and convince him that he actually does have a need for the service or product being proposed. Sometimes the unsolicited proposal is preceded by an informal inquiry of interest outlining the idea or plan. If successful, the unsolicited proposal results in what is often called a "sole-source" procurement. It is probably safe to say, though, that such noncompetitive proposals are restricted to those rather rare circumstances in which it is believed the proposing organization possesses a unique capability. Ordinarily, contracts for research and development are not let without competitive bidding.

Finally, we wish to emphasize the crucial importance the proposal bears to a company's success in meeting competition. A poorly conceived and ineptly presented proposal has an immediate and brutal effect because it fails to get a contract and thus results in less income for the company. Over a period of time, almost every employee of any organization that bids for contracts through proposals may be personally affected by the degree of skill with which the staff can write them.

With the foregoing characteristics of a proposal in mind, let's take a quick look at what happens, typically, when a proposal is prepared. Suppose a firm that manufactures jet engines for aircraft wants a new high-speed wind tunnel in which to test its engines. The firm sends out a number of "invitations to bid" to companies listed in its source file, companies whose work in this field the firm knows and respects. Without such a file, it might advertise in some of the trade and professional journals. Accompanying each invitation may be a set of specifications together with a general discussion of the firm's needs and, possibly, a formal statement of work (list of items the firm wants done). A deadline may also be indicated. The specifications would, of course, set down the requirements the wind tunnel must satisfy.

Men in the companies receiving the invitations to bid study the specifications and bid papers to determine whether to submit a proposal or not. Considerations of capability, availability of qualified staff and facilities, already committed schedules of work, and the possibility of profit are among their chief considerations. Each company deciding to bid then assigns staff members to the various tasks involved in the preparation of the proposal. Two of the most important tasks, for example, are the development of a suitable design and the preparation of a cost estimate. During this period, each of the competing companies may confer with representatives of the engine manufacturer to acquire as thorough an understanding of the problem as possible (assuming the customer is willing to grant this favor to bidders). In fact, the engine manufacturer may convene what is called a pre-proposal briefing conference for representatives of the interested companies so that a thorough discussion of the problem can be held.

As each of the competing companies makes its final decisions, it prepares its written proposal. In this proposal, the company presents its design for the wind tunnel and explains how it would do the job. If the specifications have been quite detailed, the proposal may concentrate almost exclusively on how the tunnel would be built, rather than on what kind of design it would have. In any event, by the time the proposal is completed, a great many people will probably have contributed to it in one way or another. Proposals concerned with large projects are almost always team efforts, not the work of a single author, though there may be a proposal co-ordinator or manager and an editor charged with the responsibility of making the proposal conform to a single style and format throughout. The final effort is to make sure that copies of the proposal are mailed in time to meet the specified deadline for consideration.

When the engine manufacturing firm receives the various proposals, it designates certain of its staff members to evaluate them. Usually the evaluation team will include one group to judge the relative merits of the technical proposals, another to judge the cost proposal, and still another to judge the management proposals. Out of this evaluation process there finally emerges a decision as to which company gets the contract.

Innumerable variations are found for the general situation described in this hypothetical case. The work being proposed may range from the design and production of a small, simple device to projects of enormous complexity and cost. The number of people involved in the preparation of the proposal may range from one to hundreds. And the proposal itself may be a brief letter or an elaborate set of bound volumes. See the Proposal Flow Chart (Figure 1) for an indication of what happens with an elaborate procurement cycle.

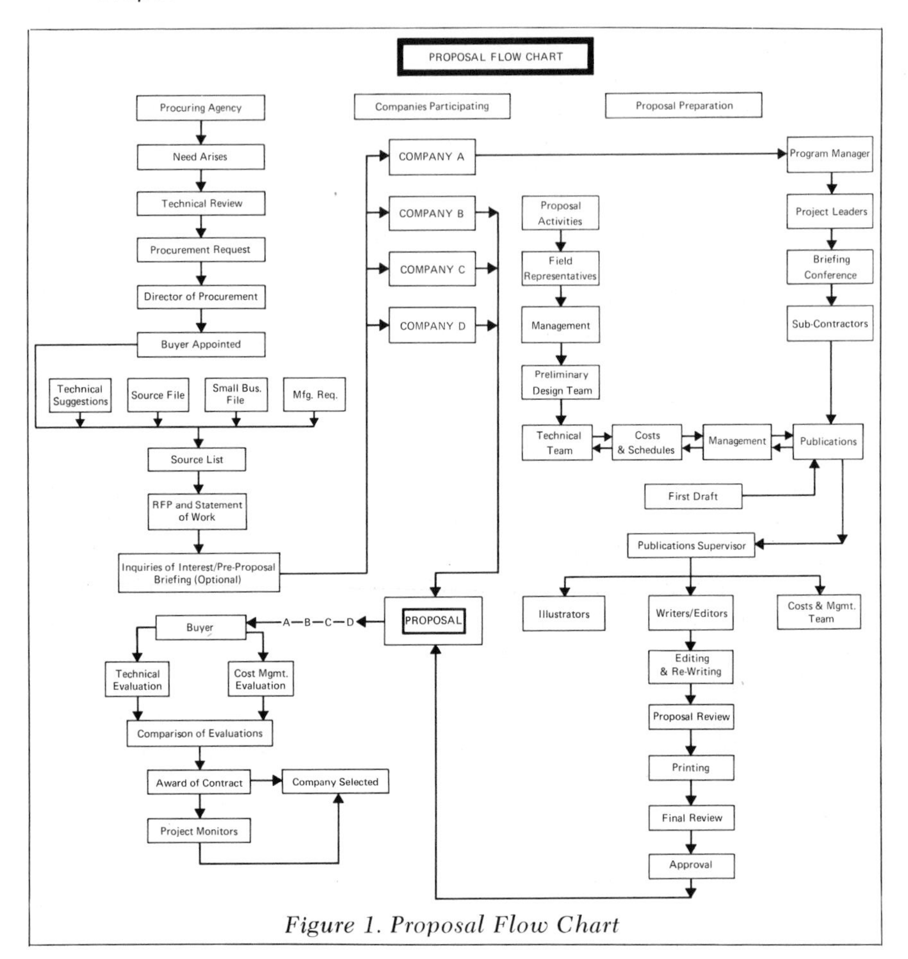


*Figure 1. Proposal Flow Chart*

You can easily imagine the importance attached to the proposal. Few experiences can be more frustrating to a company's executives than to feel sure they have developed a better technical design or plan than their competitors, only to see a competitor get the contract. The frustration is especially painful when the cost of putting out the proposal has run to tens or even hundreds of thousands of dollars.

Let us turn our attention now to a discussion of some principles and procedures that will, we hope, help you make a useful contribution to the production of effective proposals.

The process of producing a good proposal consists essentially of the following stages: (1) making a preliminary study; (2) developing a plan or making an outline, including decisions about what to emphasize or focus attention on; (3) writing a rough draft and planning illustrations and layout; and (4) reviewing and revising. We will discuss each of these stages in order and then turn our attention, in concluding this chapter, to some of the common weaknesses or faults of proposals in conjunction with a brief discussion of proposal evaluation.

## Making a Preliminary Study

In the broadest, most inclusive sense, the preliminaries to the drafting of a formal proposal begin with efforts on the part of those company representatives who have direct contact with customers. Such contacts naturally provide information as to what a customer's needs are, and also make clear to the customer what the company's capabilities are. An effective proposal may have its real origin in these relationships, long before a formal request for bids is received. Once a request is received, a decision to bid or not to bid must be made, if this decision has not already been made during the period when the company is trying to get itself on the bidder's list.

Of course many people may have a hand in making the final decision to bid or not to bid. What is the role of the technical staff in this process? They begin by studying the request for proposal, commonly referred to as the RFP or sometimes the "bid request" or "purchase request." They study the specifications, if any have been supplied, and they study any other available material which may help them define the technical problem and determine a plan of attack on it. If this work leads the technical staff to recommend making a bid, and if other departments or divisions of the company are in agreement, then members of the technical staff cooperate with representatives from purchasing and accounting to prepare a cost estimate and begin the technical proposal.

This summary of the work of the technical staff preliminary to the writing of the proposal can be broken down into the following phases or steps: (1) detailed study of the invitation to bid, of the specifications, and of any related papers or information such as notes of a briefing conference or correspondence with the procuring agency or company; (2) study of background information, such as the reports of field representatives who have called on the procuring agency or company; (3) careful analysis of the probable competition; (4) strategic evaluation of the technical design or program to be presented; and (5) preparation of a tentative schedule for completion of the various phases of proposal preparation, which should allow enough time for careful review and editing of the finished document. We will comment on each of these steps.

*Study of Bid Papers.* It is surely clear that the invitation to bid, and certainly the specifications, must be carefully studied before a successful solution to the technical problem may be devised. But it is also necessary to study these documents to determine the proper focus of attention for an effective presentation of the proposal. For instance, these documents may reveal the aspects of the problem the *customer* believes to be most important or crucial, most difficult to solve, and most urgently in need of solution. Even though the technical staff may know there are problem areas more critical or difficult than those the customer stresses, they will be unwise if they ignore the customer's convictions. Study of these documents can help the writers produce a proposal that is customer-oriented rather than author-oriented. In other words, the technical staff must satisfy themselves that they understand the problem *as the customer sees it*, since the customer's evaluation of the solution presented will doubtless be made in terms of his understanding of the problem. This initial study can be of tremendous help to the technical staff in deciding what to stress in the finished draft of the proposal.

*Study of Background Information.* As a further preparatory step in planning a presentation, it is desirable to gather as much information as possible about the customer and his needs and interests. Reports from field representatives and others who have had direct contact with the procuring agency may contain information that can be put to use. For instance, it may be possible to determine what a given company or agency's attitude is toward approaches to a solution of their problem other than that suggested in the bid papers. It may also be possible to determine what the company's attitude is toward taking exceptions to specified items and procedures, toward reliability and quality assurance programs, and so forth. It is unfortunate if a contract is denied with the remark "You proposed a fine piece of equipment but it was not what we specified," or "Your hardware looks very good indeed, but your study plan does not reflect a thorough understanding of all our problems." Thorough study of specifications, collated with background information, may not always enable the author to avoid such criticism, but it will help. In any case, the conscientious proposal writer or team cannot afford to ignore any source of information that might help produce a more effective, convincing proposal.

*Analysis of the Competition.* The value of taking competition into account in planning a proposal lies in an honest analysis of the strong points of the competitors. If a competitor is recognized as a leader in some particular technical area that is involved in the proposed design or program, it does not make sense for a writing team to plan to stress that particular area in their technical plan and in their written proposal. On the contrary, after inventorying their own company's potential advantages, they should stress those aspects of their solution in which their company has the most to offer.

Some companies maintain a file of information on other companies in the same field, information gleaned from study of journal articles, published reports, news stories, annual company reports and brochures, and the like. Classified by areas of technical competence and performance, such information can be efficiently used by a proposal writing team, particularly in the planning stage. Perhaps you can get a better feeling for the importance of analyzing the competition if you think of what you would want to stress in presenting your qualifications for a job that you know is being sought by other applicants. You would not stress your work in physics, say, if your grades were only average and your "competitors'" were excellent; rather you would stress those areas in which you were strongest.

*Strategic Evaluation of the Technical Approach.* Once the technical staff has worked out the details of the approach to a solution of the problem presented by the customer, it is wise to consider carefully which particular aspects of the proposed design or plan should receive the heaviest stress. Although you would think it only common sense to emphasize the most attractive features of a design, in practice the handling of the relative emphasis given to various features often seems rather haphazard. Examination of proposals submitted in many procurements reveals this quite clearly. A surprisingly large number give as much attention to aspects of the design that are rather trivial as they do to more important aspects. Such treatment is puzzling to the reader. We have all heard of the wisdom of putting your best foot forward; in writing proposals, this advice is not only wise but essential. Technical staff members who are intimately familiar with all aspects of a problem's solution often assume that such relative merit is self-evident, forgetting that their reader is not as intimately familiar with the technology as they are and that he needs to have his attention called to qualitative points. Sometimes it is good strategy to highlight noteworthy aspects of a design or plan in the introduction to the proposal, or in the foreword, or, possibly, on a special sheet to be placed just inside the front cover.

*Making a Schedule.* An important corollary of preliminary planning is the making of a schedule that will enable each person to carry out all steps for which he is responsible with a minimum of haste and still get the proposal out on time. Such a schedule must not only take into account all the demands made upon each person's time, but it must also allow enough time for administrative and service functions. Usually someone in management, for instance, must prepare a letter of transmittal or a foreword; the department head under whose cognizance the work comes will surely want time to review the finished document; and the publications department will need time to do a good job of editing, illustrating, and printing. It simply does not make sense to spend money and time planning a good technical design or program and then rush through the process of producing the manuscript which is designed to sell that design or program. Figure 2 is the schedule proposed by a division of Westinghouse in a

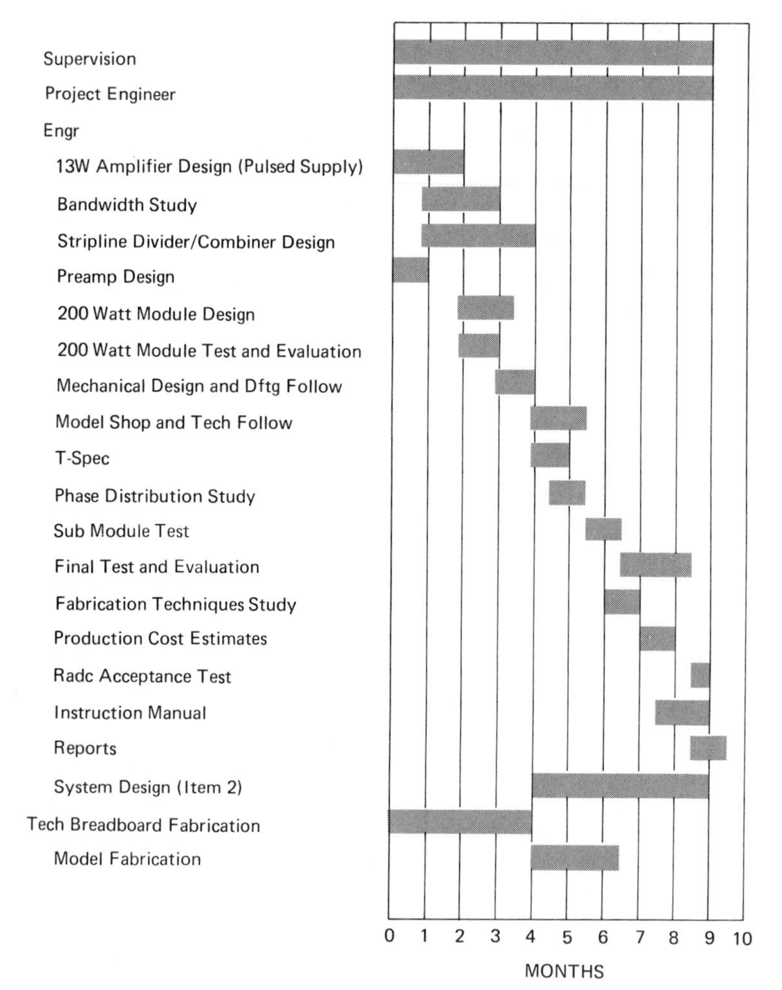


*Figure 2. Proposed Program Schedule*

proposal for the design and construction of a solid-state radar transmitter. As you see, it is concerned principally with technical work. Later in this chapter another schedule (Figure 3) appears which is concerned chiefly with the final stages of production of the printed proposal.

## Developing the Outline or Writing Plan

With the preliminaries carefully completed, the making of an outline or plan of presentation should not be particularly difficult. The important things to bear in mind about the outline are that (1) a written outline is essential as a guide for the writers to follow in developing the text, and this outline later serves a useful and necessary function as a table of contents for the reader of the proposal; (2) the outline should not only present the material in a logical order, but also in that particular logical order most likely to make a favorable impression on the reader; and (3) supplementary material should be included only as needed. An agreed-upon outline is particularly important for proposal preparation, since the writing itself is likely to be the product of several contributors, each writing up that portion of the technical approach on which he is expert and which he will, in all likelihood, be responsible for carrying out.

*The Outline as a Writing Guide and Table of Contents.* An outline of the material to be presented is usually necessary. This skeletal representation of the material offers an opportunity for checking whether all necessary information is included; it gives a visual means of determining whether the parts of the discussion are in balance and in the most effective sequence; and, of course, it serves as a convenient prod to the memory in the writing process itself. When several people are collaborating in writing a proposal, an outline becomes an indispensable help in avoiding confusion. As the writing proceeds, a need for changes in development may be seen, possibly even additions of information, and the formal outline will have to be changed correspondingly. But this is as it should be, for an outline ought never to be so binding as to prevent use of new ideas. See the discussion of planning a presentation, in Chapter 23; most of what appears there is relevant here. See also the discussion of the outline as a table of contents in Chapter 4.

The fact that the outline ultimately becomes a table of contents for the proposal as well as a system of subheads within the proposal, prompts us to offer a few words of caution here.

1. Be sure there are enough headings to reflect the content adequately. Remember that it is sometimes necessary for readers

to refer to a segment of discussion, after an initial reading; a sufficiently detailed outline, with page references, of course, makes such reference easy. We have in mind avoidance of the fault of too great an expanse of writing covered by a single outline entry—no matter how logical.

2. Be sure each outline entry is meaningful to the reader: single-word topical entries may be adequate for the writer but enigmatic to the reader. Surely a phrase like "Procedure for Carrying out Vibration Tests" is preferable to "Vibration Tests."
3. Try to keep the outline structure to three, or at most four, levels. Remember that each entry in the outline reappears as a heading in the proposal itself.

Further comments on these three points, together with illustrations, may be found on pages 83–86.

*The Outline as a Reflection of Strategy.* The word "strategy" has a double-edged meaning that invites attention to a moral—and emotional—problem often involved in the making of a decision to bid and in the writing of a proposal. On the one hand, the idea of strategy implies the situation of a man who is confronting a series of difficult decisions and is trying to work out an over-all plan, or strategy, that will be wise and effective. On the other hand, the idea of strategy also implies an effort to outwit somebody else for one's own personal advantage. We want to make clear that we are fully aware of these different implications of the word "strategy" in the subhead above. Which meaning, or implication, do we intend? The easy way to answer this question would be to adopt a simplistic attitude, like the professional moralist who is in favor of God, country, and motherhood, or like the professional cynic who insists that the idea of honesty is only a myth or is at best relative.

The trouble with these simplistic attitudes, in our opinion, is not so much that they are wrong as that they are inadequate. Let's consider an example. Let's imagine that you are an electronic engineer working for a small company that has just received a request for a proposal from a large corporation. This corporation wants a miniaturized high-frequency transmitter for use in a harsh environment; it wants the output to be extremely stable; and it wants the unit to have a maintenance-free life of 18 months at a confidence level of 95 per cent. After studying the details of these requirements, you decide that your company can produce a transmitter which would more than satisfy the stability requirement but would just barely satisfy the durability requirement. In fact, you are somewhat disturbed by the knowledge that carelessness during manufacture might well result in some units that would have to be rejected, or that would be of marginal quality in this respect—even though your records would justify asserting in the proposal that the requirement can be met.

While pondering this problem, you also give some thought to the probable circumstances in which your chief competitor finds himself. Let's assume you have good reason to believe his situation is just about the reverse of yours. That is, he can probably produce a transmitter with better durability than yours but with a less stable output.

Now what should you do? You run some risk if you recommend bidding for the contract, and then help plan and write a successful proposal in which the good points of your transmitter design are prominently displayed and fully developed and, correspondingly, in which durability is played down. Perhaps the acceptance checks run by your customer will prove that your transmitter *is* of marginal quality so far as durability is concerned—that is, there may be a large number of rejects. If so, you may be accused of having used strategy to conceal or obscure a possible weakness in your product. On the other hand, if you do not recommend bidding for the contract, you will never be certain you have not deprived your company of some much-needed income—and, after all, your transmitter *does* meet minimal durability requirements, with what you regard as a nonprohibitive number of rejects. Your dilemma is not made any the less serious when you reflect that your competitor is struggling with the same problem.

What should you do? It is difficult to say, and we shall not presume to offer a neat solution to this puzzling problem. Our purpose in introducing this particular problem has been limited to indicating the many perplexities that may present themselves in arriving at decisions about strategy when planning a proposal. And our purpose in emphasizing these perplexities has been to be as realistic as possible. We want it to be absolutely clear that nothing in this chapter is meant as a guide to outwitting people for personal advantage. At the same time, we want to be realistic in recognizing that translating good intentions into wise decisions can be very difficult.

With all these thoughts in mind, we return to our principal subject—the outline as a reflection of strategy. What is the relationship between strategy and an outline? What can you do with an outline that is strategic? The answer is that through the outline you choose what to include in your proposal and what to emphasize.

We should begin our thinking about these matters with recognition of the fact that the main divisions of a proposal are actually fairly uniform. That is, most proposals must have an introduction; a list of items and services to be supplied; a general description of the equipment to be supplied (for hardware procurements); a technical discussion or detailed description of the proposed equipment or the proposed study program; a section on packaging, perhaps, or other special considerations; and a conclusion. Frequently, appendixes dealing with such matters as reliability, quality assurance, and facilities are added. But it does not follow that the items must constitute the main divisions in every proposal, nor does it follow that these items must always appear in the order listed above.

In a given situation, for example, it might be critically important to place the list of items and services to be supplied right after the introduction, in a position of prominence; but this placement would be strategically desirable only if the customer is primarily interested in having a firm commitment as to hardware and services to be furnished, along with firm delivery dates. In another situation, the customer might be primarily interested in seeing a complete and thorough discussion of the problems outlined in the invitation to bid; in this case, it might be far better to have the introduction followed by a discussion of the key aspects of the problem, with a heading reflecting the nature of the problem.

After decisions have been made as to what must be included in a proposal, constructing an outline to reflect strategy becomes a matter of deciding on the *order* of presentation of the items to be discussed. This order should reflect your best thinking about what the customer will want discussed first. Since problems dealt with in proposals are not all alike, and since customer needs and interests are certainly not all identical, it follows that a standardized outline will not work for all proposals. What the writer should try to do is tailor-make plans to suit the particular case; what the customer reacts favorably to is a custom fit, not a ready-made job.

*Supplementary Material.* Appendixes are a useful means of presenting material that may be needed by the reader, or material that may be interesting to him but which would be awkward or unwise strategically to include in the body of the proposal. Such material may include mathematical analyses, biographical résumés of staff members who will work on the project, lists of facilities, records of successful performance by the company of work similar to that being proposed, descriptions of reliability procedures, and the like. The careful planner will see to it that such material appears in a proposal if it is needed, particularly when the customer has expressed a desire to see it.

But the careful planner will also be very cautious about overloading his proposal with boilerplate the reader has no interest in, and no need for. He will take a good, hardheaded look at this problem and limit appendix material to essentials. There is no merit whatsoever in adding to the bulk of a proposal just to make it big. Size has no necessary relation to quality or effectiveness. These questions should be asked about each unit of pre-prepared material considered for inclusion in the appendix: Does the customer *want* this information? Does he *need* it for understanding the presentation? Will it help *sell* the proposed design or plan? Any company wants to avoid the charge that they *compile* rather than *write* proposals.

## Writing the Rough Draft and Planning Illustrations and Layout

We are tempted to say that after going through all the necessary preliminaries and drafting a plan, the proposal will practically write itself; that, of course, is nonsense. But we do believe that preliminary thought and investigation followed by careful planning greatly simplify the job of producing a rough draft. With a thoughtfully prepared, sufficiently detailed outline to reflect to best advantage the technical design or plan, a pretty competent first draft can be produced by simply following that outline closely.

The main things to be especially careful about are (1) the introduction, (2) transitions between parts of your discussion, (3) correlation of text and illustrations, (4) exceptions taken to specified items or procedures, (5) the concluding section, and (6) style.

*The Introduction.* Although the introduction is the first and in many ways the most important focus of reader attention, assuming there is no foreword or introductory summary, it does not necessarily have to be written first. Many writers find it advisable to wait until the rest of the manuscript is completed before writing this critically important part of the proposal. Their reasons are simple and sound: they want to wait until the manuscript has been fully developed so they can see precisely what it is they have to introduce. By the time the manuscript is completed, furthermore, the writer may have a better knowledge of what he should emphasize in his introduction.

With one exception, which will be explained later, the problem of writing an introduction to a proposal is not fundamentally different from that of writing any other introduction. Three elements in the introduction to a proposal do deserve special mention, however. These are (1) discussing the problem to which a solution is proposed; (2) highlighting the key points of the solution; and (3) discussing solutions other than the one proposed, together with an explanation of why these solutions were rejected.

As you will remember from the discussion of introductions in Chapter 11, the principal functions of an introduction are to state the subject, the purpose, the scope, and the plan of the report. The sequence in which these functions, or parts, is taken up depends, of course, upon circumstances.

Looking back, then, at the three elements in the introduction to a proposal, listed above, we see that all three are special cases of either the statement of subject or the statement of scope. The first two, discussion of the problem and key points in the solution, are obviously part of the statement of the subject. The primary subject of virtually any proposal is a problem and its solution. It is a little harder to decide about the third element, the discussion of solutions other than the one chosen.

In any event, the strategic points to keep in mind are these. First, in presenting the subject of a proposal, it is usually crucially important to make absolutely clear your understanding of the nature of the problem and to indicate the most significant aspects of the solution you propose. Second, if you are rejecting possible solutions, it is often important to note what they are and to explain why you decided to reject them. This may be done as part of either the statement of subject or the statement of scope. And third, the introduction should never become a purely perfunctory, stereotyped statement to the effect that "this is a proposal to . . . ." Because of the title page and headings, the reader already knows what the document is when he gets to the opening sentence. And remember that first impressions are often lasting ones. The real danger is that a bad initial impression may eliminate the possibility of a good impression's being made later, since there is always the possibility that a weak introduction will cause the reader to stop reading—especially if it is a proposal that was not solicited.

We have been describing the general characteristics of an introduction to a proposal. Now we would like to comment very briefly on the exception we mentioned earlier. This exception can be explained as follows:

When you write a reply to a business letter, the way in which you set up your reply is usually much influenced by the letter you are replying to. For instance, if the letter you received contained three questions, numbered 1, 2, 3, you would probably number your answers in the same way. A similar relationship often holds true between a proposal and a request for proposal.

We are presenting this fact as part of our comments on the introduction to a proposal because it is in the introduction that the consequences are most noticeable. For example, let's think once more about the statement of the subject in the introduction. In an unsolicited proposal, this part of the introduction, together with the statement of purpose, is likely to be handled with great care and at some length. In a short proposal written in response to a request for proposal, on the other hand, there is little need for much attention to statement of subject and purpose. The reader knows what the subject is; he has a copy of the request for proposal open on his desk. What the reader will want is a formal—indeed, a legalistic—statement of what the subject is, one that will show the proposal to be "responsive" to the reader's interests. An example of such a statement will be found under the heading "Objective" in the proposal presented as an illustration at the end of this chapter.

Much could be written about other ways in which the content and organization of a proposal relate to the request for proposal. Observe, for example, the way in which the organization of part II of the proposal just mentioned, which appears at the end of this chapter, evidently reflects the request for proposal. Because each situation is unique, however, there is little point in going further into this subject.

*Transitions.* Clearly linking the parts of a discussion with transitions to produce an easily understood whole is, of course, one of the means of making a good impression on a reader. This technique is discussed in Chapter 10. Although they are always vitally important, transitions are likely to become a special problem in a proposal which is being worked on by several people. It is usually desirable to have it clearly understood that one of the people concerned will take responsibility for making sure that there are major transitions where they are needed, and that they are clearly written. If a professional editor is available, he will probably assume this responsibility.

*Illustrations and Layout.* Since most of the problems of layout and printing are dealt with by those charged with reproduction of the proposal (commonly, the publications department), we will assume that the writer will not have to solve them. This may not be strictly true, particularly for those proposals which call for a special cover and other special features in which the publications department may need the author's help. We discuss the principles of layout in Chapter 20 and they are as applicable to proposal writing as to other kinds of reports.

Illustrations are another matter, however. The author may often have to give almost as much thought to the selection and location of effective illustrations as to the text. To the information and suggestions about graphic aids presented in Chapter 21 we want to add here one observation especially relevant in the preparation of proposals. This observation is that when several people are collaborating on a proposal, it is easy for each one to become so wrapped up in his own part that he forgets to think of how his graphic aids will relate to those used by his collaborators. Imagine what an unfortunate impression would result if, for instance, Sections 2 and 4 of a proposal were lavishly illustrated with photographs, but Section 3 had only a few small graphs, even though it was obviously the best suited of all for illustration by photographs. It can also happen that *identical* illustrations appear in different sections. Consultation and planning by all concerned can easily prevent such mishaps.

The problem of providing illustrations to embellish the text, rather than to support it technically, is a difficult one to solve. Too much "window dressing" will create a bad impression. On the other hand, handsome illustrations can most certainly add dramatically to the interest of a proposal; they attract the reader's attention and do much to hold it. They may be used to advantage in breaking the monotony of successive pages of solid text. Probably the best thing the proposal author can do is familiarize himself with the artistic capabilities of the staff of technical illustrators and become familiar with available art so that, with the advice of an editor and illustrator, he can make a judicious and tasteful selection of such illustrations—within the budget and time allowed. Browsing through a generous sampling of the best proposals in the library or files would give a measure of this familiarity. It is also wise for the proposal author (or authors) to confer closely with the illustrator, explaining aims and needs. When the illustrator knows exactly what the author is trying to tell with illustrations, he is much better prepared to produce successful illustrations. Furthermore, with this knowledge, he may be able to make constructive suggestions, based on his more extensive experience and knowledge of illustrating, that will result in a more effective presentation. Finally, when the illustrator knows exactly what is wanted, he is less likely to produce an illustration that emphasizes the wrong points.

*Exception Taking, or Deviating from What the Customer Specified.* Since taking exceptions to specifications or deviating from a suggested approach involves the risk of displeasing the customer, incurring the charge of "nonresponsiveness," it is clearly necessary to exercise all possible skill in suggesting changes. Special effort should be made to demonstrate that an exception will result in a better product or program for the customer. The writer needs to call upon whatever arts of expression he possesses in phrasing an exception so that it will not sound offensive to the reader. The publications editor or proposal manager should be of help here.

Here are some general principles that should govern exception taking:

1. Give a reasonable explanation for every exception.
2. If possible, make it clear that the specification *could* be met as written but that a change is being recommended for improvement.
3. Try to relate exceptions to the customer's more important objectives: reliability, performance, ease of maintenance, weight and size, and so on.
4. If a section is devoted to exceptions, keep the discussion short and refer to pertinent detail in the technical discussion.
5. When the company does not own or cannot make available facilities for requirements such as performance tests, environmental tests, and mean-time-between-failure tests, do not state baldly that such tests cannot be made. Suggest the use of customer facilities or of subcontracting the tests. Of course, for sizable procurements with the promise of follow-on business, it may be economical to invest in such facilities.

Let's take a look now at a couple of "before and after" examples of the application of these general principles.

EXAMPLE A

*Original:* As discussed in Section II, the TR tube contemplated for this application requires a minimum of −600 V for satisfactory operation. We have not found any TR tube that functions properly at −250 V. It is therefore suggested that a higher voltage supply be provided.

*Improved:* The technical requirements specify −250 V for the TR tube. As discussed in Section II, we are proposing a TR tube that requires −600 V. Since this tube has better performance characteristics for this application than any which will operate at the lower keep-alive voltage, we propose to provide the required −600 V. The penalty in increased shielding requirements and larger power supply will be negligible, and the performance of the equipment much improved.

EXAMPLE B

*Original:* Design approval tests will be performed on the 10 prototype systems in accordance with the tests listed in Section 3 of NAVORD OS 8136, with the exception of tests listed in Items 3.2.3.5, 3.2.3.6, and 3.2.3.9. These tests are deleted as we do not have the equipment necessary to perform them.

*Improved:* Design approval tests will be performed on the 10 prototype systems in accordance with the tests listed in Section 3 of NAVORD OS 8136. We propose that tests listed in Items 3.2.3.5, 3.2.3.6, and 3.2.3.9 be conducted by our engineers at a government-owned facility.

*The Concluding Section.* Like the introduction, the concluding section of a proposal should certainly be more than a purely perfunctory, routine closing. A good, strong, final impression should be left with the reader, preferably one that re-emphasizes the strong points of the proposed design or program, and one that leaves the reader with the feeling that the bidder is eminently competent to do a fine piece of work for the customer.

Although it would be foolish to suggest a formula for writing the conclusion, since it should always grow out of the material in the proposal itself, it may be worthwhile to consider summarizing, perhaps in the form of a list, the key selling points of the technical approach. Emphasis on solidly concrete features of the proposed design or program is far more likely to leave a lasting and favorable impression than glib assurances that the bidder feels confident he can produce satisfactorily for the customer. With evidence before him, the customer becomes convinced not only that the bidder is *confident* he can produce but also that he actually *will.*

One further note: most of the proposals we have seen use the title "Conclusions" for the final section. We question the wisdom of this title because very often no "conclusions" in the technical sense are given. A title more accurately descriptive of the content would be more effective—something like "Summary of Key Features," or "Advantages of the Program Proposed."

*Style in the Proposal.* Two considerations should govern all decisions about the suitability of the writing style in a proposal: accuracy of statement and adaptation to the reader. What we want is an accurate, clear, readable style. When we speak of accuracy of statement, we must think primarily of ourselves. Do the words say exactly what we want to say? But when we speak of clarity and readability, we must think not of ourselves but of the reader. Will the words be clear and interesting to the reader?

To answer these questions correctly—and honestly—you must take as detached and critical a look at what you have written as you can. Achieving accuracy and precision of statement requires going over the text carefully and patiently to make sure the words represent what you have in mind. You must be careful and patient if you are to avoid the tendency, especially in latter stages of composition, to *as-*

*sume* that the words on paper accurately convey your thought. In fact, many writers apparently reach a stage or state of mind in which they do not actually *see* the words themselves, but rather the ideas the words are intended to express.

When you reflect on the fact, moreover, that a proposal may be read and evaluated by more than one type of reader—by nontechnical readers, for instance—you will recognize the danger of a style too closely reflecting your intimacy with the subject. In particular, you must guard against using shoptalk, jargon, slang, and highly technical terms that may be second nature to you but which may be unfamiliar to some of your readers. Further discussion of these ideas will be found in Chapter 3, Part One.

## Reviewing and Revising

The final chore in the preparation of an effective proposal is reviewing and revising the manuscript before final printing. These last opportunities to remedy defects are three: the first typescript copy of your rough pencil or dictated copy, the edited copy if you have the services of an editor, and the proof copy (perhaps the reproducible masters themselves). The latter will be provided by the people who print the final copies.

The time it takes for your rough draft to be typed will probably constitute a long enough cooling-off period so that your first formal review of the manuscript may be carried out with critical detachment. Read through the entire manuscript, preferably at a single sitting, checking it for continuity of development and completeness of technical content. Even if you have contributed only a small portion of the entire proposal, it is wise to read through the entire manuscript, unless it is extremely long, to get a sense of how your contribution fits in.

A second opportunity to review and revise will occur if you have the services of an editor. Upon his return of your edited copy, you should once again go through the text carefully. If you are not satisfied with the results of editorial changes, you should get together with the editor and work out a satisfactory version. During this second phase, you should check on the preparation of illustrations and figures to make sure they are completed, or being prepared, to determine that they are suitable, and to make certain that the text makes adequate and accurate reference to them.

Although organizations making use of editors usually assign them the responsibility of seeing to it that a title page, table of contents, list of illustrations, and the preprinted appendix material are prepared, it will do no harm for you to make sure these materials are what you want. You will want to be certain, for example, that the table of contents entries corresponds exactly with the headings and subheadings in the text, and that there is an exact correspondence between the illustrations listed and those which actually appear in the text.

Finally, remember this: although you will have another opportunity to catch errors when you read final proof copy, you should make every possible effort to have the copy exactly as you want it before it is typed for reproduction. Last-minute changes of any magnitude on reproducible masters will require unfortunate delays—and increased costs—in getting out the finished proposal in time for mailing.

The end is really in sight when you read the master copy however. This last reading is, normally, little more than a double check of the work of the people who have typed and proofread your copy. Nevertheless, this does not mean that the task should be undertaken casually. Even the best proofreader can overlook a textual error—and textual errors are what you do not want! It is best to take your time and *read*—not scan—the masters. Some indication of the time required is shown by the accompanying chart (Figure 3). Once you have done this, and have checked the corrections which have to be made, you can let the copy go to the printer, confident that you have produced a good job.

**FINAL REVIEW AND PRODUCTION SCHEDULE***

| PUBLICATIONS SCHEDULE | 1st Day | 2nd Day | 3rd Day | 4th Day | 5th Day | 6th Day (Mail Day) |
|---|---|---|---|---|---|---|
| Editing and Conferences with Author | → | 2 Days → | | | | |
| Illustrating | → | → | → | As Required → | | |
| Typing Repro Masters | → | → | → | 11 Hrs → | | |
| Makeup and Layout | → | → | → | 2 Hrs → | | |
| Author Review | → | → | → | 4 Hrs → | | |
| Approvals, Final Changes | → | → | → | → | 5 Hrs → | |
| Printing, Collating, Binding | → | → | → | → | 6 Hrs → | |
| Circulating and Mailing | → | → | → | → | → | 4 Hrs → |

*Example of an editing and production schedule chart.

*Figure 3. Final Review and Production Schedule*

## Proposal Evaluation

To many technical people, especially those whose proposals have not been successful, the process of proposal evaluation by the organization receiving the proposal remains both a mystery and an enigma. We shall not presume to explain it here, but we can call attention to some rather obvious facts and tell you what we have found reported as the principal reasons for proposal failure—and, by implication, success.

Normally, the evaluation procedure follows the pattern of the proposal itself. That is, three evaluations occur: of the technical content, the management plan, and the cost or price information. In some procurement agencies and companies, a weighted point system is assigned to each of these elements, sometimes on a 100-point basis and sometimes not. We know of one agency that set up 2500 points as the perfect score, with the largest number of points, 500, allocated to technical content and the smallest number, 200, assigned to the bidding company's interest in the job. Whatever the system, it is fair to state that the lion's share of "points" goes to technical competence. A second example of an evaluation point system is as follows:

| | *Points* |
|---|---|
| Problem Understanding | 25 |
| Soundness of Solution | 25 |
| Compliance with Requirements | 15 |
| Design Simplicity | 10 |
| Ease of Maintenance | 10 |
| Capabilities and Qualifications of Company | 15 |

The above evaluation point system applies to the technical content, chiefly; it does not include, obviously, any points for a cost evaluation. Another evaluation scheme assigns point values to the following criteria, with "technical aspects" roughly twice the value of any other single criterion.

Technical Aspects
Relative Costs
Time Schedules
Technical Competence of Bidder
Management Competence
Financial Responsibility
Interest in Proposed Work

Still another agency reports the following scheme:

Scientific/Engineering Approach
  Understanding of Problem
  Soundness of Technical Approach

Responsiveness to Requirements
Special Technical Factors (Unique Aspects)

Bidder Qualifications
Pertinent Experience
Management Organization for Project
Adequacy and Availability of Facilities
Special or Unique Qualifications
Customer Experience with Bidder (if applicable)

Although we do not have precise point values in these last two examples, which were presented in a conference on proposal writing, technical competence and soundness were given the greatest emphasis.

Whatever point system or relative value scheme is made use of, discussions of proposal evaluation make it abundantly clear that the evaluators look for evidence of the following:

1. Understanding of the customer's problem or need
2. A sound and concrete technical solution to the problem
3. Compliance with customer's requirements or specifications, or a suggestion for improvement in the product or plan if exceptions are taken to requirements
4. Recognition of the possibility of solutions other than the one proposed, together with knowledge of their strengths and weaknesses.
5. Clarity of presentation
6. Realistic and reasonable pricing
7. Financial responsibility
8. Sound and intelligent management planning
9. Adequacy of qualified personnel and facilities
10. Realistic time schedules for proposed work
11. An effective communication plan
12. Effective quality and reliability procedures

Although the faults of proposals are, largely, implicit in a listing of strengths, the following are the most frequently heard criticisms:

1. Failure to understand customer's problem
2. Failure to explain, adequately or satisfactorily, deviations from specifications or requirements
3. Overemphasis on company product or products (too much space devoted to successful performance on other contracts and too little to concrete solution of *this* problem)
4. Oversimplified technical treatment
5. Overoptimistic estimates of performance and schedules
6. Inadequate test information

7. Too many vague generalities and sweeping statements
8. Misinterpretation of specifications
9. Too much "window dressing"—too much attention to attractive packaging of the proposal
10. Imbalance in presentation: too much space devoted to problems of interest to bidder but not to customer
11. Uncertainty of tone: overcautiousness about ability to carry out program
12. Ineffective presentation: wordiness, obscurity, irrelevancies, weak or nonexistent transitions
13. Unrealistic cost proposal
14. Weak performance record on past contracts
15. Insufficient detail in technical approach and management plan; vagueness about assignment of expert personnel

We realize, of course, that the listing of good and bad points about proposals will not ensure that you can produce a winning proposal, but we hope that together with the preceding discussion these listings will be helpful.

## Conclusion

Every proposal presents its own special problems and requires its own special solutions, not only in technical content but also in technique of presentation. Remember these essential steps, however: collect and study all available pertinent information as a preliminary to working out a plan of presentation; write your rough draft and then alter and work on it until you are satisfied; finally, review and revise your copy with all the ruthlessness you can muster, and encourage the editor—and anyone else who will help—to do the same.*

# ILLUSTRATIVE MATERIAL

The following proposal was prepared by the staff of the Atlantic Research Corporation for submittal to the Bureau of Aeronautics of the United States Navy. As you will see from the table of contents, we have included the first three main sections only. Section IV gives a detailed breakdown of estimated costs in terms of direct labor, over-

*The literature on various phases of proposal writing is becoming quite extensive. A recent article in *STWP Review* lists some 84 items, chiefly magazine articles, although two complete books are devoted to the subject. See Bibliography.

head, materials, communication and travel, and special capital equipment. Section V tells, by name, who will supervise and direct the work. Section VI itemizes the special pieces of capital equipment that will be needed to carry out the project. The last section describes briefly some projects the company has carried out in which experience applicable to the proposed work was gained. Detailed contract experience, as well as personnel résumés, is given in the appendix.

Examination of the text of this proposal reveals that Section III accounts for about half of the total number of pages in the report, and the reason for this proportion is quite clear. Since this is to be a research project, a principal strategy in the preparation of the proposal was to persuade the reader that the staff of Atlantic Research Corporation had a thorough grasp of the problem and would be able to work effectively toward its solution.

### Design and Development of an Aviation Fuel Contamination Detector*

Table of Contents



*Quoted by permission of Atlantic Research Corporation, a division of The Susquehanna Corporation, Alexandria, Virginia.

## Design and Development of an Aviation Fuel Contamination Detector

### *I. OBJECTIVE*

The ultimate objective of the program will be to furnish the Bureau of Aeronautics with a prototype model of an instrument system which will determine accurately and reliably the degree of contamination in aviation fuels and which will meet the service conditions and service requirements set forth in the General Information section of Request for Proposal No. PP-412-125-59.

### *II. BREAKDOWN OF THE WORK*

The breakdown into three phases, including the eight items to be delivered, given in the Request for Proposal is suitable for this development project, and the work proposed here would follow the specified breakdown.

*Phase I—Feasibility Study*

The objectives of the feasibility study are: (1) to select the one, or perhaps two, physical or chemical phenomena most likely to be useful as the basis for the desired instrumentation system; (2) to examine the selected phenomena in some detail to determine the feasibility of making use of them for the desired measurements; and (3) to outline the experimental program required for the instrument development.

To achieve these objectives the work would include, but not necessarily be limited to, the following: (1) the collection of detailed information about the contaminants as found in the practical situation; (2) an examination of published literature to obtain as much information as possible about the physical and chemical properties of the contaminants and the fuels; (3) analyses of the various physical phenomena by which the presence of particulate contaminants suspended in aviation fuel can be detected and measured, in terms of the known properties of the fuels and contaminating materials; and (4) an examination of any instrumentation system proposed for development with regard to safety, reliability, and accuracy of operation in shipboard use.

It may turn out that not enough is known about the contaminants as found in fuel lines or that the parameters needed for an effective quantitative analysis of the performance of a proposed system are not reported in the literature. In this event, some experimental work may be required before the above-stated objectives can be achieved in full.

*Phase II—Developmental Model*

The objectives of Phase II are to design, develop, fabricate, test, and deliver a developmental model of the contamination detector.

The work of Phase I will have outlined the experimental program, and the precise nature of the work cannot be specified in advance of the study. In a general way, however, the work may be outlined as follows: (1) basic experimental investigation of the one, or perhaps two, phenomena chosen as the basis of the instrumentation system, to determine the effects of various parameters and to obtain an estimate of the signal output likely to be achieved in the final system; (2) tests of a measurement system, simulating in the laboratory the practical situation by using artificial contaminants and a small-scale flow system; (3) laboratory tests of a measurement system using actual contaminated fuel; (4) design, fabrication, and test of the developmental model.

We anticipate that the investigation of the basic phenomena and the design of the sensing element will require considerable work, but that the design of the associated electronic gear to provide the required display and alarm functions can make use of established techniques.

*Phase III—Prototype Model*

We assume that the developmental model will be given service tests and that these tests will suggest desirable improvements. The work of Phase III will involve, then, the design and test of a modified system, the design of the final system for production, and the fabrication and delivery of the prototype model.

## *III. PROPOSED METHODS OF SOLUTION*

### A. *General Discussion*

Under ideal conditions, there would be no great difficulty in making continuous measurements of the concentration of particles of a given material suspended in a fluid. The conditions presented by this problem are far from ideal, however. A multi-component system is involved in which the relative and absolute concentrations of the components, the particle sizes, and the particle-size distribution all vary. It seems unlikely, therefore, that the measurement of a single quantity, an absorption coefficient for example, will meet the requirements. It is more likely that several sensing elements will be required whose outputs will be electrically combined to give a single display. Another possibility is that a single sensing element can be used to make a specified set of measurements under different conditions, the set of measurements being repeated several times a second to give effectively on-line operation.

A further complication may be presented by the fact that different lots of aviation fuel can vary somewhat in physical properties, even though the applicable specifications are met, and that this variation might be displayed by a given measurement system as a contamination. In this event, a comparison system might be adopted in which the output of a sensing element in the main fuel line is compared with the output of a similar element in an uncontaminated fuel sample and the difference in output displayed as contamination. The uncontaminated sample would be prepared continuously from the main fluid stream by filtration, with appropriate safeguards to insure that the reference sample is actually uncontaminated.

A rapidly moving stream of fluid may contain bubbles of air or vapor which could be falsely reported as contamination. Even turbulence might have a similar effect on certain measurement systems, particularly if an instrument of high sensitivity is used. If these effects are troublesome, it may be necessary to sample the main fluid stream continuously, evaporate the liquid in the sample, and measure the solid content of the resulting aerosol. A separate measurement of liquid water content would be required in a system of this kind.

In summary, we envision a multi-element sensor with the outputs combined to give a single display and the possibility that special continuous sampling techniques may have to be adopted to eliminate perturbing factors. Although we do not wish to minimize possible difficulties which may arise in this development, it does appear that the problems which can be foreseen have solutions which can be adapted to instrumentation requirements. The most important part of the project is the selection of the principle on which the sensing elements will operate so that an accurate, reliable, and safe instrument will result.

*B. Principles of Operation*

Our initial study of the problem has indicated a number of phenomena which should be investigated further in Phase I of the project. The discussion here is intended only to illustrate the possibilities, without implying that one of these may be successful or that these are the only phenomena to be investigated.

*1. Attenuation of Sound:* A study of the transmission of ultrasound in aqueous suspensions, recently reported, shows that the attenuation is large in the frequency range from 15 mcps to 30 mcps for aqueous suspensions of lycopodium spores (representative of organic matter) and of quartz sand (representative of inorganic matter). The mean diameter of the lycopodium spores was about 31 microns and of the quartz sand was about 2 microns. The attenuation was found to be strongly dependent on the number of particles per unit volume of suspension. Suspensions in which the particle concentrations were one or two parts per million by volume were tested, and in this concentration range the attenuation coefficient was approximately a linear function of the number of particles per unit volume. Considering a path length of one centimeter, the intensity of the beam was reduced by more than ten percent when the number of particles per unit volume was doubled.

The attenuation of sound depends on the temperature, the frequency, the particle size, and the nature of the particle, as well as on the particle concentration. These are complicating factors which must be considered in detail in designing an instrumentation system based on sound absorption.

The temperature range of the fuel in the practical situation may be small enough so that its effect can be neglected. If the effect cannot be neglected, a comparison system may solve the problem. Alternatively, a measurement of the temperature and a corresponding adjustment of the output meter of the instrumentation system may be operationally acceptable.

The effects of frequency and of particle size are inter-related. If the wave-length of the sound is short compared to the particle size, the particles contribute to the total attenuation principally by scattering. If the wave-length of the sound is long compared to the particle size, the particles contribute to the total attenuation principally by frictional effects. In any event, the attenuation will depend on the particle size distribution, and we would expect that measurements would have to be made at more than one frequency, either by using a broad-band noise generator or by sweeping a single frequency over a given frequency band.

Figure 1 illustrates the principle of a possible system using a sweep frequency generator to drive a crystal sound-generator. The signal is received by a second crystal, after transmission through the fuel, and is amplified, detected, and displayed by appropriate electronic circuitry. A control system can be provided so that the amplitude of the detected signal in the pure fuel is independent of frequency or has any other desired frequency dependence. Introduction of particles into the system will markedly increase the attenuation in a certain frequency range, depending on the relation between the wave-length of the sound and the particle diameter. It may be possible to adjust the frequency range of the sweep and the frequency response of the system so that the output will be a measure of particle concentration, independent of particle size distribution. [Note: Figure 1 is not shown.]

The experimental conditions of the referenced work are close enough to those expected in measuring fuel contamination, and the reported effects are so large, that we regard the use of sound attenuation as the most promising principle of operation of all those we have examined.

2. *Light Scattering:* The light scattered by a suspension of particles is determined by the number, size, shape, and refractive index of the particles, the refractive index of the suspending fluid, and the wave-length of the light. In principle, then, light scattering can be used to measure particle concentration.

Light scattering phenomena are similar in some respects to sound attenuation phenomena, and much of the discussion of sound attenuation given above applies in a general way also to light scattering. In this case, also, one cannot expect a single measurement, of turbidity for example, to be adequate because of the effects of particle size distribution and of different materials. The additional measurements required might be made at different frequencies or at different angles with the incident beam. There is also the possibility of using polarized light.

Our preliminary examination of the problem leads to the following tentative conclusions for planning purposes: (1) a light scattering system can probably be devised; (2) a light scattering system will have all the complications of a sound-absorption system plus the additional instrumentation complication of combining an optical system with an electronic system; (3) the light scattered by some of the contaminating particles will not be great, because the refractive index is too close to that of the fuel, and sensitive electronic equipment will be required. For these reasons we regard a light scattering system as less likely to be successful than a sound attenuation system and therefore would give priority to investigation of the latter in this project.

3. *Electrical Properties:* Neither sound-attenuation nor light scattering effects depend on the volume of the contaminating particles in any simple way, and it would be worth some effort in Phase I of this project to examine phenomena which do depend on volume. If one can be found which would give a sufficient output, the instrumentation would be much simplified.

The dielectric properties of a composite medium consisting of particles of one material dispersed throughout another material depend on the volumes and dielectric properties of the two materials, but do not depend on the particle size distribution. The measurable properties include dielectric constant, resistivity, and loss factor. If the concentration of one component is a few parts per million, the change in dielectric constant and resistivity is usually of this same order of magnitude. In certain cases, however, the loss factor is markedly changed with the addition of contaminants.

We expect the addition of particulate contaminants to a fuel would cause only small changes in dielectric properties and that such measurements would be difficult to make on a routine basis. Nevertheless, the advantages are great enough to warrant investigation.

## SUGGESTIONS FOR WRITING

1. Write a letter-form proposal of a subject to be investigated for your library research paper (assuming one is required). Your proposal should contain a clear definition (at least as clear as you can make it) of the specific topic to be researched, some explanation of the scope of the investigation you plan, and an explicit account of the sources of information you will make use of. Your proposal will be more effective if you accompany it (perhaps as an attachment) with a tentative bibliography of the sources of information you have located through consulting the card catalogue, the guides to periodical literature you have examined (be sure to name them and state the time period covered—for instance, *Applied Science and Technology Index* 1960–1970), and any other publications you expect to consult. Since you are writing a proposal, you might wish to comment on the value the information you will collect may have, not just for you, but also for the reader. Remember, one important purpose of a proposal is to convince the reader that a project is worth carrying out. Other items

that might well be discussed would include the estimated time schedule for completion of various phases of the research project, the amount of illustrative work that may be needed to support the textual presentation, an account of relevant experience you have had in carrying out such projects, etc. Obviously, a cost proposal would scarcely be relevant.

2. Write a short letter proposal for improvements in the conduct of some course you are taking. Address it to the head of the relevant department (not to the teacher of the course).

# 16
# Forms of Report Organization

## Introduction

Many firms have found that their written reports are most efficient and satisfactory when they are organized according to a prescribed form. The form may range from a general one consisting of three or four divisions to the kind of detailed sheets used in colleges for laboratory reports. The method of organizing the components of a form varies according to the materials concerned and the preferences of the firm. In this chapter we shall introduce a few of these special forms beginning with the highly generalized type and concluding with the highly detailed. Since format is discussed in Chapter 20, we will not take it up here.

## Generalized Forms

The principle underlying the policy of prescribing certain major divisions for the organization of a report is usually to make the report convenient to use. Convenience is often achieved by organizing a report in a different way from what would naturally be suggested by the subject matter. In a report on a series of tests, for example, it would seem natural to present the results of the tests toward the end of the discussion, whereas convenience is often served by presenting them near the beginning: some readers will want to know only the significant results without having to examine the rest of the report. Furthermore, a report may go to several readers, some of whom are not technically trained. This situation may be met by providing a preliminary nontechnical statement followed by what amounts to a restatement of the same material but in more detail and with the addition of technical material.

The General Motors Institute, for example, recommends the following organization:

1. The statement of the purpose of the report
2. A summary of the findings, conclusions, or recommendations
3. Supporting expansion of the steps which lead to the findings, etc.
4. Evidence in support of findings, etc.

In this organization, parts 1 and 2 are stated in a form intelligible to the nontechnical reader, part 3 is a technical discussion of method and interpretation of results, and part 4 is a presentation of data in support of 2 and 3.

A similar purpose may be seen in the form used at Battelle Memorial Institute:

1. Covering letter (one page, nontechnical)
2. Introduction (half-page to one page, nontechnical summary)
3. Technical summary
4. Body
5. Section on "Future Work"
6. Appendices

Part 5 of this form applies primarily to progress reports, a type of report that is especially important in a research organization like Battelle.

The National Advisory Committee for Aeronautics prefers a more conventional organization:

1. Summary
2. Introduction
3. Symbols [They say that all symbols should be defined.]
4. Description of apparatus
5. Test procedure
6. Precision [Statement of probable accuracy of measurements, where pertinent.]
7. Analysis and discussion
8. Conclusions
9. Appendix
10. References
11. Illustrative material

In a form used by a large construction company that prefers to remain anonymous, the major components of the form are these:

1. Purpose and scope
2. Summary
3. Conclusions, recommendations
4. Text
5. Appendix

Two general comments should be made about the forms that have been shown.

In the first place, almost all firms state that such forms should not be regarded as absolutely binding, but should be modified by the writer if circumstances require. The implication is, of course, that circumstances won't usually require much modification. The flexibility of the forms may be seen in the use of the appendix, a division that appears almost universally. The appendix is primarily intended to contain data that supports the discussion and the conclusions in the body of the report. The decision about how much of the data should go into the appendix and how much should be introduced directly into the discussion must be governed by the particular problems in each individual report. In effect, the requirement of an appendix is a recognition of the principle that the discussion should not be cluttered up with unnecessary details, but that the details should be available to prove the soundness of the discussion. The forms may be regarded as inflexible, on the other hand, in view of the fact that major divisions are prescribed and that a young engineer will naturally be reluctant to make any modification of these major divisions.

In the second place, what about the relation of these standardized forms to the whole problem of types of reports? If all the reports issued by a given firm can go into one or two standardized forms, does it follow that one form is actually suitable for several types of reports? Frankly, we are raising this question simply because we thought you might be puzzled about it if we didn't, and not because we believe that it matters. One firm may use a single form for several types of reports; another may require two or three forms. And that is about all we need say here—except to suggest that you notice how easily the Battelle form, in particular, would accommodate both types of reports discussed in previous chapters: progress and recommendation reports.

There is one other commonly used form to be mentioned before we turn to the highly detailed variety. That is the memorandum report. The form used looks like this:

To: [addressees' names]
From: [author's name, and title after]
Subject: [subject in all capitals]

Usually the sender puts his handwritten initials or full name at the bottom of the sheet. Sometimes other items are printed on the page, like Subject ____________, File ____________, Project No. ____________, or any other information that will prove useful. Not uncommonly the paragraphs are numbered, often with underscored headings to identify the subject matter. See Figure 4, Chapter 18, for an illustration of headings. The memorandum report is essentially a rather informal communication between acquaintances, often employees of the same firm, about a project with which each is familiar. As with all reports, the content and style are determined by the relationship of sender and recipient.

## Detailed Printed Forms

A detailed, printed form is often a great convenience in making routine reports. Thousands of such forms are in daily use. Figure 1 is representative of the type. Also shown are illustrations of forms for keeping a record of report production (Figures 2 and 3).

It might seem that in a form report there would be no problem of reader analysis; but not so. It is wise to think about such matters as symbols, abbreviations, systems of units in stating values, probable accuracy of measurements, and sampling techniques. Don't use a symbol that your reader won't recognize. Remember that an abbreviation that looks clear as crystal to you may be puzzling to somebody who is not intimately acquainted with what you have been doing. Or, if you have taken measurements in the British gravitational system only to discover that everybody else in your organization is using the metric absolute, you'd better convert your values. In brief, give some attention to the needs and knowledge of your reader.

The same attitude of consideration for the reader should lead you to think carefully about what help you can give him in the "Remarks" or "Comments" section, if such a section is provided. If certain measurements, for instance, were taken under conditions of unusual difficulty, a short explanation might relieve your reader of undue concern over slightly erratic results.

It is interesting to think about a form report from the point of view of the man who has to prepare one. Here is a comment made by the engineer at Westinghouse who was given the assignment of preparing the form on the facing page —Edward T. Shaw.

> When I was given this assignment, I quickly learned that I had to know specifically what information about a malfunction or a failure is needed by the engineers in the plant where the equipment is designed and made before I could hope to organize a suitable and usable form. In addition, I had to know who would be reading the report. Therefore, I had first of all to concern myself with the purpose of the report; after

that, the content; and, then, the readers. And, with nearly 60 plants involved, this was not a simple matter. But I was the principal beneficiary. I learned a lot more about reporting than I knew before I was given the assignment. For this reason, it seems to me that if a student has to dig deeply—as I did—into purpose, content, and readers of a report before he can plan a form, he too will learn a lot more about reporting than he knew before he began the assignment.

ADVICE OF TROUBLE
FIELD REPORT
SERVICE ORDER

ELECTRIC SERVICE DIVISION
FIELD RELIABILITY ANALYSIS FEED BACK
WESTINGHOUSE PROPRIETARY

TO (LOCATION) | (RELIABILITY MGR.) | CONTROL ACCOUNT | JOB NO.

(SERV. OFF.) | (MGR.) | RESPONSIBILITY | SERVICE ESTIMATE | R.R. NO.

(SALES OFF.) | (SALESMAN)

(WORKS) | (ACCTG. MGR.) | CUSTOMER NAME

PRELIMINARY APPROVAL | TROUBLE AT (NAME OF PLANT) | LOCATION

TO BE FILLED IN BY SALES ENGINEER

APPARATUS (PRIME) | REPORT TO | TITLE | TELEPHONE

APPLICATION | APPARATUS SOLD TO | LOCATION

STYLE NO. | DWG. | SUB. NO. | ITEM | SERIAL NO. | G.O. (INCLUDE SUFFIX) | S.O.

**1. CUSTOMER COMPLAINT** - ACTION DESIRED (USE EXTRA SHEET WHEN NECESSARY)

DATE OF TROUBLE OR FAILURE | TIME IN SERVICE EST. | ACT. | DATE WARRANTY STARTED | SHIPPING DAMAGE - TERMS FOB DEST. ☐ FACTORY ☐ | CLAIM ENTERED YES ☐ NO ☐ BY

WILL THERE BE A CHARGE CUSTOMER ORDER FOR THIS WORK? YES ☐ NO ☐ | SEND RECOMMENDED BILLING TO SALESMAN FOR DISPOSITION ☐ | CUSTOMER ORDER NO.

CREDIT APPROVAL (BY & DATE) | BKG. DIST. | SALES OFFICE | WESTINGHOUSE SALESMAN | DATE WRITTEN

ANALYSIS FEED BACK

ACCURACY OF APPARATUS IDENTIFICATION WILL BE VERIFIED BY ENGINEER | TO BE FILLED IN BY INVESTIGATING ENGINEER RETURN PROMPTLY (WITHIN 2 DAYS AFTER ARRIVAL) ➡ | DATE REPORTED ON JOB

PHYSICAL LOCATION OF TROUBLE | PANEL NO. | CABINET NO. | CAGE NO. | LOCATION NO. | SCHEMATIC

ENVIRONMENTAL CONDITIONS WET | DRY | DIRTY | DUSTY | CLEAN | CHEMICAL ATMOSPHERE ☐ TYPE | AMBIENT TEMP. MAX. | MIN.

**2. PROBLEM AS FOUND** (GIVE COMPONENT N.P. DATA WHEN REQUIRED)

PROBLEM DUE TO CHECK APPROPRIATE BLOCK(S) | A. DESIGN | B. MFG. | C. COORDINATION | D. PKG. | E. BRACING | F. TRANSP. | G. IDENT. | H. INSTR. MATL. | I. STORAGE | J. INSTALL. PROD. | K. OPER. PROB. | L. MAINT. | M. APPLI.

**3. ACTION NEEDED TO CORRECT TROUBLE** - IS JOB COMPLETE YES ☐ NO ☐ | EST. MAN HOURS TO FINISH

**4. MATERIAL REQUIRED:** YES ☐ NO ☐ REQ. NO. | DATE OF FINAL GREEN SHEET

CITY OFFICE | INVESTIGATING SERVICE ENGINEER | DATE | REPORT NO. OF | PRELIM. | FINAL

TO BE FILLED IN BY DIVISION RELIABILITY MGR.

CORRECTIVE ACTION (SEND XEROX COPY TO SERVICE OFFICE)

RELIABILITY MANAGER | DATE CLOSED

RECORD OF SPLIT DECISION
CONTROL ACCT. | COST
TOTAL COST CLEARED
FINAL APPROVAL | DATE

FORM 35661 B

USE FORM 35662 WHEN ADDITIONAL SPACE IS REQUIRED FOR COMMENTS

*Figure 1. Printed Service Report (Westinghouse Proprietary Material, Reprinted by Permission of Westinghouse Electric Corporation)*

Thought of in the terms of what this engineer has to say about the preparation of a form report, the importance of reader analysis in using it becomes clearer.

**REPORT PRODUCTION RECORD**

Report Title ____________________

____________________

____________________

Author(s) ____________________

Classification ____________________ Report No. ____________________

Contract No. ____________________ Mailing Deadline ____________________

Rough Draft Rec'd ____________________ Completed ____________________

Rough Typed Draft Proofed ____________________

Rough Draft Checked by Author ____________________

- Title Page ____
- Foreword ____
- Abstract ____
- Table of Contents ____
- Equations Checked ____
- Rough Illustrations Prepared ____
- Tables Checked ____

Illustrations Received ____________________ Completed ____________________

Cover Completed ____________________

Edited ____________________ Date ____________________

Editor-Author Conference ____________________ Date ____________________

Final Copy Typed ____________________ Date ____________________

Final Copy Proofed ____________________ Date ____________________

Proof Copy Checked by Author ____________________ Date ____________________

Approval Signatures ____________________ No. Copies Needed ____________________

Assembly and Binding ____________________ Date ____________________

Quality Checked ____________________

Mailed ____________ Logged ____________ Filed ____________

Remarks ____________________

____________________

*Figure 2. Report Production Record*

PRODUCTION RECORD

Publication No. ______________ Due: _____ Date _____ Hour

W.O. ________ E.O. ________ (Classification) Editor ______ Engineer ______

Title ________________________________

Copies Required:

_____ Mailing (Complete)
_____ Without Appendixes
_____ In T/I Binder (Complete)
_____ Tissues in T/I Binder
_____ Total Copies

Consists of: ____ Sections, ____ Appendixes
Dividers: ☐ All Sections ☐ Appendixes Only

| Appendix: | Title: | Classification: |
|---|---|---|
| A | | |
| B | | |
| C | | |
| D | | |
| E | | |
| Inserts: | | |

Special Instructions: ________________________________

Reproduction Class: Brief Class Explanation:

☐ A Justowriter–Film–Plates–Two Sides–Special Cover
☐ A-1 Justowriter–Film–Plates–One Side–Regular Cover
☐ A-2 Justowriter–Film–Plates–Two Sides–Regular Cover
☐ B IBM Executive–Masters–Two Sides–Special Cover
☐ B-1 IBM Executive–Masters–One Side–Regular Cover
☐ B-2 IBM Executive–Masters–Two Sides–Regular Cover
☐ C Justowriter–Xerox onto Masters–One Side–Regular Cover
☐ D IBM Executive–Albanene–Blueline–T/I Cover
☐ E (Special): ________________________________

Text Rough Draft:

| Operation | Received | | Completed | | Initials |
|---|---|---|---|---|---|
| | Date | Hour | Date | Hour | |
| Editing | | | | | |
| Typing | | | | | |
| Proofreading | | | | | |
| Rechecking | | | | | |
| Project Reviewing | | | | | |

☐ Approved for Final Reproduction with Changes as Marked

Final Reproduction:

| Operation | Received | | Completed | | Initials |
|---|---|---|---|---|---|
| | Date | Hour | Date | Hour | |
| Typing | | | | | |
| Proofreading | | | | | |
| Correcting | | | | | |
| Pasteup | | | | | |
| Prod. Check | | | | | |
| Edit. Check | | | | | |
| Signatures | | | | | |
| Edit. Recheck | | | | | |
| Prod. Recheck | | | | | |
| Filming | | | | | |
| Xeroxing | | | | | |
| Printing | | | | | |
| Collating | | | | | |
| Quality Check | | | | | |
| Binding | | | | | |
| Delivery | | | | | |

Illustrations: Total ______ Date Needed ______

| Fig. | Size | HT | LN | OK | Film | Xerox | Pg. No. |
|---|---|---|---|---|---|---|---|
| 1 | | | | | | | |
| 2 | | | | | | | |
| 3 | | | | | | | |
| 4 | | | | | | | |
| 5 | | | | | | | |
| 6 | | | | | | | |
| 7 | | | | | | | |
| 8 | | | | | | | |
| 9 | | | | | | | |
| 10 | | | | | | | |
| 11 | | | | | | | |
| 12 | | | | | | | |
| 13 | | | | | | | |
| 14 | | | | | | | |
| 15 | | | | | | | |
| 16 | | | | | | | |
| 17 | | | | | | | |
| 18 | | | | | | | |
| 19 | | | | | | | |
| 20 | | | | | | | |
| 21 | | | | | | | |
| 22 | | | | | | | |
| 23 | | | | | | | |
| 24 | | | | | | | |
| 25 | | | | | | | |

Return to Editor:

☐ Copies Requested
☐ Excess Printing (Including Tissues)
☐ Original Art (Collated)
☐ Negatives (Collated)
☐ Plates or Masters (Collated)

Record Reviewed and Filed:

________________
Production Editor

*Figure 3. Production Record Form*

## Conclusion

The few special forms of organization which have been shown in this chapter provide only a glimpse of the multitude of varieties in existence. It quickly becomes apparent, however, that there is nothing really new in these forms for the person who has a knowledge of the fundamental skills of technical writing.

The content of this chapter will be given added meaning if you will turn to Appendix C, where you will find reproduced instructions issued to employees by the Research and Development Department of the Sun Oil Company.

## SUGGESTIONS FOR WRITING

Devise a prepared format for keeping a record of the progress you make in the preparation of your library research report, making provision for a record from the time you submit a topic for your instructor's approval until you submit your completed report.

# 17
# Oral Reports

## Introduction

The purpose of this chapter is to make a few practical suggestions about talking with people. For the most part it will be concerned with talking rather formally to an audience, but some attention will be given also to conferences. This chapter is not a substitute for a course in speech, nor for reading a good textbook on speech* – both of which we strongly recommend. This chapter is merely a brief introduction to a broad and important subject. Emphasis will be given to speech problems especially common in the technical field.

Most of what has been said earlier in this book about the organization and language of technical writing applies to speaking on technical subjects as well. The discussion that follows will be confined to factors that appear only because the form of communication is oral, rather than written.

## Making a Speech

To be an effective speaker you must know how to use your voice properly and how to maintain a good relationship with your audience. These subjects aren't as formidable as they sound. Actually, the chief need of the novice speaker is simply the application of

*Harry E. Hand, *Effective Speaking for the Technical Man: Practical Views and Comments* (New York: Van Nostrand-Reinhold Company, 1969), is a good example.

common sense to his problems—and practice. In addition to the subjects mentioned, we will comment on transitional material, graphic aids, and the question period at the end of a talk.

*The Voice.* It is impossible to become a polished speaker without making speeches. Practice is unquestionably the most important single element in acquiring skill. Advice on how to speak is often ineffective until practice begins to lend meaning to it.

Fortunately, there is one aspect of speech making that each of us practices every day, at least to some degree. We all talk. We all say words. So we might as well practice saying words in a way that is pleasant to hear and easy to understand. Here are four suggestions that are helpful, whether you are talking to one person or a hundred.

1. *Relax.* Tenseness causes the muscles in the throat to constrict and raises the pitch of the voice. Your lungs are a pair of bellows forcing air through the vocal cords. The force is applied by muscles in the abdomen. When you are relaxed and speaking naturally the vocal cords vibrate easily.
2. *Open your mouth.* Speaking with your mouth insufficiently opened is like putting a mute on a trumpet. This fault is one of the commonest causes of indistinct speech. If you find you're having trouble being heard, you'll probably feel that you look ridiculous when you first start opening your mouth wider. Look in a mirror. Watch other people.
3. *Use your tongue and lips.* We remember a student who announced he was going to explain how to graft ceilings. It sounded ominously political, until it turned out he meant "seedlings." You can't say a "d" or a "t" without using the tip of your tongue. Nor can you say "b" or "p" without using your lips. Repeat the alphabet and notice the muscular movements required for the different sounds. It's a mechanical problem. As with opening the mouth, you may feel foolish if you start using your tongue and lips more than you have been doing. Of course you may sound foolish, too, if you overdo it. Make the sounds clearly, but not affectedly. Listen to and watch a good TV announcer.
4. *Avoid a monotone.* It's hard for a speaker to interest an audience, or a companion, in a subject in which he doesn't sound interested himself, and there is nothing interesting in a monotonous drone. Enthusiasm is naturally shown by a variation in the pitch of the voice to match the thought being expressed. See how many shades of meaning you can give the following sentence by varying the pitch of your voice: "You think he did that?"

Some other suggestions that are also related to the problem of using your voice to best advantage are these:

1. Pronounce syllables clearly. Don't substitute "Frinstance" for "for instance." Be sparing of the "I'm gonna because I gotta" style of pronunciation.
2. Give a little attention to the speed with which you talk. Moderation is a good principle: neither very fast nor very slow.
3. Try to talk along smoothly, with fairly simple sentence structure, and without repeated "and-uh's," or habitual pauses between groups of words. It is probably best not to think much about sentence structure in your first few speeches. Concentrate on what you have to say and keep going. But you can practice good sentence structure every day in conversation.
4. Speak loudly enough so that everyone you are addressing can hear easily, but don't blast people out of their seats.

*Your Relationship with Your Audience.* The audiences you can expect to address as part of your professional work will be made up of people who are seeking technical and economic information, not a show. Typically, you may expect to address fellow members of professional societies, fellow employees conferring on special problems or meeting on special occasions, prospective clients, and so forth. Aside from reports made in college classes and seminars, the young technical man is likely to do his first speaking before a chapter of a professional society.

With such audiences your relationship should be unaffected and unassuming, but at the same time confident and businesslike. You should by all means avoid anything approaching what is sometimes called florid oratory. Say what you have to say as directly and simply as possible.

As for posture, the best advise is to be natural—unless nature inclines you toward sprawling limply over the table or lectern. Stand up straight, but not stiff, and look at the audience. If you feel like moving around a bit, do so; but don't pace, or walk away from a microphone if a public address system is being used. If you feel like emphasizing a point with a gesture, go right ahead, but don't make startling or peculiar flourishes that will interest the audience more than what you are saying. In general, it is wise to move slowly. Don't do anything (like toying with a key chain) that will draw attention away from what you are saying.

Above all, act like a human being, not a speech-making automaton. Try to convey to the members of your audience a feeling of interest in your subject; show that you enjoy talking with them about it. A particularly useful device is to bring in occasional references to personal incidents involving yourself or your co-workers, incidents that have some relation to your subject and may be used to illustrate a point. People are always interested in other people, and an appropriate personal anecdote may warm up and give life to an otherwise dull body of information.

What can be said about preparing for such a performance? You can choose among three basic possibilities. You may read your speech from a manuscript, you may memorize the speech from a manuscript, or you may deliver it "off the cuff," using a few notes if necessary.

The last method, with or without notes, creates an impression of spontaneity and naturalness that is greatly to be desired. The use of notes is not a significant barrier to this impression and is a considerable support to self-confidence. Very often, however, custom calls for, or sanctions, the reading of a paper. This method is especially desirable when the material to be presented is complex, as it is likely to be in meetings of professional societies. The possibility that you would need to commit a speech to memory, word for word, is very remote.

If the speech is given extemporaneously and notes are used, it is generally wise to put them on small cards, to type them or write them clearly, and to indicate only major headings. Too much detail in the notes might result in confusion. You might lose your place.

The initial preparation of the speech, whether notes are used or not, is like the preparation of any report. That is, an outline should be made first (some differences in content will be noted later). If the speech is to be memorized, or read, the outline is used as a guide in writing the manuscript. If the speech is to be delivered more spontaneously, the writing step is omitted and the outline becomes a guide for practice (to a friend or relative) and a basis for the notes.

Naturally, you will want to learn all you can about your subject. Make it a point to know more about every phase of it than you expect to reveal. This extra information is like armor between you and the fear of running out of something to say when you get up to speak.

It may be helpful to read something aloud, in private, and at your normal speaking rate, to count the words per minute and thus estimate the number of words you'll deliver in the time allotted to your speech. But remember that almost everybody uses more words to cover a given subject when speaking than when writing. Don't underestimate the length of your talk and keep your audience longer than they expected: they won't like it.

Finally, there is the problem of nervousness. You may feel about your first few speeches as crusty Dr. Johnson did about women taking up preaching: it's not a question of doing it well but of doing it at all. Nervousness is best regarded simply as a nuisance that will diminish with experience. Most people never do get over feeling a little trembly when they first arise to speak. There are two sources of comfort with regard to this matter. One is that you are almost certain to find that after you have been on your feet a few minutes, the going is easier. Speak slowly at first, and pause for a good breath now and then. The other comfort is that your nervousness will be less apparent to the audience than you think. We once sat in the front row of an audience to which a young engineer was making one of his first speeches. We were thinking what a fine job he was doing, and what composure he had, when we just happened to notice that the knees of his trousers were vibrating at what we roughly estimated to be ten cycles a second. He made an excellent impression, and it is doubtful that anybody else in the room knew that his knees were shaky. Speak whenever you have the chance; experience will put you at ease.

Maintaining an effective personal relationship with an audience is exceedingly important in making a speech, but so is maintaining clarity. In this connection, transitional devices deserve a comment.

*Introductions, Transitions, and Conclusions.* Two problems faced by a speaker which are not faced by a writer are that an audience cannot be expected to give unwavering attention to what the speaker says, and that the audience cannot turn back to review an earlier part of a speech. Consequently, the speaker is under a heavy obligation to provide clear introductions, transitions, and conclusions. There is nothing new in principle here, and you need expect no special difficulty if you give careful thought to the matter. Sometimes, if you are using notes, it is helpful to indicate points at which transitions are needed. A glance at the headings on the card will supply its content.

A third problem that should be mentioned is the possible need for a more dramatic introduction in a speech than would seem necessary in a written report. We said earlier that, for the kind of audience you are likely to have, you should be supplying information rather than putting on a show. That statement holds true; nevertheless, it is almost inevitable that a speaker will find it desirable to use certain devices to heighten interest (still far short of putting on a show) that would seem out of place in a written technical report. One such device that has already been mentioned is simply an attempt to make the whole delivery animated and enthusiastic; a second is the use of personal anecdotes; a third is the use of graphic aids to lend drama and emphasis to your discussion; a fourth is the use of a dramatic introduction or conclusion.

Such an introduction is more easily illustrated than discussed abstractly. We suggest you turn to Mr. Galt's speech, *Cold Facts,* in Appendix B, and compare the introduction to this speech with the introduction to his article on the same subject. It is worth noting here, however, that caution is always necessary in an attempt to be dramatic. If you feel uncertain of success, don't try. Observe Mr. Galt's elimination of the joke about the peanuts that appeared in an early version of his talk.

*Graphic Aids.* Four principles to remember about the use of graphic aids are these:

1. Use graphic aids if you can—so long as there is no special circumstance that would make them inappropriate. There are almost unlimited possibilities as to types: graphs, tables, flowsheets, objects that can be held up by hand or specially mounted, slides, moving pictures, sketches on a blackboard. If you draw sketches on the blackboard do it beforehand if possible. Try not to have to let the audience sit in silence while you draw, but don't talk to the blackboard as you draw.
2. Don't use too many graphic aids. If you keep popping up with new gadgets the total effect may be spoiled.
3. Make sure that all your graphic aids are properly located and are big enough to be seen by everyone in the audience.
4. Keep your graphic aids simpler than would seem necessary in a written report on the same subject. And don't use any aid (like certain types of graphs) that some members of your audience won't understand.

*Answering Questions.* You may be asked to answer questions after you finish your speech. Naturally, the best preparation for this part of your performance is to acquire such a thorough knowledge of your subject that you can answer any question promptly and precisely. Needless to say, such omniscience lies beyond the reach of most of us. What, then, is to be done?

In the first place, prepare yourself as thoroughly as you can, and then try not to worry about the question period. Chances are it won't be half the ordeal you might imagine. If you don't know the answer to a question, say so. A simple statement to the effect that you are sorry but you just don't know the answer is preferable to an attempt to bluff or to give an evasive answer.

In the second place, be considerate of the questioner. He may be a little nervous himself, and may ask a foolish question or put his question more sharply than he really intended.

In the third place, don't try to answer a question you don't understand. Ask politely for a restatement.

In the fourth place—and also in connection with the preceding comment—make sure that the audience has heard and understood the question before you answer it. If a chairman is running the meeting, he may take care of this problem, repeating questions when necessary to clarify them or to make sure everyone has understood. If he does not, or if there is no chairman, you should assume the obligation yourself. You can start your reply with some such statement as "If I understand correctly, you are asking whether . . . ."

Finally, it may be necessary for you to bring the period to a close. If a specific length of time has been allotted for questions, the time limit should be respected. You should, however, try to gauge the feeling of the audience. If there is reason to think that most of the audience would like to continue, you can suggest that the time is up but that anyone who cares to may stay and go on with the discussion. In any case the audience as a whole should not be kept against its will merely because two or three persons persist in raising questions. You can usually achieve a graceful halt by declaring that time will permit only one more question.

## Conferences

A large portion of almost every professional man's time is taken up by conferences (estimated at 12 percent by one corporation). These conferences may involve only two people, or many people. They may vary from nothing more than informal conversations to highly formal group proceedings. Your preparation for, and conduct in, a conference deserve serious thought.

1. Try to formulate the purpose of a conference ahead of time. Is the purpose to clarify a problem? To single out feasible courses of action? To make a final decision on a course of action? It is easy to permit the words "Let's get together and talk things over" to lull one into a passive state of mind in which problems that should have been thought out carefully beforehand are not even recognized until the conference is in progress. This is a waste of time and energy.

2. Try to formulate your own objectives before you go to a conference. Your chances of making a significant contribution are very much greater if you know your own mind before discussion begins than if you drift into the meeting like a boat without a keel. On the

other hand, you should go equipped with a rudder as well as a keel so that you can change direction if the conversation opens up facts and points of view that had previously escaped you. Don't be stubborn.

3. Estimate the attitudes of the people in the meeting. This is not a new problem: it is simply the principle of "reader analysis" carried into the conference room.

4. Take some time to speculate on how things are likely to go. Try to think of the conference as a structure. A skilled chairman can lead a group of people through a series of deliberations with an ease and clarity little short of astonishing when viewed in retrospect. He can do this—for one reason—because he is thinking of the situation as a whole and not letting progress bog down in irrelevancies. He knows where he's going. Other members of the group cannot direct the discussion quite as freely as the chairman, but they can nevertheless accomplish much by well-timed suggestions.

5. A last bit of advice is that you give some attention to your oral delivery as you engage in discussion, according to the principles suggested earlier. In some respects, more skill and flexibility are required in the conference room than on the lecture platform. The situation is less under the speaker's control, and he must adjust himself quickly as it changes. Voice control is particularly important. We are all familiar with complaints about people whose voices are so loud in conversation they can be heard in the next block. It is probably true that certain types of people are actually offended by being addressed in an especially loud voice; but some psychologists assert that there is also a type of personality that is offended by an especially soft voice. People with this type of personality, it is said, tend to feel that anyone who addresses them in a soft voice must dislike them or he would "speak up."

At any rate, remember that your voice is an important part of your personality, and in the close quarters of a conference it should be used with care. If you avoid either roaring or whispering, enunciate clearly but not affectedly, and pronounce your words without slurring syllables, you need have no worry.

## Summary

The best advice we can give you is to take a course in speech, and to speak before groups of people whenever you can. Meanwhile, you can help yourself by making sure that you are enunciating distinctly, with adequate movement of mouth, lips, and tongue; that you are varying the pitch of your voice effectively; and that you are pronouncing words without an irritating or confusing slurring of syllables.

You can watch the technique of speakers you hear. Do they use their voices well? Do they interest you in their subjects, and seem interested themselves? Is their posture suitable? Are their speeches well organized? Are introductions, transitions, and conclusions clear, so that you don't get lost? Have they employed graphic aids to best advantage? Can they handle a series of questions smoothly? Whenever you get a chance to speak, practice these techniques yourself.

In conferences and discussions in which you take part, you can practice formulating purposes and deciding upon your own objectives. You can also try to guess the attitudes of the other participants and to predict the probable course of the discussion. Make mental notes on the chairman's handling of the general course of the discussion. Finally, use your voice effectively.

## SUGGESTIONS FOR SPEAKING

1. Bring to class an article from a magazine or professional journal and give a brief analysis of its construction. Discuss the introduction, particularly subject, purpose, scope, and plan. Write the main headings in the organization of the article on the blackboard, and comment on the logic of the organization. Discuss the use of transitions. Discuss the conclusion or summary. Don't choose a complex article, nor one more than 3000 words long. Time: three to five minutes.
2. Give a short talk based essentially on one of the special techniques discussed in Section Two of this book. For instance, describe a simple device like a miniature flashlight. Time: three to five minutes.
3. Take one aspect of your library research report as your topic, and discuss it in detail. Time: ten to fifteen minutes.
4. Present a persuasive argument in favor of a course of action or a certain way of doing something. Time: three to five minutes.

# 18
# Business Letters

You will probably have to do a lot of letter writing. Most professional men do. And the more successful you are, the more correspondence you are likely to have to carry on. This chapter may be taken as a guide to the form and layout of letters, to the handling of style and tone, and to the organizing of a few selected types of letters.

There are a great many details and refinements in the art of letter writing that lie beyond the scope of this chapter. In the future, as correspondence assumes an increasingly important place in your work, you will find it useful to consult such books on letter writing as are listed in Appendix A. For the present, this chapter should be adequate.

## The Elements of a Business Letter

The elements, or parts, which normally appear in a letter are the heading, the inside address, the salutation, the body, the closure, and the signature. See Figure 1. Additional elements which appear in some letters are the subject line, the attention line, and notations about enclosures, distribution, and the identity of the stenographer. We shall discuss each of these elements before commenting on their over-all layout and appearance on the page.

---

```
                    1201 Linwood Avenue
HEADING ——→         Peoria, Illinois 61650
                    February 16,19--

Wakey Products, Inc.
1410A Grand Avenue      ←—INSIDE ADDRESS
Detroit, Michigan 48239

Gentlemen:  ←—SALUTATION

I would appreciate it if you would send me your
catalogue of home movie equipment, as advertised in
your Circular 33-C.

If you handle stereoscopic cameras and equipment, I
would also be grateful for information about what you
have. I am especially interested in securing a
projector.
BODY OF THE LETTER
                    Yours truly,
                      COMPLIMENTARY CLOSE

                    Richard Roe
                      TYPED SIGNATURE
```

---

*Figure 1. Elements of a Modified Block Business Letter*

*The Heading.* The heading of a letter includes the sender's address and the date. Business firms ordinarily use stationery with a printed heading containing the name of the company and its address, and frequently other information—the names of officials, the telephone number, the cable address, the company motto. When letterhead stationery is used, therefore, the writer need add to the heading only the date, either directly beneath the printed heading or to the right of center so that it ends, roughly, at the right margin.

If you write a business letter on stationery without a letterhead, you will need to put down, at the right and in order, your street address, the name of the city and state in which you live, the zip code, and the date of the letter, as in the following example:

4516 Ramsey Avenue
Austin, Texas 78731
October 15, 19__

Note that the zip code number appears after the name of the state.

*The Inside Address.* The inside address includes the full name and business address of the person written to, just as it appears on the envelope. Particular care should be exercised to spell the addressee's name correctly, and courtesy demands that his name be prefaced with "Mr." or an appropriate title. Business titles, by the way, should not precede a name; they may appear after it, separated from the surname by a comma, or on the line below. Compare the following illustrations:

Mr. John C. Doe, President
American Manufacturing Company
110 First Street
Houston, Texas 77021

Dr. John C. Doe
Director of Research
Wakey Products, Inc.
1410A Grand Avenue
Detroit, Michigan 48239

In writing the name of the company or organization, take pains to record it just as the company does. For instance, if the company spells out the word "company" in its correspondence, you should spell it out too, rather than abbreviate it. This is simple courtesy.

If you must write a letter to a company but do not know the name of an individual to whom to address it, you may address the company or a certain office or department of the company. Deletion of the complete first line in either of the examples above would leave an adequate address. When a letter is officially addressed to a company but the writer wishes some particular individual or office of the company to see the letter, he may use an "attention line." Placed a double space below the inside address, or below and to the right of the inside address, this line has the word "Attention," or the abbreviation "Att.," followed by a colon and the name of the proper person or department, as shown here:

Wakey Products, Inc.
1410A Grand Avenue
Detroit, Michigan 48239

Attention: Head, Drafting Department

Gentlemen:

At least a double space should be left between the heading and the inside address. Further comment on this point will be found below in the section "Form and Appearance."

*The Salutation.* The salutation or greeting is located a double space below the last line of the inside address and flush with the left-hand margin. In formal business correspondence, "Dear Sir" is always acceptable in greeting an individual man. The greeting "Sir" should be reserved for very formal letters, and the even more formal "My dear Sir" can probably be dispensed with altogether; to most persons, it has a stilted, artificial sound. More informal than "Dear Sir" and more suitable when you are not acquainted with the individual you address is "Dear Mr._________________." The latter greeting is used more than all others, with the possible exception of "Dear Sir." In addressing a company, or a group of men, use "Gentlemen." When writing to a woman or a group of women, use the equivalent of the forms just noted (Dear Miss_________________, Dear Mrs._________________, Dear Madam, Ladies).

Remember that the only acceptable mark of punctuation following the greeting is a colon. The comma is satisfactory in personal letters, but not for business letters. Too often we see an even less satisfactory mark—the semicolon. It is always incorrect.

*The Body of the Letter.* The body of the letter is, of course, its message, or what you have to say to the addressee. In a general way, we can say that the body of most letters is made up of three parts: (1) the introductory statement identifying the nature of the business the letter is about or the occasion for it, along with references to previous correspondence if appropriate or necessary; (2) the message proper; and (3) the closing paragraph, often a purely conventional statement. The body of the letter begins a double space below the salutation.

*The Complimentary Close.* The complimentary close is the formal way of signalizing the end of the letter. See Figure 2. It is ordinarily a conventional expression which should correspond in formality with the greeting. Standard closings are as follows:

> Yours respectfully, or Respectfully yours
> Yours truly (not Truly yours), Yours very truly, or Very truly yours
> Yours sincerely, or Sincerely yours
> Yours very sincerely, or Very sincerely yours
> Cordially yours

The first of the closings listed, or a variant, "Respectfully submitted," is proper for letters of transmittal to superiors, letters of application, or for any letter in which you wish to show special respect to the addressee. "Cordially yours" is suitable only when you are personally acquainted, on a basis of equality, with the person to whom you are writing.

The usual practice calls for a comma after the closing. Only the first word of the closing should be capitalized. Although many letter writers like to place the closing so that it ends in alignment with the right-hand margin, accepted practice approves of its being placed anywhere between the middle of the page and the margin, a double space below the last line of the text.

*The Signature.* Directly below the complimentary close and aligned with it appears the typed signature of the writer of the letter. The typed signature should be placed far enough below the closing so as to allow plenty of space for the handwritten signature. Four to six spaces are about right.

Often the writer will need to include his business or professional title ("Chief Engineer," for instance) and sometimes the name of the company or department of a company for which he is writing the letter. The business title is placed either above or below the typed signature. The use and location of the name of the company or department depend upon circumstances. The name of the company or department should appear below the signature only if it does not appear in a printed heading. But there is one exception. If you use a business title, like "Manager," which indicates your relationship to a department or section but not to the entire company, then the department or section should be stated after or below the business title even if the department is also identified in the heading. You will almost certainly have company letterhead stationery for official correspondence, but you may not have a departmental letterhead. The name of the company may appear *above* the signature, however, if you wish to emphasize the fact that you are speaking only as an instrument of the company and not with personal responsibility. The examples in Figure 2 illustrate various forms.

Cordially yours,

*John C. Doe*

John C. Doe

Yours very truly,

*John C. Doe*

John C. Doe, President

Yours sincerely,

*John C. Doe*

John C. Doe
Chief Technical Advisor

Yours very truly,

*John C. Doe*

John C. Doe
Chief Technical Advisor
Research Division

Very truly yours,
AMERICAN MANUFACTURING CO.

*John C. Doe*

John C. Doe, President

*Figure 2. Forms of Signatures*

*Miscellaneous Elements.* Several other items may be necessary or useful parts of a business letter. They include a notation identifying the stenographer, an indication of enclosures, a distribution list for copies of the letter, and a subject line. The stenographer identification consists of the sender's and the stenographer's initials, separated by a colon or a slant line. This notation is placed at the left margin, either directly opposite the typed signature or two spaces below. If there are enclosures to the letter, the abbreviation "Enc." or "Encl." is typed just below the identification notation. Many writers indicate the number of enclosures in parentheses after the abbreviation, as "Encl. (4)." If copies of the letter are distributed, the phrase "Copies to" or the abbreviation "cc" is typed at the left margin, below the identification notation, and below the enclosure notation, if there is one, and the names of those receiving a copy are listed below it. If a subject line is used, it appears either just below or below and to the right of the inside address. Most of the items discussed above are illustrated in Figure 3.

## Form and Appearance

Although the content of a letter is of first importance, attractive form is also necessary if the letter is to be effective. Good appearance requires that the materials used for the letter be of good quality, that margins and over-all layout of the letter on the page be pleasing to the eye, and that the spacing and arrangement of the elements be in accord with accepted conventions of good taste. And the letter must be neat. A typical business letter is shown in Figure 3.

The paper chosen for business correspondence should be a high quality white bond, 8½ by 11 inches in size, and of about 24-pound weight. If the letter is typed, as business letters are, the typewriter ribbon should be new enough so that it will make firm, easily legible letters; if it is handwritten, black, blue, or blue-black ink should be used. Other colors of ink are not generally considered in good taste. Carbon copies should be made with new carbon paper on good quality onionskin paper.

Attractive appearance calls for a minimum margin of at least one inch on all sides. Margins will have to be increased all around, of course, for letters which do not occupy a full page. Although an experienced stenographer can estimate accurately from shorthand notes about how wide the margins should be set, the inexperienced letter writer will probably have to type a trial effort or two before attractive placement can be achieved. The letterhead, by the way, is ignored in determining over-all layout of the letter on the page. It is permissible

WAKEY PRODUCTS, INC.

*Education Department*
*General Office Newark, New Jersey*

1410A Grand Avenue
Detroit, Michigan 48239
October 17, 19--

American Manufacturing Co.
110 First Street
Houston, Texas 77021

Subject: Training Films
Attention: Mr. Richard Roe

Gentlemen:

In response to your letter of October 12, I am glad to say that we have several training films now available that would be suitable for the needs you described. I am enclosing two pamphlets which will give you an idea of the contents of these films. I am also requesting a representative from our Dallas office to call upon you within the next week.

Very truly yours,

Joe C. Ashford

Joe C. Ashford, Manager
Education Department

JCA: wk
Encl. (2)
cc: Mr. Joseph Smith

*Figure 3. A Typical Business Letter*

to allow a somewhat narrower margin at the top of the page than at the bottom, about a 2:3 ratio being acceptable. This means that the center point of the letter may be slightly above the actual center of the page.

Balanced margins on the left and right sides of the page are desirable, but it is impossible to keep the right margin exactly even all the way down the page because of the necessity for dividing words at the ends of lines, or the necessity for not dividing words. Words must be divided between syllables or not at all. In general it is best to avoid divided words as much as possible. A dictionary should be consulted to find the correct syllabication of words if you are uncertain of syllable division.

In letters that are more than one page long, you should write the name of the addressee, the page number, and the date on page two and any additional pages. This notation appears just below the top of the page (one acceptable form is shown below). The text begins two or three spaces below the notation if it occupies the full page, and about eight spaces below if it occupies only a portion of the page.

Mr. John C. Ashford —2— November 23, 19__

With the exception of very short letters, you should single-space each of the elements of the letter and as a rule double-space between elements and between the paragraphs of the body. This means that the lines of the heading, inside address, and so forth, are single-spaced but that there is double spacing between the heading and the inside address, between the inside address and the salutation, between the salutation and the opening paragraph of the body, between paragraphs, and between the text and the complimentary close. The rule is not a hard and fast one, however; for pleasing proportions on the page, you may need to triple-space, or more, between the heading and the inside address and between the last line of the body and the complimentary close. Quite short letters may be double-spaced throughout.

The two commonly used styles of arrangement for the elements are the straight or full block form and the modified block form, and the only difference between the two is that paragraphs in the body are indented in the modified block form and not indented in the full block form. Full block form is illustrated in Figure 1. Figure 5 is a modified block form.

In the past it was common practice to use staggered indention for the elements, so that a heading, for instance, would look like this:

1919 South Second Street<br>
    Phoenix, Arizona<br>
        September 23, 19__

This style is scarcely ever used now. Another style seen now and then is called the "left-wing" form: in it each element, including the heading and closing, is begun flush with the left margin.

Although the block form or the modified block form is entirely satisfactory for most formal business correspondence, there is a simpler and more convenient form for interdepartmental and personal communication within an organization as well as for much intercompany and interagency communication. In fact, this form is often used for short, informal reports such as the monthly progress report. The memorandum form employs the principal features of the military correspondence form: "To," "From," and "Subject" headings printed on the stationery. In many organizations, stationery of different colors is used, each department having a different color. Much less formal than the conventional letter, the memorandum is usually headed with a date, often expressed in abbreviated form such as 10/21/70, and simply signed or initialed at the end. Neither a salutation nor formal closing is needed. See Figure 4. Two details are worth noting: the use of all capitals for the subject (or "title") and the use of underscored side headings.

## Style and Tone

In determining the style and tone of a business letter, you should keep three facts in mind: it is a personal communication, it serves as a record, and it is usually brief.

Since a letter is a personal communication, it should be characterized by courtesy and tact. In a sense, a letter is a substitute for a conversation with the person you are writing to, and you should therefore try to be as polite and considerate in your letter as you would be in dealing directly with the addressee. This consideration of the addressee, commonly called the "you attitude," will suggest that your letter should not only be perfectly clear in meaning but also free of any statements which might antagonize or irritate the reader. The "you attitude" thus has two aspects: the general one suggested by the phrase "tact, courtesy, and consideration"; and a mechanical aspect.

To be truly considerate of the reader, you need to grasp his point of view. Try to anticipate questions he might ask and to estimate his reaction to your statements. You should examine your sentences to see if they are free of ambiguities, free of words the reader might not understand—in short, to see if the letter will say *to the addressee* what you want it to say. It is important to try to read it from the point of view of the addressee because you know what you intended to say and are likely to take it for granted that the words express your intention unless you examine critically everything you have written.

In a more mechanical sense the "you attitude" means substituting the second person pronoun ("you") for the first person ("I" or "we"). Use of the second person has the effect of keeping attention centered on the reader rather than the writer and thus helps to avoid any impression of egotism. Suppose a writer has made the following statement in a letter:

I have noticed that your shipments to us have consistently been delayed. We are inconvenienced by these delays and request that you investigate the matter at once.

---

MEMORANDUM

20 June 19___

To: John Ableman, Chief
Engineering Department

From: Jim Moore

Subject: RENOVATION OF MODEL SHOP AREA

In response to your request, I have studied the Model Shop Area (see earlier memoranda), and I have a number of recommendations to make for its renovation.

Drains and Flooring

Since the Model Shop will probably have a pilot finishing area, the open drains in the finishing area should be left intact by being covered with boiler plate mounted flush with the floor. This plating would support the machinery now in the Model Shop and make it possible to convert the area for pilot finishing or etched board work. The floor should be covered with tile similar to that in other areas of engineering.

Ceiling Insulation

Acoustical insulation should be installed in the ceiling to absorb noise from the machinery. The type used in the most recent annex to our building is quite effective.

Wall Construction

Since this area is temporary, the walls shown on the layout (see attachment) should be constructed of plywood or masonite.

Welding Booth

Two special requirements need to be met for efficient welding booth operation: (a) duct work and blower (a hood is available) for venting fumes from the welding area, and (b) a 440-volt bus capable of supplying 90 amperes for the Heliarc welder.

Conclusion

So that we may function as efficiently as possible, I respectfully suggest that you approve these recommendations as soon as convenient so that we may get to work on the remodeling.

Jim

Jim Moore
Supervisor, Model Shop

Attachment

---

*Figure 4. Modified Military Form Memorandum*

This rather blunt statement could have been better phrased as follows:

> We are curious about the cause of the rather consistent delay in the receipt of shipments from your company. You will understand, of course, that delay in receiving these shipments is the cause of considerable inconvenience to us, and we are sure you will want to correct the situation as soon as possible.

Perhaps we should add that personal pronouns are entirely suitable to personal communication. Do not hesitate to use "I" or "you" when it is natural to do so. Expressions like "the writer " instead of "I," or "it will be noted" instead of "you will notice" are out of place in correspondence. Be direct and natural.

In addition to being considerate, courteous, and unaffected in style, the letter should be concise. Since it makes no sense to waste a reader's time with nonessential discussion, a letter should be held to a single page if at all possible. But do not make the mistake of believing that brevity alone is a virtue in letters. Too much brevity makes for an unsatisfactory tone. When carried too far, it results in a curtness and bluntness which can be irritating. A further danger of carrying brevity too far is that it may result in a lack of clarity and completeness.

The fact that letters are filed for reference makes it necessary that a letter be clear and complete, not only at the time of writing but also at any later time at which it might prove necessary to look at the letter again. This need for clarity of reference is one reason why most letters of reply begin with a concrete reference to the date and subject of previous correspondence. It also explains why phrases like "the matter we corresponded about last month," are not satisfactory.

## Types of Letters

In books devoted exclusively to letter writing, a great many types are discussed which we shall not have space for here. Should you need information about such types as claim and adjustment letters or sales letters, consult one of the volumes listed in Appendix A. Our discussion will be limited to five frequently used types of letters: inquiry, reply, instruction, transmittal, and application.

*Letters of Inquiry and Reply.* A good letter of inquiry should (1) identify the nature of the inquiry at the very beginning, (2) state the reason for the inquiry if it is not obvious, (3) clearly and explicitly phrase the inquiry to make reply as easy as possible, and (4) close with an appropriate and courteous statement.

Since anyone reading a letter naturally wants to know immediately what it is about, the writer should state in his opening sentence what he is writing about. This does not mean that the salutation must be followed immediately by the specific inquiry or inquiries. It means that the purpose of the letter should be identified immediately as an inquiry about a specific subject. Thus you might begin, "I am writing this letter to inquire whether you have any new perfomance data for release on the ramjet engine you are developing." This statement could then be followed by concrete, specific questions.

Explaining the reason for making an inquiry is not absolutely necessary unless a response to the inquiry you are making constitutes the granting of a favor. It is always courteous, however, to explain why the inquiry is being made, and when you are asking a favor, it is a courtesy that should not be neglected.

It is exasperating to receive an inquiry phrased in such general terms that no clear notion of what is wanted can be determined. Too often one sees statements like this: "Please send me what information you have on television antennas." The writer of this request probably wants far less information than it seems to call for. Actually, his inquiry would have had more meaning had it been rephrased:

1. What types of television antennas do you manufacture?
2. Can you send me installation instructions for the types you manufacture?

Concrete, specific questions make a reply easier to write. Questions do not necessarily need to be numbered and listed as above, but such a form is perfectly satisfactory and is desirable when several questions are asked. Remember that vague, general requests present an impossible problem to the man who receives the inquiry. We recall a student who wrote a request to a research organization for "any information you have about new aircraft designs" being worked on by the organization. He did not realize that a literal granting of his request might result in his receiving a truckload of reports!

The problems of concreteness and courtesy are both well illustrated by the letter of a graduate student of chemical engineering who wrote for information to be used in a report for one of his courses. The letter he wrote is quoted below, the only change being the deletion of names.

Dear Sir:

It is requested that literature concerning the history and background, wage and benefit plans, and general research policy of [your] Corporation, and a current statement to stockholders be sent to me at the above address.

> This information is to be used in a report assigned in a graduate chemical engineering course. Additional information which you may have available will be appreciated.
>
> This literature will be made available to the University Library after I have finished with it.
>
> Any information you supply will be greatly appreciated.
>
> Very truly yours,

The first paragraph of the reply received by the author of the above letter went as follows:

> Dear Mr._______________:
>
> We are glad to comply with your rather blunt request that we supply you with a great deal of not altogether inexpensive material to be used in a report in a graduate chemical engineering course. Since you are going to give it to the library after you have finished with it, we are less critical of the tone of your letter than we might otherwise be. I speak this way to bring to your attention something which may be useful to you later on, since believe me your letter of March 29 could be couched in more gracious terms.

The unfortunate tone of this letter of request is due primarily to the use of the passive voice. The phrase, "It is requested that . . ." is ungracious. Perhaps, "I am writing to ask if you could help me . . ." would have been more cordially received. A comment on the kind of report that had to be written would have helped the reader decide what materials should be sent. The letter should have been rephrased throughout, particularly to indicate the writer's realization that he *is* asking for a great deal of help.

Custom suggests that letters of inquiry, especially those in which a favor is asked, close with a statement showing that their writers will appreciate a reply. This is adequately illustrated in the letter quoted above. But remember that good taste suggests that you avoid ending your letter with "Thanking you in advance, I remain, . . . ." If you are in a position to do so, it may be appropriate to offer to return the favor.

In writing a reply to a letter of inquiry, keep these two points in mind: (1) begin your letter with a reference to the inquiry, preferably both by date and subject; and (2) make the reply or replies as explicit and clear as possible. If the inquiry contained itemized questions, it is a good plan to itemize answers. Naturally, the reply should be courteous.

*The Letter of Instruction.* When instructions are to be issued by letter, you will find the following plan of organization suitable:

1. The opening paragraph of the letter should explain the situation or problem which necessitates issuing the instructions.
2. The body of the letter should contain the detailed instructions.

Common sense will tell you that these instructions should be clear and definite. Vague and ambiguous instructions often defeat their own purpose by confusing and irritating the reader and by making it less likely, consequently, that they will be satisfactorily carried out.

3. The conclusion of the letter should suggest any action, other than carrying out the instructions, that should be taken. This may be a request for a report, a conference, or the like.

*The Letter of Transmittal.* The letter of transmittal is a communication from the writer of a report to its recipient. In a general way, it serves about the same purpose for a report that a preface does for a book. Although letters of transmittal are often sent through the mails separately from the report itself, they may be bound in with the report, following the title page. We shall discuss the five primary functions of a letter of transmittal in the sequence in which they usually appear.

1. The letter typically opens with a reference to the occasion of the report or an explanation of why the report is being submitted. There may be a reference to a contract or other authorization of the work being reported on.

2. The letter should state the title of the report being transmitted. Both of these first two functions are illustrated in the following sentences:

> In response to your request, dated October 26, 19__, I have investigated the possibilities for a new plant location in the Southwest. The accompanying report, entitled *Advantages and Disadvantages of Five Southwestern Cities as Sites for a New Assembly Plant,* is an account of this investigation and the conclusions it led to.

3. The second paragraph of the letter of transmittal should explain the purpose and scope of the report (unless the opening paragraph has already done so). Beyond this, it should be devoted to any comments about the report that the writer feels should be made to the addressee. It might happen, for example, that the writer had on a previous occasion said that the report would contain information that was finally omitted. This omission should be explained.

Do not hesitate to duplicate in the letter of transmittal those elements that also appear in the abstract or the introduction to the report. A statement of purpose, for instance, almost always appears in all three places.

4. If the writer has received assistance in carrying out the work with which his report is concerned and feels that this assistance should be acknowledged, the letter of transmittal is a place in which he can name and thank those who helped him.

5. Customarily the letter closes with a statement expressing hope that the report is satisfactory.

The five functions that have been mentioned are illustrated in Figure 5.

---

5127 Clearview Street
Austin, Texas 78730
November 20, 19--

Professor John C. Doe
Department of English
University of Texas
Austin, Texas 78712

Dear Professor Doe:

The accompanying report entitled Joistile-Concrete Beams is submitted in accordance with your instructions of October 10.

The primary purpose of the report is to present information about joistile-concrete beams for use in floor and roof construction. An effort has been made to cover the subject thoroughly, including the development of joistile, its use and application in beams, investigations and tests of both joistile and joistile-concrete beams, and general specifications for the construction of tile-concrete beams and floor slabs. The section on investigations and tests is limited to the most important and pertinent tests and results.

I wish to acknowledge the information and assistance given me by Mr. James Wood of the Greek Key Ceramics Association.

I sincerely hope that this report will meet with your approval.

Respectfully yours,

*Edward Donaldson*
Edward Donaldson

---

*Figure 5. Letter of Transmittal*

*The Letter of Application.* Most students feel that no form of the business letter is as important to them as the letter of application for a job. We shall therefore discuss this type of letter in considerable detail.

Since the amount and variety of information an employer will want to know about you is great, it is ordinarily impracticable to include it all within the framework of the conventional business letter. We will, therefore, consider that form of application which is made in two parts: (1) a data sheet or qualifications record, and (2) an accompanying letter. We shall discuss the data sheet first since what is said in the letter will depend, to some extent, upon what is included on the data sheet.

*The Data Sheet.* The data sheet contains about the same information called for on most printed application forms, organized in four sections:

A. Personal data
B. Education
C. Experience
D. References

The form of a data sheet is illustrated in Figure 6.

---

DATA SHEET

Malcolm Richards
5127 Clearview Street
Austin, Texas 78730
Telephone: 472-8509

Personal Data, May, 1969

Age 25
Height 6 ft, 2 in
Weight 175 lb
Married, no children

Health excellent
Eyesight excellent
Hearing excellent
Draft Status: IV-A

Education

Leaton High School (Texas), 1962
Senior Classification in Petroleum Engineering in the University of Texas: Grade point average, 2.0 of 3.0.
Degree sought: B.S. in Petroleum Engineering
Important Courses (Semester hours shown):
Mathematics (15)
Chemistry (16)
Fluid Flow (3)
Petrophysics (2)

| | |
|---|---|
| Engineering Mechanics (9) | Petroleum Production (3) |
| Engineering Drawing (4) | Petroleum Technology (3) |
| Physics (8) | Technical Writing (3) |
| Reservoir Engineering (6) | Mechanics of Drilling (4) |

Percentage of Expenses Earned: 30%

Experience

Summer of 1968
"Roughneck" on oil drilling rig under Mr. Frank A. Thomas, Marshall Drilling Co., Sarnia, Texas

Summer of 1967
Assistant "Shooter" on explosive truck under Mr. James Stone, United Geophysical Co., San Antonio, Texas

Summer of 1966
Laborer with Gunn Supply Co., under J.A. Gunn, San Antonio, Texas

References (by permission)

Mr. Joseph Wood, President
Chamber of Commerce
Sarnia, Texas 78999

Mr. James Stone, Manager
United Geophysical Co.
San Antonio, Texas 78298

Professor Howard Bitt
Department of Petroleum Engineering
College of Engineering
The University of Texas
Austin, Texas 78712

*Figure 6. Data Sheet*

The personal data section contains enough information to enable an employer to get some idea of what you are like individually. Relevant headings may include age, place of birth, health data (height, weight, eyesight, and hearing), marital status and dependents, recreational interests, organization and society memberships.

Accompanying this data and placed in the upper right-hand corner of the sheet should appear a photograph. Of regulation size (any commercial photographer will know about this), the application photograph should be a straightforward, unretouched, serious pose. The "portrait" type of picture, such as you may have had made for gifts, is not appropriate. The photograph is of interest to any employer because he wants to know what you look like. We believe one should accompany every application. We do know, however, of one interesting instance in which an applicant purposely left his out. As a matter of fact, he made a point of referring to its omission in his letter and explained it by remarking that he was undoubtedly the ugliest man living. Since the prospective employer was impressed by the man's qualifications and curious about this comment on appearance, he invited him to an interview and hired him. He said later, though, that the man's remark was correct—he was the ugliest man he'd ever seen.

In an application from a man with several years' experience, the section on education will need to contain only the bare facts: schools attended, dates of attendance, and degree (or degrees) awarded. Experience, or what the applicant has done, will be of chief interest to an employer. But for the young man, especially the one just completing his training, the section on education is more important. Education is what the young man has to offer. He should make the most of it in his application by giving full and complete details about all of it that is pertinent to the job applied for. He should state in this section the name of his high school and the date of graduation; the name of his college and the date, or expected date, of graduation; the degree he has, or is seeking; his rank in his class; his major and minor fields; and a list of the courses pertinent to the requirements of the position sought. It may be desirable to give the number of credit hours earned in each course. Some applicants prefer to classify courses as basic, specialized or technical, and general.

Some applicants feel that giving an employer the information that they have a B.S. degree is enough. But since programs differ a good deal from one college to another, it is better to call attention to the courses you feel help qualify you for the job. Do not, of course, name those subjects which cannot possibly qualify you for the particular position you are trying to get. You will not need to list the same courses in every application you make, though there will naturally be a considerable amount of duplication. You may offer to send a complete transcript; some employers will want one.

Although the young graduate may have had little or no experience related to the kind of professional work he is seeking, he should nevertheless record the facts about jobs he has held, including part-time and summer jobs. Many an employer will be interested in finding out how well an applicant has discharged responsibility and how diligent and cooperative he has been in carrying out duties and working with other people, for he may believe that a man's attitude toward his job and the people with whom he works is not likely to change, regardless of the nature of the work.

List the jobs you have held in reverse chronological order: most recent job first, etc. For each job give (1) the dates of employment, (2) the kind of work done, (3) the name of the company employee qualified to evaluate your services, (4) the name of the company or organization, and (5) its address. Be sure to record the name of your superior and the name and address of the company accurately so that an inquiry will be certain to reach its destination.

One often hears, "It's not *what* you know that gets you ahead; it's *who* you know." We do not subscribe to this cynical remark, but neither do we wish to minimize the importance of having influential people back you up in your application. Often the recommendation of a man whose word is respected is the deciding factor in getting a job. This is perhaps especially true when an employer is considering a number of applications from graduates who do not have professional experience. The academic records alone might not provide a clear basis for a choice. We believe you should devote careful thought to this matter of references and make the most of your opportunities.

At least three and perhaps as many as five references should be listed. One of these should be an employer, if possible; one should be a person who has known you personally a long time and who can therefore vouch for your character; and the others should be those of your teachers who can vouch for the quality of your work as a student. Be sure to get permission from each one before giving his name. It's a good plan to tell each reference, at the time you ask his permission, something about the job you are applying for so that he can write a better letter of recommendation for you, one in which he can emphasize qualifications which are pertinent to the job you are after. Be very sure that you show your references the courtesy of spelling their names correctly, and for your own sake be sure to give an address by which they can be reached.

One final word about references. If you are especially eager to get the job you are applying for and feel that you can presume upon the kindness of some of your references, ask them to write unsolicited letters of recommendation, to be sent so that they will be received shortly after your application has been received. This support for an application may be quite effective, and it is comforting to know that a recommendation has been made. A prospective employer may not take the initiative by writing for information about you. We know of one young applicant who went a step further; he asked one of his references to put in a long-distance telephone call in his behalf. He got the job, too.

*The Letter.* The letter acompanying a data sheet has four principal functions: (1) making reference to your source of information about the opening, (2) explicitly making application for a job, (3) elaborating on pertinent qualifications, and (4) requesting an interview. These functions are illustrated in Figure 7.

---

5127 Clearview Street
Austin, Texas 78730
May 14, 19--

Mr. M. A. Lindstrom, Head
Personnel Department
Rhode Island Oil Company
Sarnia, Texas 78999

Dear Mr. Lindstrom:

I talked to your Austin representative, Mr. Clapper, and he informed me that it is a policy of your Company to employ college students of petroleum engineering each summer as assistant gaugers. I should like to submit my application for that position.

While working with the Marshall Drilling Company, I became thoroughly familiar with the location of wells on your Chalker and Sarnia fields. In fact, I happened to be working with the crew that drilled the first deep dual-completion well on the Dole lease last summer. The course which I completed this year in Petroleum Production covered production gauges and their functions and operation quite thoroughly.

Please refer to the enclosed data sheet for details of my education and experience and the names of persons who have consented to express an opinion about my ability and character.

I shall be glad to go to your Sarnia office any

Saturday on a few hours' notice. Unfortunately, my schoolwork makes it virtually impossible for me to go during the week. I shall be available to begin work June 1.

Yours respectfully,

Malcolm Richards

Malcolm Richards

---

*Figure 7. Letter of Application*

If some person, like a company representative or employee, has told you of an opening, the best way to begin your letter is by making reference to that person by name. Seeing a familiar name, the employer is likely to continue reading and to give your application consideration. If your source of knowledge is an advertisement, refer to that. If you do not know whether an opening exists or not, you may begin by mentioning your interest in getting into the particular kind of work done by the company, or your desire to be associated with the company. Openings which stress an interest in a particular company must be tactfully written so they will not sound as if flattery is being employed to gain a sympathetic hearing.

Explicitly stating that you are applying for a job is more than a conventional formality; it permits you to state exactly what work it is that you are after. Applying for a specific job is always better than just asking for employment, and this is particularly true when you are applying to a large corporation in which many technical men are employed in all sorts of jobs, jobs which do not always bear a direct relationship to the academic training of those who fill them. This emphasis on making application for a specific position may seem to ignore the fact that many graduates are put into a training program upon first being employed by a large organization. If you know that the company you are applying to puts all newly hired men into a training program, make your application for a place in that program and also state what your professional interest is.

The first paragraph of the letter, then, contains at least the second and perhaps both of the following two elements: (1) a reference to your source of information about the job that is open, and (2) a statement about what job you want. The next paragraph (or paragraphs) is the hardest to write—and the most important. It is the real body of your letter; it is here that you distinguish your application from others.

In it you may single out for detailed discussion something from your training or experience that particularly qualifies you for the job you are trying to get.

Remember that the data sheet gives the bare details. In an application to a company for a position in a tool and die design department, a mechanical engineer, for instance, may have listed a course in machine design on his data sheet. But merely naming the course does not tell what the engineer did in the course, or whether he did well or poorly. Suppose he had undertaken several projects and completed them successfully, and suppose he had learned that this experience will help him if he gets the job. He will be wise, then, to provide the employer with the details about this course and the projects he completed. Similarly, merely naming a job on the experience record scarcely does more than suggest the duties, skills, and responsibilities that the job demanded. The letter gives you an opportunity to submit full information about those aspects of your training and experience which best fit you for the job you are after.

You must, of course, carefully analyze the job's requirements and measure them against your own qualifications before selecting something to give details about. It is this elaboration of selected details of your qualifications which makes your letter more than a letter of transmittal for your data sheet. If there is a parallel—even a distant one—between what you demonstrate here that you are familiar with and the work the employer has available, he will probably conclude that you are a promising candidate for the job he has open. One final caution: do not state that you feel you are fully qualified for the job unless you supply enough information to support such a claim. And if you present the support, it is hardly necessary to make the claim. After all, very few graduates are fully qualified for any job until they acquire some experience.

Since the immediate objective of a letter of application is an interview, you will want to close your letter with a request for one. You should take pains to phrase this request so that it will convey the fact that you want the interview, not merely that you are willing to have it. Accommodate yourself to the employer's convenience so far as possible. Suggest that you will be glad to appear at a time convenient for him. If there are restrictions on your time which make it impossible for you to be interviewed at certain periods, be sure to state what they are and explain them.

If time and distance make it impossible for you to go to an employer for an interview, explain the facts, express your regret, and suggest an alternative, such as being interviewed by a company employee in your vicinity.

If you receive no response from your application within a reasonable time, it may help to write a follow-up letter. In this letter you may inquire whether your application has been received and thus remind the employer of it. If possible, it is a good idea to add some new support to your application. This may be presented as a sort of afterthought. Sometimes the follow-up letter is just what is needed to secure you that extra consideration that results in a job.

## Conclusion

If you hired a man to act as your personal representative you would want him to be pleasing in appearance, businesslike and alert in manner, and intelligent in speech. A letter is your personal representative. Don't be satisfied with inferior specimens. This chapter will give you a start toward good letter writing. A special book on the subject will help you further. Intelligent practice will help you most of all.

### SUGGESTIONS FOR WRITING

1. If you are writing a library research report for your course in technical writing, you might attempt to supplement the library materials available by writing to several companies for information not obtainable in the library. Be sure, if you write such an inquiry, to explain why you are making it.
2. Write a letter of application, together with a data sheet. To get the most benefit from this exercise, you should make the application as realistic as possible: aim it at a job your present experience and training qualify you for. A summer job is a suggestion.
3. Write a letter of complaint to a manufacturing company about a defect in a mechanical device you have purchased. Then assume you are that employee of the company to whom the letter has been routed for reply and write an appropriate letter in response to the complaint.
4. When you have a report to write, prepare a letter of transmittal to accompany it.

# 19
# Writing for Professional Journals

Publication of articles in professional journals may benefit you in many ways. Such publication is likely to increase your circle of professional acquaintances; it is certain to put an example of your work in the hands of leaders in your field; and it will be a strong stimulus toward mastery of your area of specialization. It may also have direct effect on your advancement, for many firms strongly encourage their employees to publish. (See the accompanying chart from a General Motors discussion of research publication.)

Publication of semitechnical articles in popular magazines is also financially attractive. Technical journalism of this sort interests only a small minority of scientists and engineers, however, and for that reason we shall not discuss it here.*

The professional journals—by which we mean loosely any journal published for trained specialists—do not ordinarily pay for contributions. Nevertheless, it is by no means always easy to place an article with them. Usually these journals have many more articles submitted to them than they can possibly publish. You should not be surprised to have your offering rejected or returned with a request for revision. You should not assume that a rejection means you have written a poor

*If you are interested in this field, see Helen M. Patterson, *Writing and Selling Feature Articles*, 3d ed. (Englewood Cliffs, New Jersey: Prentice-Hall, Inc., 1956). See also F. Peter Woodford, *Scientific Writing for Graduate Students* (New York: Rockefeller Press, 1968) and John Mitchell, *Writing for Technical and Professional Journals* (New York: John Wiley, 1968).

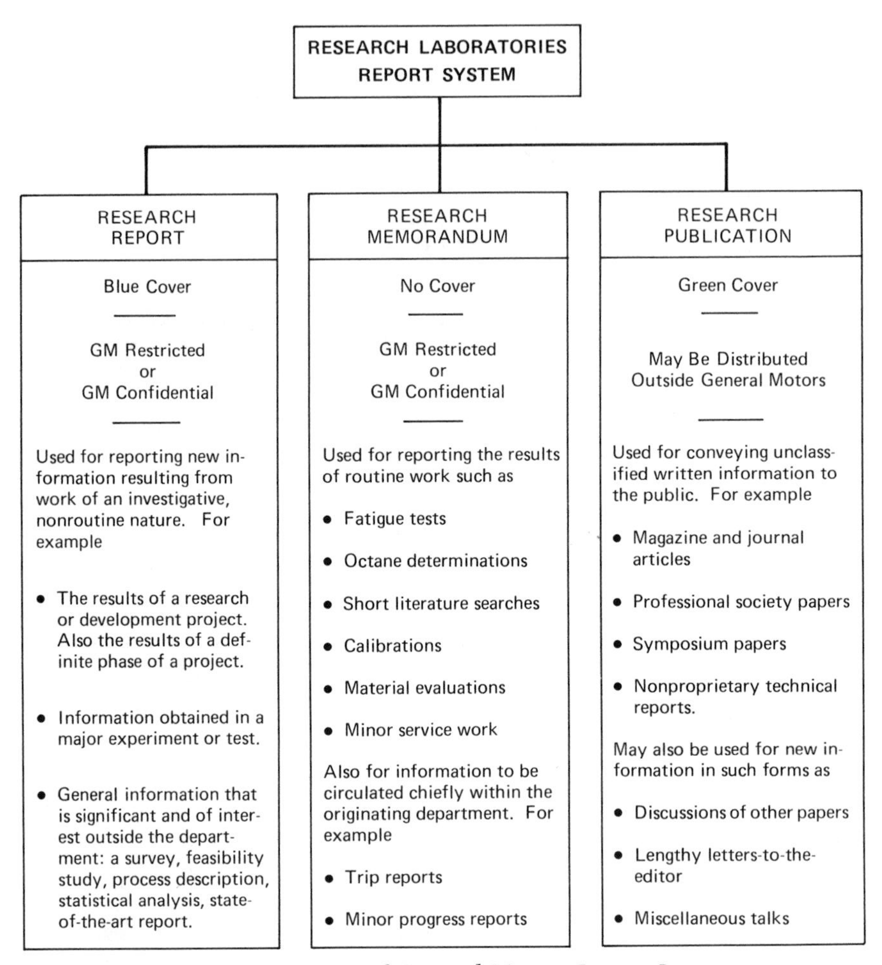


*Figure 1. Diagram of General Motors Report System*

article; an article may be turned down for a number of reasons which have nothing to do with its quality. Your article may be concerned with a subject which the editors feel has already been given all the space in recent issues that they can devote to it. It may be presented in a way that the editors feel would not interest their readers, although it might interest other readers (in this case they may suggest another journal or journals to you); or it may be an article that at another time would have been accepted but which is now rejected simply because the editors have on hand a large number of unprinted articles of high quality. And frankly, an article from an unknown author is less likely to be accepted than one bearing the name of a man with a nation-wide reputation. But don't assume that a man without a reputation should abandon hope: nothing could be farther from the truth. If you have something significant to say, and say it clearly, you are almost certain to have your work published. Don't forget the virtues of patience and common sense.

In this chapter we shall discuss the problems of choosing a subject, selecting a journal to send your article to, writing the article in a suitable style, and putting the manuscript into the proper form.

It is often difficult to say whether, in practice, a person selects a subject because he wants to write an article, or wants to write an article because he has a subject. Ideally, the latter would always be true. Publication would be considered only when a person's thinking and research had developed facts or theories which he realized might be of value to other workers. A famous fictional presentation of this ideal, and a satire of its opposite of seeking fame through shoddy, pretentious, over-hasty publication, is Sinclair Lewis's *Arrowsmith*. But, as Lewis said, human motives are seldom unmixed. Granted honesty and sincerity, there is little point in thinking much about whether desire for publication or interest in a subject comes first. In any case, there isn't much we can say here about your personal interests.

The only advice we can give here about how to find a subject, in contrast to evaluation of a subject, is to read widely in your field, to acquire a wide acquaintance among your colleagues, to attend meetings of professional societies, and everywhere to use your imagination and to be critical. Be slow to assume that an explanation is correct or that an accepted method is the best method. Out of such an attitude will come new ideas for research and publication.

When you have an idea that looks interesting, work at it. We once heard a well-known physicist say that one of the chief differences between creative and noncreative men in his field was that the noncreative men simply failed to develop their ideas. A good deal of determination and some stubbornness are called for. Do your own thinking, and allow in advance for opposition to new ideas. But be sure to distinguish between boldness in conceiving new ideas and carelessness in developing them. You should be patience itself in calculating, testing, checking, and criticizing an idea once you have gone to work on it.

It is wise to keep a file of possible subjects for investigation. Here are some questions to ask yourself in evaluating subjects for your file:

1. Is development of the idea within your present ability? Of course you will want to add to your knowledge and skills, but don't take on too much at once.

2. Is equipment available to you for the work that will be required?
3. Do you feel a real interest in the subject? Don't let circumstances coax you into work you don't care for when you could be doing something you'd like.
4. Will the subject open up further possibilities of research and publication, or is it a dead end?
5. Is the subject in a field that has received little recognition? Or is the subject in a field that is overworked? Either possibility may mean difficulty in getting your article published and the merits of your work recognized.
6. Will work on the subject contribute to your ability and success in the kind of career you desire? An article on the fuel cell would probably do little to further the career of a civil engineer.

Such considerations as these are very much worth your attention before you commit yourself to any project that will take more than a few days of your time.

Having decided to write an article, you must begin thinking about where to send it. The first step is to find out what journals publish material of the kind you will have. Of course you should be acquainting yourself with such journals anyhow; familiarity with them is an important part of your professional equipment. (For quides to help you locate the journals in your field, see Chapter 22.) A second step is to analyze the journals you have decided are possible targets. Can you find any articles in them, dating from the past year or two, which are on a subject comparable to yours? You needn't feel that you should find exactly the same subject, and of course you shouldn't expect to find an article that says approximately the same thing that you are going to say. This possibility of duplication, incidentally, brings up another matter. You should be very careful to look at every article that has been published anywhere on your specific subject, no matter how tedious the hunt may be, to make sure that you are not merely repeating somebody else's work, as well as to inform yourself fully on your subject.

When you have a list of the journals that show an interest in the kind of subject you have, it is wise to make your next step a conference with a man who has had considerable experience with professional publication. You may acquire invaluable information about editorial whims, possible places to publish the article, and the like. It is not uncommon for a beginner to achieve his first publication through the friendly help of an older, well-known man whose recommendation carries weight. If you work for a company that has a technical editor, or a staff of technical writers, it may be that you can get help from them. Many companies encourage their technical staff to seek such help.

Next, you should analyze the style of the journal you have chosen. This analysis should cover two elements—literary style and physical format.

Literary style is perhaps less important in a professional journal than in a popular journal, where appeal to a large, untrained audience demands a vivid presentation. Nevertheless, it is well worth your time to see if there are any special preferences or prejudices in style and general attitude that examination of numerous issues of the journal will reveal. Is the treatment theoretical or practical? Speculative or down to earth? Informal and colloquial, or formal and restrained? You will probably find considerable variety even within a single issue, but usually a distinctive tone will become evident as you read through several issues. Try to get the feel of the journal, and write your article accordingly. Remember that editors are human beings, and the problem of reader analysis is essentially the same in writing for an editor as in writing a routine report in college or on the job. All you know about the editor, however, may be what you can infer from analysis of the articles he has chosen for publication. As a matter of fact, articles are usually read by several people, usually two or three besides the principal editor who is responsible for the final decision.

Often it is a good idea to send a query to the editor of the journal you have chosen, asking if he thinks the idea you have for an article is promising. Here is what the editor and publisher of *Automotive Industries* has to say to prospective contributors on this point.

> Before preparing the final text, send AI editors a 150 word outline of your proposed text, stating the subject, the sources of information used, the problem that will be discussed, the illustrations available, the data or charts available, the benefits developed for readers—and your own name, address, occupation, title and business affiliation. Graduate engineers or scientists are also expected to mention their educational attainments and outstanding career achievements.
>
> When the editors have approved the outline, the next step is to accept and *incorporate in your outline*, the suggestions made by the editors.*

*Automotive Industries* is a trade journal. That is, it publishes articles on the subject of what might be called technical news about new techniques and new products. Another kind of journal is that which is devoted more definitely to published articles describing original research. An example is *Review of Scientific Instruments.* Generally

*Hartley W. Barclay, "Letter to Contributors of Technical Articles to *Automotive Industries*" (no date).

speaking, editors of trade journals are likely to be more receptive to queries about how promising an idea for an article is than are editors of journals publishing reports of original research.

Analysis of the physical format (form of footnotes, subheads, and the like) preferred by a journal is a simpler task than analysis of the literary style. Very often specific directions are available, and when they are not, the form of articles printed in the journal serves as a model. Some professional societies that publish journals issue pamphlets giving instructions on form. Some journals regularly print short statements about form. Examine closely the journal you are interested in, and if you find no hint of directions to be followed, pick out two or three articles and use them as models. Note particularly such matters as use of subheads, footnote and bibliographical forms, whether or not an abstract is used (and if it is, what type it is), types of illustrations, how numbers are written, and what abbreviations are used.

Your article should be typed, preferably with pica type, and double-spaced. You should make at least one carbon copy to keep for yourself. Some journals request that two or three copies of a manuscript be submitted so that the several editorial readers can read the article simultaneously. Good clear carbons are acceptable, and—in the carbon copies—illustrations can be roughed in or, if that is not feasible, a brief explanatory note can be substituted. Clear photocopies are also acceptable.

Here are some general suggestions about manuscript form.*

1. Use good paper of standard size (8½ by 11 inches).
2. Leave a margin of 1¼ inches at the left and 1 inch on the other three sides of the page.
3. Type your name and address in the upper right-hand corner of the first page.
4. Type the title of the article about halfway down the first page. Underneath, type "by" and underneath that your name, triple-spaced, like this:

```
ELECTRONICS


    by


John Warren
```

*For more detailed instructions see Sam F. Trelease, *Scientific and Technical Papers.* (Baltimore: The Williams and Wilkins Company, 1958.)

The empty space in the top half of the page is a convenience to the editor for making notes.

5. In the upper right corner of each page after the first page type the title of the article, followed by a dash and the page number. If the title is long, use an abbreviated form of it.
6. There are various ways of handling illustrations, but if you have no specific directions the following will be satisfactory. First, be sure you have put a clear title on every illustration (photographs, drawings, charts, graphs—and also tables) and, if there are several illustrations, a figure number. Next, instead of putting the illustrations into the text, collect them in an envelope at the end of the manuscript. To show where they go in the body of the text, write the figure number and the title in a blank space left in the appropriate place in the text. Finally, add to the collected illustrations a typed list, on a sheet 8½ by 11, of figure numbers and titles (this sheet should not have a page number). Further suggestions on illustrations will be found in Chapter 21, and additional suggestions on manuscript form in Chapter 20. Incidentally, remember that reproduction of illustrations is expensive, and professional journals do not always have an abundant supply of money.
7. Proofread your manuscript with painstaking care, particularly tables and graphs. This job is dull and time-consuming, but it is imperative that it be well done. Ask a friend to read aloud from the rough draft while you check the final copy. A page on which numerous corrections must be made should be retyped. If there are no more than two or three corrections on a page, however, it is permissible to make them neatly between the lines.
8. Mail the manuscript flat. Enclose some kind of stiffener, like heavy cardboard, if there are illustrations that would be seriously damaged by folding. It's a good idea to mark the envelope "Do Not Fold." Include in the envelope a self-addressed stamped envelope to bring back the manuscript if it is rejected.
9. Resign yourself to a long wait. You may learn the fate of your manuscript in six weeks, but it may take six months. If you've had no word in six months, an inquiry would not be out of order.

If your manuscript is rejected, mail it out again at once. But there are two things that should be done first. One is to make sure that the manuscript looks fresh. An editor is never flattered by a suspicion that his office isn't the first your manuscript has visited. The second thing is to consider whether any changes should be made in the article to adapt it to the policies and attitudes of the journal you now have in mind. You should be as careful about this on the second, third, or fourth mailing as on the first.

Of course your manuscript may be accepted the first time out, or it may come back with a request for revision. Whether or not to revise as requested is a matter to be settled between you and your conscience. Chances are that the editor is right. If you think he is wrong, and feel strongly about the matter, it may be better to seek publication elsewhere. In any case, don't be fussy about little things. Few editors can resist changing a few commas, at least.

When your manuscript has been accepted, there will probably arrive, in due time, some "proof sheets" or "galley proof." These are long sheets of paper on which the printed version of your article appears. Your job is to proofread these sheets and return them to the editor. You should again get someone to help you.

Corrections on proof sheets should be made with standardized "proofreader's marks." These marks, together with directions for their use, can be found in most good dictionaries. With a few exceptions, corrections should be made only of errors that the printer has committed, because of the expense of resetting type. On the other hand, if you discover that you have overlooked errors in grammar or in facts, you should certainly correct them.

You may or may not later on receive the corrected proof sheets to examine. If you do, the checking process should be conducted as meticulously as before, but—with very rare exceptions—only printing errors should be corrected.

Writing an article for publication in a professional journal is fundamentally like any technical writing. The principles of reader analysis, logical organization, and clarity of expression must be observed. There are some special problems: selection of a subject, choice of a journal to submit your article to, handling of the manuscript, and correction of the proof sheets. But these are all problems that can be solved by a methodical approach.

## SUGGESTIONS FOR WRITING

The most obvious suggestion about writing for professional journals is to write an article and send it to a professional journal. If you have an idea for such an article, fine. Give it a try. Some preliminary work might prove helpful. Select a few journals that seem appropriate and study them. Write out what you find concerning the kind of articles they print, the length of the articles, the readers they aim at, the format of the articles, instructions they provide for contributors, and anything else that is pertinent. It would be a good idea to send a query to the editor of the journal you select as your first target concerning his possible interest in your subject; request any suggestions or instructions he might care to offer. With this kind of investigation completed, you can write the article itself with far greater assurance than if you simply sit down and start writing.

If you are not interested in publishing an article, the sort of investigation just described can prove highly valuable as a means of getting well acquainted with those professional journals in your field you will want to make it a habit to read regularly. The advice of an experienced member of your profession would be helpful in selecting the journals.

# SECTION FIVE

# Report Layout

*The two chapters in this section discuss the format of reports and graphic aids. Format includes such "mechanics" of report preparation as the arrangement of a title page or the placement of subheads. Graphic aids refers to any nontextual device included in a report: a photograph, a table, or a chart. The subject matter of the two chapters is similar in that both deal with the problem of visual effect.*

# 20
# The Format of Reports

## Introduction

If you were to make a careful survey of the format of reports prepared by a representative number of companies, you would observe two facts: (1) although all companies do not use the same format, the differences are likely to be minor ones of detail; (2) all companies and organizations agree that attractive format is necessary.

While accuracy and clarity are always of paramount importance, remember that a report makes an impression on its reader even before he has an opportunity to determine whether its contents are accurate and clear. A well-known engineer once told a story of visiting an industrialist's office and seeing the industrialist pick up a handsomely bound report just as it was delivered to his desk, leaf through it, and remark that it was a fine job of engineering report writing. He hadn't read the report; he made this judgment solely on the basis of its appearance.

Common sense will tell you that it pays to make your reports look good. The question is not whether attractive format is desirable, but what *is* attractive format. The following pages, therefore, will be devoted to a discussion of typescript standards and the form of the elements of a report, plus some notes on the relationship of form to organization and style. We want to say, before presenting the "rules" which follow, that no body of rules exists for report format which could be regarded as authoritative the country over. The ones we present are representative of good practice, however, and will be acceptable whenever you do not have other instructions.

## Typescript Standards

When you prepare a typewritten report, you will have to make decisions regarding the choice of paper, width of margins, spacing and indenting, and paging.

*Paper.* Reports should be typewritten on white paper of high quality, preferably 20-pound bond, 8½ by 11 inches in size. Second sheets, if carbon copies are to be made, should be of high quality, rag content white bond, about 13-pound weight. A good quality paper is essential if a neat, attractive copy is to result; inking and erasures, for instance, require good paper. Always use white paper unless you have instructions to the contrary. Some companies use colored sheets to identify certain types of reports or reports from certain departments.

*Margins.* Margins for the typewritten report should be approximately as follows:

| *Left Side* | *Top* | *Right Side* | *Bottom* |
|---|---|---|---|
| 1½″ | 1″ | 1″ | 1″ |

Since the left-hand margin must be wide enough to allow for binding, up to 2 inches may be needed, depending upon the nature of the binding. No reader likes to be forced to strain the binding or his eyes in order to read the words on the bound side of the sheet. The right-hand margin cannot, of course, be kept exactly even on account of the necessity of dividing words properly, but an effort should be made to keep a minimum margin of ¾ inch.

Where quotations are introduced into the text of a report, an additional five spaces of margin must be allowed on the left side and approximately that on the right.

*Spacing and Indenting.* The text of a report should be double-spaced throughout, except as noted below:

1. Triple- or quadruple-space below center headings.
2. Single-space and center listings (if items are numerous, number them).
3. Single-space long quotations—those which run four or more lines in length.
4. Triple-space above and below quotations and listings.
5. Single-space individual footnotes more than a line long; double-space between notes.

6. Single-space individual entries in the bibliography; double-space between entries; use hanging indention in bibliographical entries of more than one line in length.
7. Single-space the abstract if space demands it; otherwise, double-space it.
8. Usually, single-space material in the appendix.
9. Double-space above and below side headings.

The customary indention at the beginning of a paragraph is five spaces. An additional five spaces (or more if necessary for centering) should be allowed before beginning a listing.

*Paging.* Use Arabic numbers in the upper right corner, except for prefatory pages and the first page of the body, and pages which begin new divisions. The number should be in alignment with the right-hand margin, at least two spaces above the first line of text on the page, and about 3/4 inch down from the top edge. The prefatory pages of a report—title page, letter of transmittal, table of contents, list of figures, and abstract—should be numbered with lower-case Roman numerals centered at the bottom of the page, about 3/4 inch from the bottom edge. It is customary to omit the numbers from the title page and the letter of transmittal, although these pages are counted; thus the table of contents becomes iii. In the body of the report, it is customary to omit placing the number 1 on the first page, since the title there obviously identifies it as page one. As for pages which begin main sections of the report, it is probably best to place the number in the bottom middle of the page. Pages of the appendix are numbered as in the body, in the upper right corner. No punctuation should follow page numbers.

## Formal Report Format

By formal report we mean the conventional "full dress" report, with all or nearly all of the parts which will be described below. Informal reports do not possess all the parts usually included in the formal report and thus present a somewhat different problem so far as format is concerned. Informal reports will be considered later in this chapter.

*The Cover.* Ordinarily you will not have to worry about making up a cover for your report, for most companies have prepared covers. These are made, usually, of a heavy but flexible paper, with a printed heading naming the company and the division, and with a space for information about the report itself. This information consists of (1) the title, usually prominently displayed in underlined capital letters,

(2) the report number, and (3) the date. Sometimes this information is typed on the cover itself; sometimes it is typed on gummed slips and pasted on, and frequently a window is cut in the cover so that the title block on the title page will show through. In any case, the title should be clearly legible. Triple spacing between the lines of two- and three-line titles is advisable.

Occasionally the name of the client to whom a report is submitted and the name or names of its authors may be found on the cover, but as a general rule only the three items of information mentioned above are recorded. These serve to identify the report for filing and reference; additional information may lessen the prominence of these important facts and detract from the attractiveness of the layout.

If prepared covers are not supplied, you can use a plain Manila folder or one of the readily available pressboard binders. Many companies have covers made of special stock, with an identifying picture or symbolic device.

*The Title Page.* Besides duplicating the information found on the cover, the title page gives a good deal more. The most significant additional information presented here is the name of the person or persons who prepared the report and an identification of them as to position in the company or organization. In addition to authorship, the title page of reports from many industrial concerns provides space for the signatures of those who approve, check, and (sometimes) revise the report. Some companies provide space for "Remarks" of those who check the report. Finally, it is not unusual to find a notation of the number of pages of the report. The accompanying illustration (Figure 1) is fairly typical. Note that the title appears in underscored capital letters, centered about one-half of the way down from the top of the page.

Should you be required to write a report for a company which does not provide a prescribed form for the title page, you will find the model given as Figure 2 satisfactory. Note that it contains four elements attractively grouped and spaced. The title appears in the upper third of the sheet, underscored and centered, with triple spacing between the lines. Centered on the page appears information about the recipient of the report. On the bottom third of the page appears the reporter's name and professional identification; the last entry is the date of submission. In centering material on the page, do not forget to allow about half an inch for binding.

*The Letter of Transmittal.* Since we discuss the letter of transmittal at length in another place (see Chapter 18), we simply want to

REPORT NO.

NO. OF PAGES

**McDONNELL AIRCRAFT CORP.**

LAMBERT — ST. LOUIS MUNICIPAL AIRPORT
ST. LOUIS, MISSOURI

**ENGINEERING DEPARTMENT**

AERODYNAMICS REPORT

FOR

McDONNELL MODEL 56 AIRPLANE

SUBMITTED UNDER CONTRACT

PREPARED BY APPROVED BY

CHECKED BY

DATE

REVISIONS

PAGES AFFECTED REMARKS

*Figure 1. Title Page, Example 1 (Used by Permission of the McDonnell Aircraft Corporation)*

A Report

on

<u>COMBATING THE STALL PROBLEM</u>

Prepared for

The Director of Research
Wakey Products, Incorporated
Detroit, Michigan

by

Richard Morrison
Aeronautical Research Assistant
September 10, 1970

*Figure 2. Title Page, Example 2*

point out here that this part of the report should be meticulously accurate in form and layout. Although it usually appears immediately after the title page, some companies require that it appear as the first item in the report, just inside the cover (or even stapled onto the outside of the cover). Sometimes the letter of transmittal does not form a part of the report at all but is sent separately through the mails. And sometimes the functions of the letter of transmittal are performed by a foreword.

*The Table of Contents.* The table of contents of a report is an analytical outline, modified in form for the sake of appearance. It serves as an accurate and complete guide to the contents of the report. The entries in this outline also appear in the text of the report as headings; thus a reader may easily refer to a particular section or subsection of the report. Every heading in the outline must appear in the text as a heading or subheading. It is not necessary, however, that every subheading in the text appear in the outline.

Except for the Roman numerals for main division headings, the conventional outline symbols (A,B,C, . . . for subdivisions) are omitted in the table of contents, indention alone being used to show subordination. Omitting the capital letters and Arabic numerals results in a neater page. But although this is majority practice, it is by no means unanimous. Some companies retain all of the conventional outline symbols; some omit all of them. Whether they are retained or not, it is a good idea not to clutter up a table of contents with minute subdivisions: three levels are enough (this is not intended to suggest, of course, that your *plans* should not be detailed). Examine the accompanying examples in Figures 3, 4, and 5.

You will note that all of the specimens have this in common: they provide plenty of white space so that the prominently displayed headings may be easily read and so that the page as a whole presents a pleasing appearance. For the best layout of the page, follow these suggestions:

1. Center and underscore "Table of Contents" at the top of the page. Use either capitals or lower-case letters.
2. Triple- or quadruple-space below the centered "Table of Contents." Double-space between the major items in the contents. If there are numerous subtopics, they may be single-spaced.
3. Begin items preceding Roman numeral I flush with the left margin. These items include the List of Illustrations, Symbols, List of Figures, and Abstract.
4. Indent second-order headings five spaces and third-order headings ten spaces.

Table of Contents



*Figure 3. Table of Contents, Example 1*

Table of Contents



iii

*Figure 4. Table of Contents, Example 2*

ABC LABORATORIES

PRODUCTS APPLICATION DEPARTMENT
GREASE AND INDUSTRIAL LUBRICANTS TESTING GROUP

DETERMINATION OF SUITABLE ROTOR BEARING
GREASE FOR USE IN THE BROWN MAGNETO

INDEX



INDEX TO TABLES



INDEX TO FIGURES



*Figure 5. Table of Contents, Example 3*

5. Use a row of periods to lead from the topic to the page number at the right margin.
6. After the last Roman numeral entry, list items in the appendix. Place the word "Appendix" flush with the left-hand margin, as shown in Figure 4, and indent the individual entries. The bibliography comes first. If nothing besides a bibliography is to be appended, do not use the word "Appendix": place the word "Bibliography" where the word "Appendix" would otherwise appear. If several appendixes appear, it is common practice to label them "A," "B," "C," and so on, and to give each a title written in all capital letters, or underscored—as with other section titles.

*The List of Figures.* If a report contains a half-dozen or more illustrations, drawings, exhibits or the like, an index to them should follow the table of contents. Usually called "List of Figures," this page gives the number, title, and page reference of each figure in the report. See Figure 6.

The actual layout of the page is simple. Center the title at the top of the page (allowing for top margin) and underscore it. Triple- or quadruple-space before beginning the list. Figure numbers should be aligned under the word "Figure" and followed by periods. The initial letter of each important word in the titles of figures should be capitalized. Page numbers should be aligned at the right margin, with a row of periods connecting title and number. Double-space between entries, but single-space an individual title requiring more than one line (and remember that a line should not be carried all the way over to the right margin). This spacing will allow for plenty of white space—a requirement for a neat, attractive page. See Figure 6.

There are, of course, some variants of this form. Some companies like to classify nontextual material so that there is, besides a list of figures, an index to tables and perhaps even a list of photographs. Separate pages are not needed for these individual listings unless the length of the listings requires them. When the table of contents is quite short and there are few illustrations to list, both the table of contents and the list of figures may be placed on one page (Figure 5). Informal, short reports containing fewer than five or six illustrations usually omit a formal list. Custom in an organization will dictate whether omission of the list of figures is permissible.

*The Abstract.* The word "Abstract" should be centered and underlined at the top of the page. Allow triple or quadruple spacing after this title and then double-space the text of the abstract itself,

List of Figures



iv

*Figure 6. Layout of List of Figures*

maintaining the same margins used for the body of the report. In some cases, where space is at a premium, the abstract may be single-spaced, but double spacing is better.

You may have observed that the term "Abstract" is not universally used; some companies call this part of the report a "Digest," some a "Summary," and some an "Epitome." Whatever it is called, format requirements are the same. For what goes into the abstract, see Chapter 4 on outlines and abstracts.

*Headings.* The topical entries of the outline table of contents also appear in the text of a report as headings which identify the individual portions of the subject matter. All of the entries in the table of contents should appear as headings in the text, and they should appear exactly as they are worded in the table of contents. The headings serve as transitional devices and enable a reader to find a specific part of a report's discussion with ease. We are concerned with the form and location on the page of the three types: main or center headings, and two types of subheadings.

*Main or Center Headings.* Main headings name the major divisions (Roman numeral divisions) of a report. Written in either lower-case letters or capitals, a main division heading is underlined and placed in the center of the page, with the Roman numeral preceding it, as "II. *Circuit Elements and Transmission Lines.*" See Figure 7. In formal reports, it is customary to begin new divisions of a report on a new page, just as a new chapter in a book begins on a new page. The centered title should stand a minimum of three lines above the first line of text of the division or the first subheading.

*Subheadings.* Usually only two levels of subheading are needed beyond the main headings: that is, headings corresponding to capital letter and Arabic numeral divisions, respectively, in the outline.

The capital letter or second-order headings should be placed flush with the left margin. Underline each word separately. Use lowercase letters, but capitalize the initial letter of each important word. Double-space above and below the heading, and do not put any text on the same line as the heading. Don't put any punctuation after the heading.

The Arabic numeral, or third-order, headings should be handled exactly like the second-order headings with three exceptions: (1) indent the heading five spaces; (2) put a period after the heading; (3) start the text on the same line as the heading. See Figure 7.

If it is necessary to use fourth-order headings, treat them like third-order headings but number them with Arabic numerals.

A variant custom of identifying or numbering sections and subsections is the use of the Arabic decimal system. With this system, the first main division topic is headed with "1," the second with "2," and so on. Subdivisions are headed "1.1, 1.2," and so on. Sub-subdivisions may be headed "1.1.1, 1.1.2," and so on. This system can be-

II. Circuit Elements and Transmission Lines

In order to understand the problems and principles of the operation of carrier circuits, it is first necessary to understand the characteristics of circuit elements and transmission lines at both power and carrier frequencies. The most common elements are resistors, capacitors, inductors, transformers and sections of transmission lines.

Circuit Elements

Different circuit elements have separate and distinct properties. Because of this difference, resistors, inductors, capacitors, and transformers will be considered separately.

Resistors. A resistance is an energy-absorbing element. Although the value of the resistance of a material does not vary with frequency, the effective resistance of a resistor or section of wire varies because of the skin effect, or the movement of the current to the outer edges of the conducting resistance. Another important characteristic of a resistance is the phase relation between the current through and the voltage across a resistor. The current is directly proportional to the applied voltage and there is no time lag between a change in voltage and a change in current.

Inductor. An inductor is a circuit element that has an impedance to the flow of current but absorbs no energy. When the voltage across an inductance is

*Figure 7. Layout of Headings*

come cumbersome (indeed, we once saw a long report with one obscure heading prefaced by eleven digits); but it is used by a good many organizations and it does lend itself to convenient reference. A correspondent, for example, may refer his reader to "section 4.2.1" of a document. Figure 8 shows a suggested breakdown for a section on equipment installation, together with the decimal numbering we have been talking about.

---

CHAPTER 2

INSTALLATION

2.1 *Introduction*

2.2 *Unpacking and Inspection*

2.3 *Preparation*

or

2.3 *Installation*

2.3.1 Mechanical

2.3.2 Electrical

2.4 *Cables*

2.5 *Accessories*

2.6 *Fuses*

---

*Figure 8. Illustration of Decimal-System Breakdown of Topics*

*Quotations and Listings.* Formal quotations are single-spaced, indented five spaces from the left margin and approximately the same number from the right. Quotation marks are unnecessary; single spacing and extra margin adequately identify the material as a quotation. If the quotation does not begin with the first word of a sentence in the original, the omission of words should be shown by a series of three periods, and any deletion within the quotation should be similarly indicated. Triple-space above and below the quoted matter.

Informal, short quotations a sentence or less in length should be run in with the text. As Gaum, Graves, and Hoffman say, "Every quotation, therefore, must be so set off from the text that its nature is unmistakable."*

Formal listings, such as a numbered list of the parts of a device, are mentioned here because they are handled very much like formal quotations: indented an extra five spaces and single-spaced. The list of rules below, under the next heading, illustrates the form.

*Equations and Formulas.* If you find it necessary to present equations in the text of a report, the following "rules" should be observed:

1. Center each one on a separate line.
2. If more than one line in length, the equation should be broken at the end of a unit, as before a plus or minus sign.
3. Place all of an equation on a single page if possible.
4. Allow three to four spaces above and below or even more if it is necessary to use symbols of more than letter height: $\int$, for example.
5. Use no punctuation after the equation.
6. Number equations consecutively in parentheses at the right margin.
7. If necessary, define symbols used.

Study the following illustration, adapted from a Civil Aeronautics Authority report**:

> It is shown that the panel penetration velocity, where failure occurs in the butyral plastic interlayer, varies approximately as the logarithm of the plastic thickness. This can be expressed by the equation
>
> $$T = Ke^{v/c} \qquad (1)$$
>
> where
>
> $T$ = thickness of vinyl plastic in inches,
> $v$ = penetration velocity of windshield panel in mph,
> $K$ and $c$ = constants.

*Carl G. Gaum, Harold F. Graves, and Lyne S. S. Hoffman, *Report Writing*, 3d ed. (Englewood Cliffs, New Jersey: Prentice-Hall, 1950), p. 162.

**Pell Kangas, and George L. Pigman, *Development of Aircraft Windshields to Resist Impact with Birds in Flight*, Part II, Technical Development Report No. 74 (Indianapolis, Ind.: Civil Aeronautics Administration Technical Development), p. 13.

## Informal Report Format

The terms "informal report" and "formal report" are vague, and are descriptive of a tendency rather than of an exact format. In general, form reports, letter reports, and reports designed for circulation only within an organization are called informal. The term usually denotes a short report, say fewer than ten pages, but company practice dictates whether the format of a formal or informal report be used. You will probably have no trouble finding out which you should use in a specific situation.

A typical informal report has no cover, no letter of transmittal, no title page, no table of contents, and no list of illustrations. If there is an abstract, it appears on page one, preceded by the title and the author's name and followed immediately by the text. The text is usually single-spaced. An example of an informal report is given in the Illustrative Material for Chapter 14.

With the exceptions just noted, the suggestions for the format of a formal report also apply to an informal one.

The format of a letter report is simply the format of a business letter except that headings may be used in the text after the first paragraph. The system of headings previously described is satisfactory. Besides the conventional block form letter, however, a modification of the military letter form is frequently used for informal reports (see Figure 4 in Chapter 18). This form calls for "From," "To," and "Subject" caption lines and numbered sections or paragraphs. One reason the military form is especially favored for interoffice or interdepartmental memoranda is that the forms may be conveniently printed.

## A Final Note: Relation of Format and Style

There is a problem as to whether a well-planned format can perform certain functions that are usually performed in the text. For example, does a table of contents in a report make it unnecessary to say anything about plan of development in the introduction? Does an abstract make it unnecessary to mention scope in the introduction? Does the use of a system of subheadings make transitions unnecessary?

The popularity of form reports clearly indicates that format can take over certain textual functions if we stretch the term "format" to include the detailed headings printed on a form report blank. In fact, the form report, which is an extreme case of the development of format, indicates both the potentialities and the limitations of the principle of assigning textual functions to format. An intelligently designed form report blank, when filled out by an intelligent man, is highly efficient. That is its strength. Its weaknesses are two. First, it can deal with only a limited number of situations. When something unusual happens, the report writer starts adding explanatory notes. The more initiative he has, and the more unusual the situations he encounters, the less useful the form report becomes. Second, the form report is impersonal. It gives the writer almost no opportunity to make himself felt as a human being. Perhaps it will help you to understand what we mean if you will try to imagine yourself attempting to present, in a form report, a persuasive statement of the advantages of a device you've just invented!

The point of these remarks is this: Yes, sometimes a table of contents makes a statement of plan in the introduction unnecessary, and sometimes a subhead is a sufficient transition; but the further you go toward letting the format take sole responsibility for such functions, the closer you are getting to the form report, which is efficient within a limited range but is neither particularly pleasing nor persuasive. In short, we urge that you recognize the many advantages of a clear and attractive format, but we also urge that you avoid letting it lull you into writing a careless text. Eight times out of ten, you should write a transition even where there is a subhead; you should state the plan of development even when there is a table of contents; and you should clarify the scope even when there is an abstract. If you really want to be understood, try to communicate with every means at your command. To be even more specific about one question that often arises, we might add: Do not be concerned about duplication in the content of the letter of transmittal, the abstract, the table of contents, and the introduction. Such duplication is entirely acceptable and is the common practice.

## SUGGESTIONS FOR WRITING

We have offered a number of suggestions in this chapter for handling various "formal" aspects of report presentation. We have also stated or implied that there is nothing absolute about our suggestions because practice throughout the technical and scientific community is not standardized: different organizations and companies have developed or evolved different formats to suit their own needs. With this in mind, we make the following suggestions for writing.

1. With the help of some of the publications discussed in the chapter on finding published information, locate the library's file of reports. Check out some of these and examine the format (or formats)

used. Write a report to your instructor in which you tell how the formats examined differ from that specified for reports in your technical writing course. The material in this chapter can serve you as a guide to what to look for.

2. Write a brief description of the prescribed format (and organizational plan) for laboratory reports in one of your science or engineering courses. Write the description as if for a student who will be taking the course some time in the future.

3. If there is a research laboratory associated with your school in which government contract research is carried out, see if you can find out what report format researchers are obliged to follow in reporting on their research. Write a brief account of what you find for your technical writing instructor. (You can consult the faculty and staff directory to find the telephone number and address of such research facilities.)

# 21
# Graphic Aids

## Introduction

In this chapter our general purpose is to introduce the extensive and important subject of graphic aids. More specifically, our purpose is to discuss some of the commoner varieties and functions of graphic aids and to consider elementary problems in their construction, exclusive of problems associated with their reproduction. Because the subject of graphic aids is so extensive, we strongly urge you to consult the pertinent volumes listed in Appendix A.

The graphic aids discussed in this chapter are charts, drawings and photographs, and tables. The term "chart" covers a broad field, however, which will actually occupy most of our attention.

Before entering into a discussion of the particular types mentioned, we must note two problems in the selection and use of any graphic aid: (1) differentiating between dramatic emphasis and communication, and (2) establishing the proper relationship between the graphic aid and the text.

All graphic aids communicate facts to the reader, but some communicate with much more precision than others. This difference can easily be seen by comparing a curve carefully plotted on coordinate paper with the pictograph often found in newspapers. You might imagine, for example, that a newspaper has indicated the number of workers in a certain industry by a series of drawings of identical overalled men, each man representing 5000 workers, except the last man, who is worth only 3000 and consequently lacks part of the left side of his anatomy. Such a pictograph may be dramatic, but it is not precise. A curve plotted on coordinate paper, on the other hand, can be fairly precise in communicating information. For a technically trained reader, it may also be dramatic, but the dramatic element is a secondary, rather than a primary consideration. This difference between precise information and dramatization, qualified by reference to the intended reader, should always be noted in selecting a graphic aid.

Our second general problem is how to establish the proper relationship between the graphic aid and the text. Practically, this usually means deciding how much to say about the graphic aid, and deciding where to put it.

Our experience has been that writers often go to extremes in deciding how much to say. One writer will repeat in words practically everything that is shown in a graphic aid, and another will not even note that he has used one. If you question the second man, he will tell you that it's all there in the graph, why should he have to talk about it? You will have to make up your own mind as to which of these offenders is the worse. We suggest that you note your reactions on this point as you read various technical materials. You will probably find yourself most nearly satisfied when the following three practices are observed:

1. If a graphic aid has some bearing on a conclusion to be drawn, no matter how simple, a reference is made to it in the text. An aid used solely for aesthetic or "dramatic" purposes need not be mentioned.
2. The significant points shown by an "informational" graphic aid are commented on in the text, but minor details are not mentioned.
3. Some directions are given on the reading and interpretation of a complex graphic aid. What "complex" means depends on the reader.

Finding the most effective location for a graphic aid is usually a simple matter. Informational aids that have a direct, immediate bearing upon conclusions or arguments presented in the text are usually located as close as possible to the pertinent portions of the text. Informational aids of a more general, supporting character are put in an appendix, unless they are so few in number as to offer no serious interruption to the reading of the text. Aids used to dramatize are placed at appropriate points in the text. In general, graphic aids that belong in the text are likely to represent derived data; in the appendix, original data.

If the aid is small enough, it may be placed on a page on which text also appears. Usually it has a border. Larger aids should be put on a separate page. In a typed manuscript they may be bound on either the right or the left edge. If comments on the aid are pretty well concentrated on one page, the aid should be bound on the right edge so that it may face the comment. If there are several pages to which the aid is pertinent, it may be wise to bind it on the left edge and locate it near the beginning of the comments (or place it in the appendix). A page occupied solely by a graphic aid is given a page number if it is bound on the left edge but is not given a page number if it is bound on the right edge.

In connection with the preceding discussion, you may find it helpful to study the use of the tables in the second report quoted at the end of Chapter 9, and the figures in two of the reports quoted at the end of Chapter 6. Some eminently practical suggestions about use of graphic aids may be found in Appendix D.

## Charts

*Introduction.* Charts, or graphs, are a means of presenting numerical quantities visually so that trends of, and relationships among, the numerical quantities can be easily grasped. Although a chart does not, in most respects, permit as accurate or detailed a presentation of data as a table, it has the advantage of making a significant point more readily and in a manner that is more easily remembered. The basic kinds of charts are the line or curve chart, the bar or column chart, and the surface chart. Additional varieties are the circle, or "pie," chart, the organization or line-of-flow chart, and the map chart. Each of these varieties will be discussed. First, however, we must review briefly some elements of chart construction. The elements to be discussed are the scales, the grid, the title, the scale captions, the source reference, and labels or a key.

Figure 1 illustrates the fundamental parts of a chart. Although it is a line chart, it could be easily converted into a bar or column chart by filling in a column from the base line up to the value for each division of the horizontal (abscissa) scale, and to a surface chart by shading the area beneath the line connecting the plotted points.

Most charts have only two scales, a horizontal (often called the abscissa) and a vertical (or ordinate). Typically, an independent variable is plotted on the horizontal scale, and a dependent variable on the vertical. Thus, if we were graphing the temperature rise of an electric motor, we would plot time on the horizontal scale and temperature on the vertical. Ideally both scales should begin at zero, at

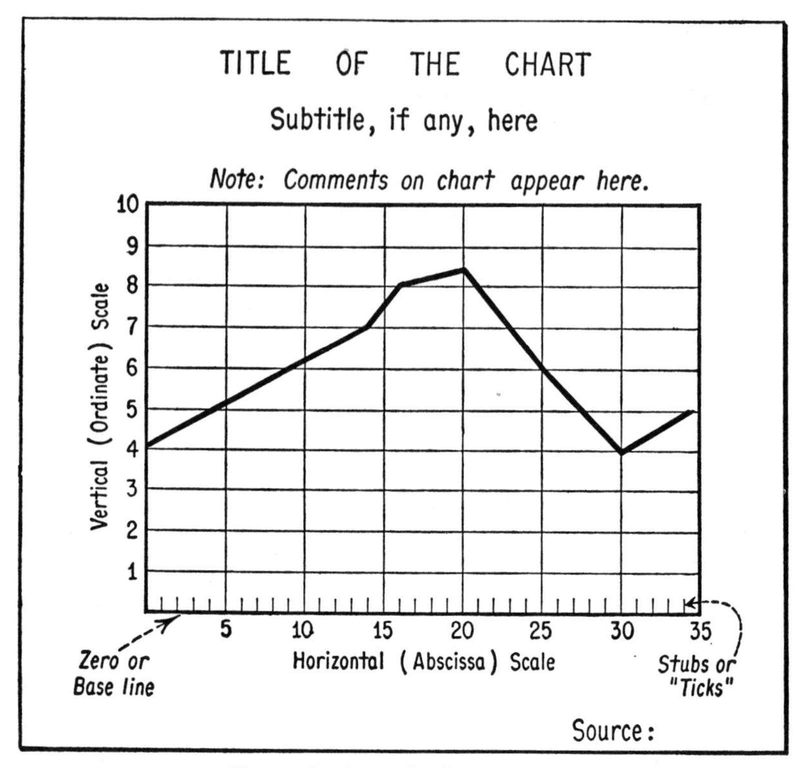


*Figure 1. Typical Chart Layout*

their point of intersection, and progress in easily read amounts, like 5, 10, 15, 20. Failure to observe either of these last two principles increases the possibility that the reader will misinterpret the chart. Often, however, the scales cannot be started at zero. Suppose that values on the vertical scale, for instance, begin at a high numerical range, as in plotting temperature changes above 2000 degrees Fahrenheit. It would be impractical to begin the vertical scale at zero if intervals in the scale beyond 2000 are to be small. In such a case, it is occasionally desirable to give the base line a zero designation and place a broken line between it and the 2000-degree line to indicate the gap in the numerical sequence of the scale.

Much of the visual effectiveness of a chart depends upon the proper slope or height of the line or bar or area plotted. The idea of movement and trend is emphasized by steepness and minimized by flatness. The U.S.A. Standards Institute suggests that an angle of slope over 30 or 40 degrees in a curve is likely to be interpreted as being of great significance, whereas an angle of 5 degrees would be regarded as of little significance. It is often difficult to satisfy all the ideal requirements: that is, the proper slope or height, an easily read scale, ample room for scale captions, and a little space between the highest point of the curve or bar and the top of the grid. (These last two points, not previously mentioned, are illustrated in Figure 1.) Sometimes it is desirable to use the long dimension of the coordinate paper for the horizontal scale, to meet the above requirements. If this method still does not solve the problem, larger paper should be used, and a fold or folds made so that the folded chart, when bound into the text, will come somewhat short of the edges of the pages of the text. If you construct a grid yourself, you should if possible use "root-two" dimensions (ratio of about 1:1.5) for the rectangle formed by the grid. Such dimensions are aesthetically pleasing. (In a precise root-two rectangle, the long side is equal to the diagonal of a square made on the short side.) However, this advice must be qualified by observance of pleasing proportions between the shape of the grid and the shape of the page.

In general, you should use coordinate paper with as few grid lines per inch as the necessary accuracy in reading will permit. The purpose of the chart—the degree to which it is informational—and the probable error in your data determine the accuracy with which it should be readily possible to read the chart. Sometimes the use of stubs or "ticks," as shown in Figure 1, provides a good compromise between the precision afforded by numerous grid rulings and the clarity and force of fewer rulings. In a bar chart, the grid normally has only horizontal rulings if the bars are vertical, and vertical rulings if the bars are horizontal.

The title of a chart may be placed either at the top or the bottom. Usually, but not invariably, it is placed outside the rectangle enclosing the grid. If there is a figure number, it should appear either above or to the left of the title. In using 8½ by 11 co-ordinate paper you will often find it necessary, because of the narrow margins, to draw the axes an inch or so inside the margin of the grid to provide space on the grid itself for the title, the scale numerals, the scale captions, and the source reference if there is one.

The scale captions need no particular comment except that they should be easy to understand. Sometimes the whole effect of a graphic aid is spoiled by one ambiguous scale caption. See Figure 1 for illustration of the placement of captions. Don't forget to note units, like amperes or milliamperes, where they are necessary. The scale numerals or values are written horizontally if space permits.

Source references for graphic aids are written generally in the same manner that text source references are (see Chapter 23). More abbreviation is permissible in the reference to a graphic aid than in a footnote reference, however, because of the need to conserve space. Any abbreviation which will not confuse the reader is acceptable. The placement of the source reference is shown in Figure 1.

It is often necessary to use labels (Figure 2) or a key to identify certain parts of a chart, such as bars or curves representing various factors or conditions. Labels often appear in a blank area with a "box" or border around them, but this is not always possible or necessary. If you are using commercially prepared coordinate paper, it may be helpful to put a box around the label even though there is no white space left for it. Labels for bars may be written at the end of the bar or, if there is no possibility of confusion, along the side. In circle, or "pie," charts, the labels should be put within the individual segment or alongside the "slice" in easily readable form. A "key" or "legend" is simply an identification of symbols used in a chart (see Figure 2, top and lower right). Another element occasionally found is a note, usually in a box on the grid, about some aspect of the chart.

We turn now to consideration of the types of charts.

*Line Charts.* Of all charts, the line chart (Figure 1) is the most commonly used. Simple to make and read, it is especially useful for plotting a considerable number of values for close reading or for plotting continuous data to show trend and movement. It is usually not as good as the bar chart for dramatic comparisons of amount. For making comparisons of continuous processes, however, the use of several curves on the same chart (Figure 2) makes the line chart superior to the bar chart. An illustration of this point might be seen in a chart in which the plate current of a triode tube is plotted on the vertical scale against the grid voltage on the horizontal. For different values of plate *voltage*, the relationship between the plate current and grid voltage is different; therefore, if one curve is drawn for each of several different values of plate voltage, the result of such changes is very effectively shown.

In a multiple-line chart of the kind just mentioned, labels are often written along the sides of the lines, without boxes. When lines intersect, the lines may be broken in various ways (dotted and dashed) to help in differentiating them; or colors or symbols with an accompanying key may be used (Figure 2). Particularly when lines intersect, you should be careful not to put too many lines on a chart, nor too many within a small area of the chart. The latter problem can of course be alleviated somewhat by the use of an appropriate scale on a large sheet of paper. If comparisons are to be made between different charts, the scales used on the charts should be identical.

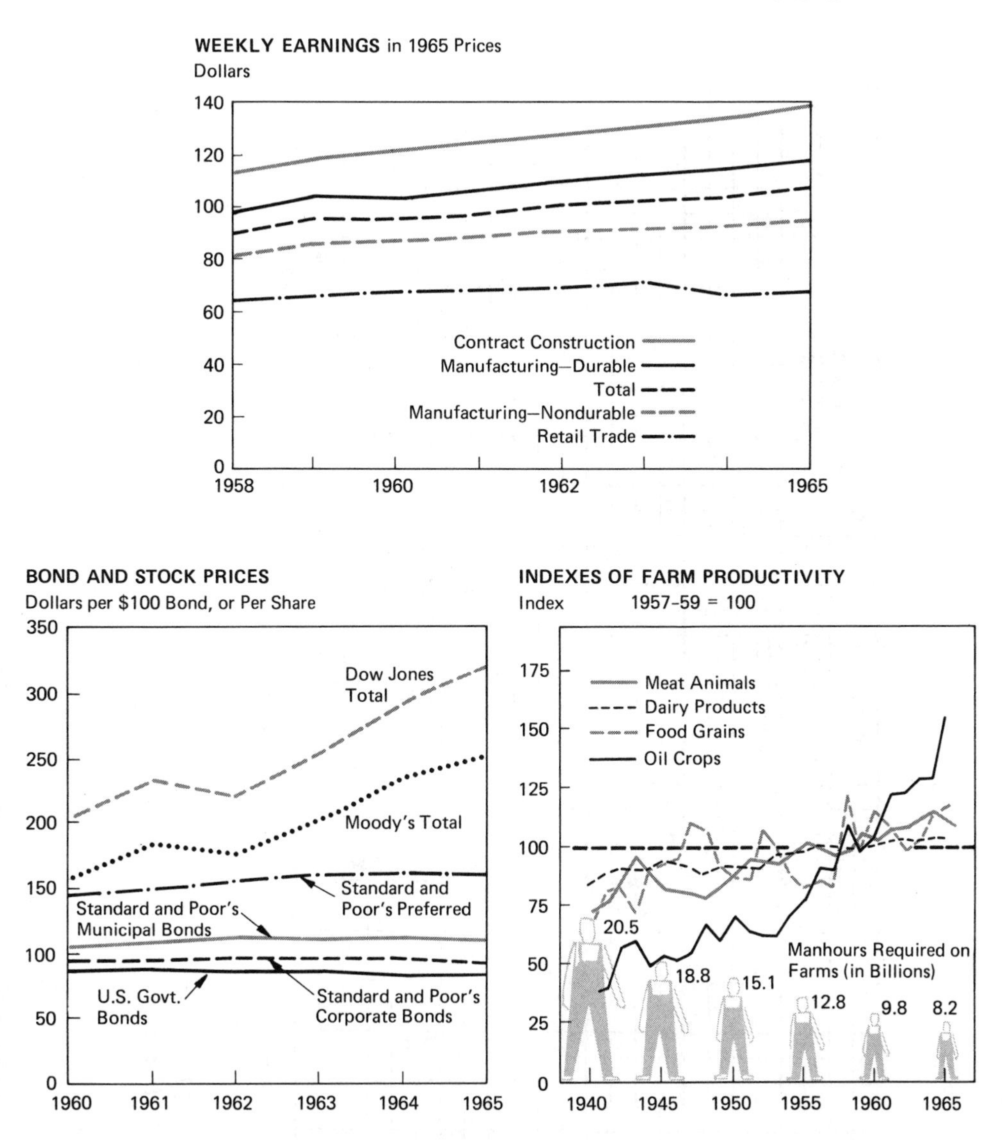


*Figure 2. Some Typical Multiple-Line Charts with Accompanying Keys. Source: U.S. Bureau of the Census,* Pocket Data Book, USA, 1967 *(Figs. 10, 20, 26).*

Another problem in either single- or multiple-line charts is whether the line connecting points plotted should be drawn straight from point to point or smoothed out (faired) (see Figure 3). If you are showing the trend of a continuous process, like the temperature

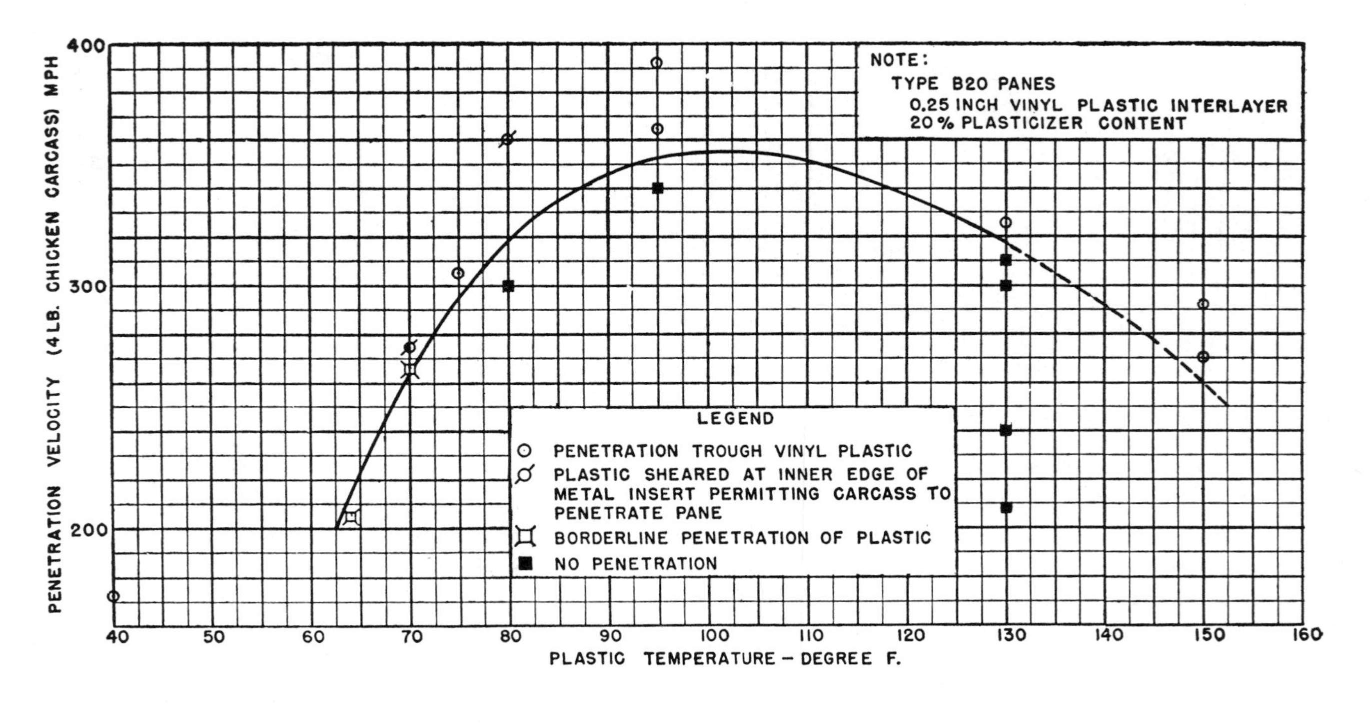


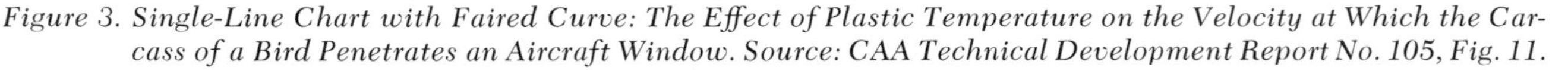

*Figure 3. Single-Line Chart with Faired Curve: The Effect of Plastic Temperature on the Velocity at Which the Carcass of a Bird Penetrates an Aircraft Window. Source: CAA Technical Development Report No. 105, Fig. 11.*

rise of a motor, it is usually desirable to make a faired curve; but if the process or change is not continuous, fairing the curve may be misleading. For example, if you were plotting an increase in student enrollment in a certain university for successive years, and your data showed enrollments of 10,000, 10,200, 14,000, and 14,300, a fairing of the curve would obscure a significant fact, the sharp increase in the third year, and would also falsely imply that the enrollment was rising steadily throughout each year. Incidentally, where precision is necessary, a point should be plotted by making a very small dot and then circling it lightly with a pencil so that it can later be found easily.

The foregoing discussion has been concerned only with the simplest and commonest of line charts. There are a great many possible variations of elements, including the use of special grids like the logarithmic and semilogarithmic, that it is important to know about. Again we urge you to consult the books listed in Appendix A.

*Bar Charts.* Bar or column charts represent values or amounts by bars of scaled lengths. They are useful for showing sizes or amounts at different times, the relative size or amount of several things at the same time, and the relative size or amount of the parts of a whole. In general the bar chart is preferable to the line chart for making dramatic comparisons if the items compared are limited in number. Arranged vertically (these are often called "column" charts), the bars are effective for representing the amount of a dependent variable at different periods of time; arranged horizontally, the bars are effective for representing different amounts of several items at one time. See Figures 4 and 5.

Although the bars of a bar chart may be joined, it is more common practice to separate them to improve appearance and increase readability. The bars should be of the same width, and the spacing between them should be equal. The proper spacing depends upon keeping the bars close enough together to make comparison easy, yet far enough apart to prevent confusion. Another convention of bar chart construction is that the bars are arranged in order of increasing or decreasing length. This convention applies to charts in which each bar represents a component; it does not apply, of course, to those representing a time series.

Portions or subdivisions of an individual bar may be used to represent components or percentages. A single bar so subdivided is usually called a 100 per cent bar chart. Shading and hatching differentiate the portions, along with labels or a key. Darker shadings (often solid black) are used to the left of horizontally placed bars,

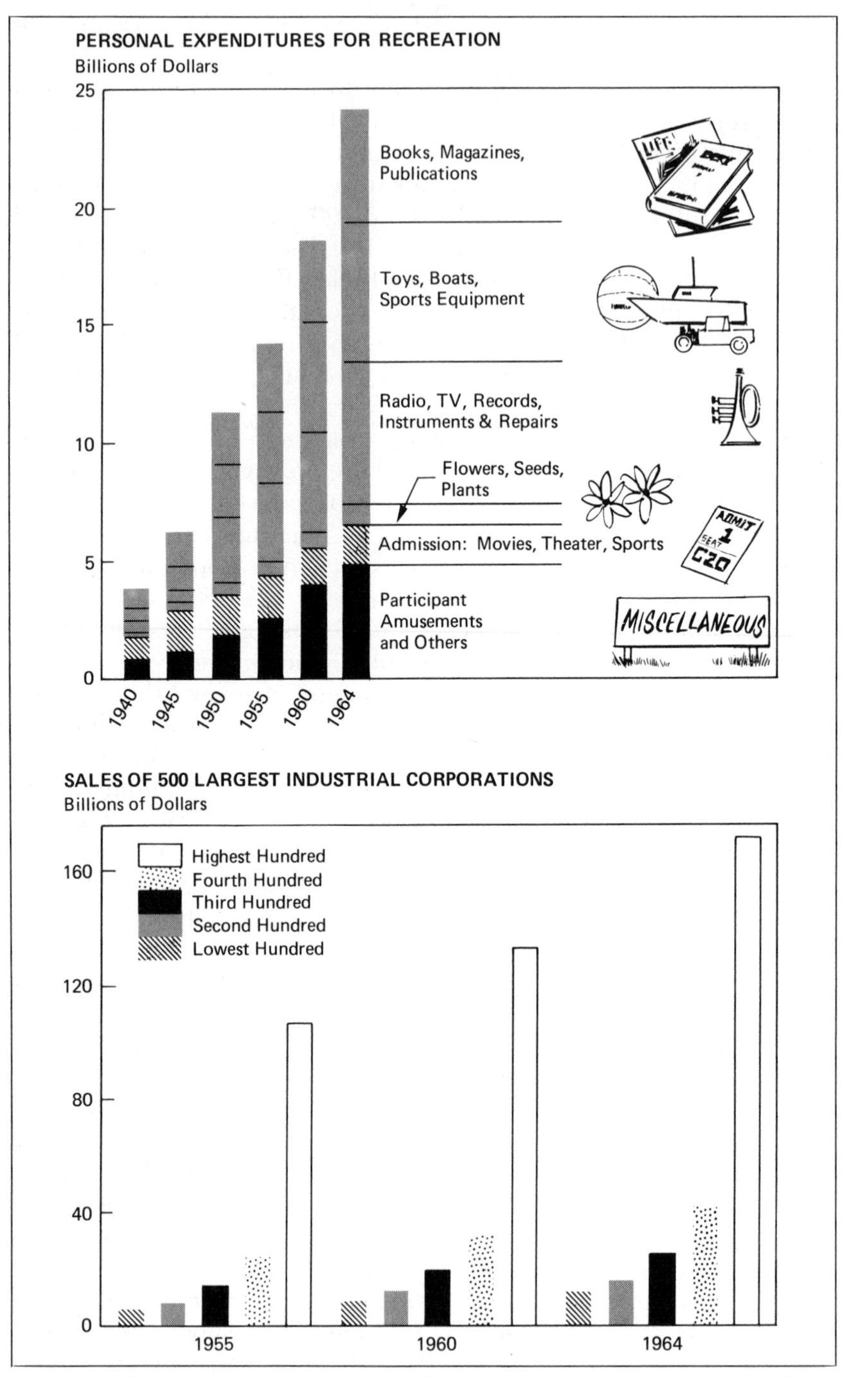


*Figure 4. Two Bar Charts. Source: U.S. Bureau of the Census,* Pocket Data Book, USA, 1967 *(Figs. 18 and 21). The lower chart is reprinted from* Fortune Directory *by Special Permission; © 1967 Time Inc.*

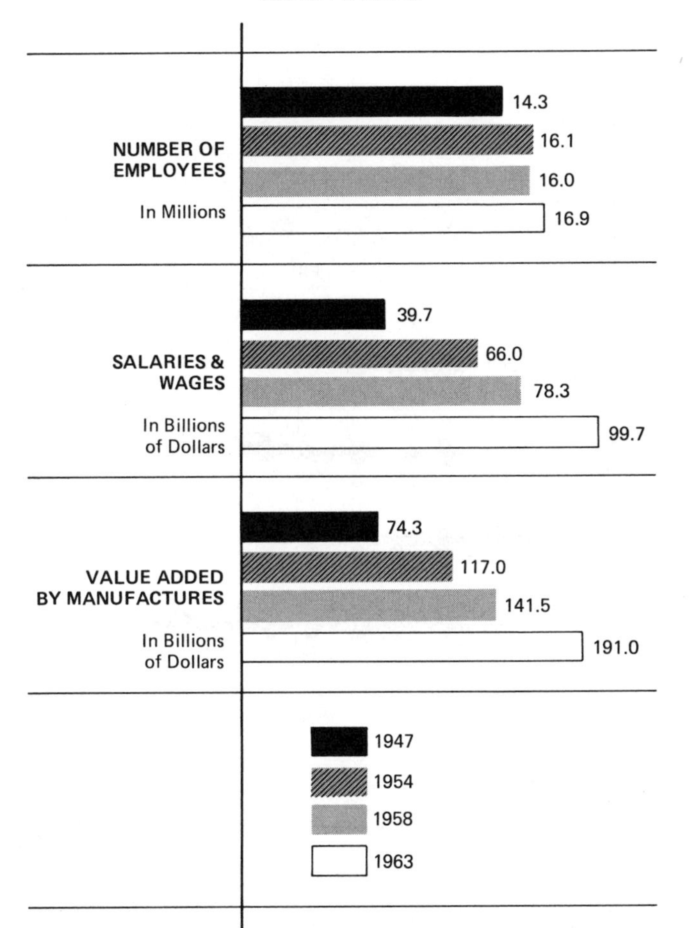


*Figure 5. Horizontal Bar Chart. Source: U.S. Bureau of the Census,* Pocket Data Book, USA, 1967 *(Fig. 42).*

with lighter shadings or hatchings being used for successive divisions to the right; on vertical bars, the darker shadings are used at the bottom. Colors may be used instead of shadings and hatchings.

One of the more interesting developments in bar chart making is the pictograph, mentioned earlier in this chapter (see Figure 6). The pictograph substitutes symbolic units, like the figure of a man or the silhouette of a ship, for the solid bar. The purpose of the pictograph is to increase interest and dramatic impact. The difficulties of preparing this kind of chart make it impractical for most technical reports, but when a report is to be distributed to a large audience of laymen,

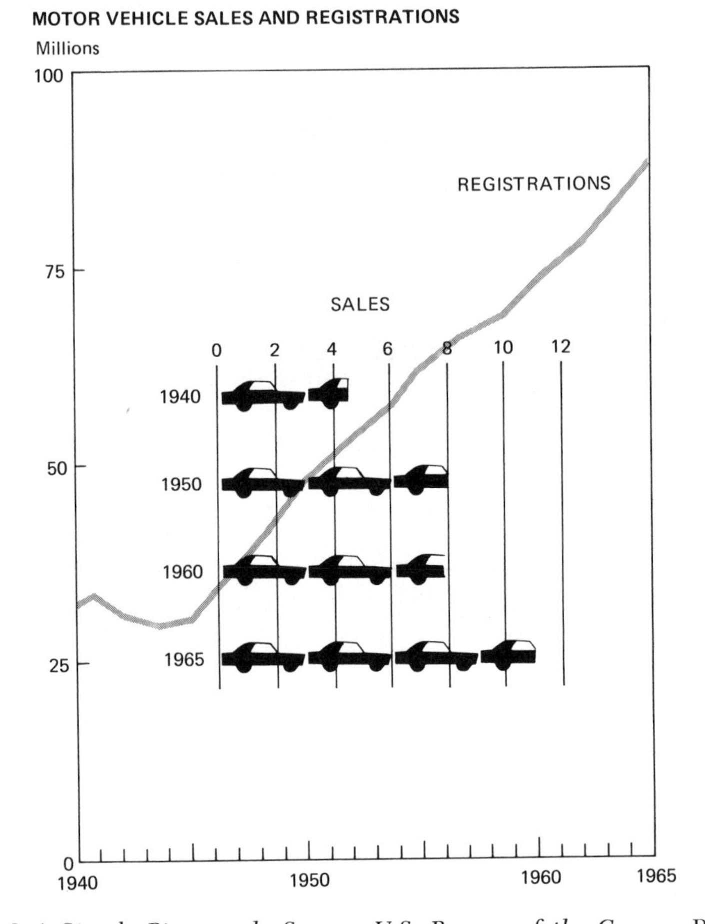


*Figure 6. A Simple Pictograph. Source: U.S. Bureau of the Census,* Pocket Data Book, USA, 1967 *(Fig. 45).*

and when professional help is available for preparing the illustrations, the pictograph may prove highly desirable.

*Surface and Strata Charts.* A single-surface chart is constructed just like a line chart except that the area between the curve line and the base or zero line is shaded. Multiple-surface or strata charts (sometimes called bands or belt charts) are like multiple-line charts with the underneath areas shaded in differentiating patterns or colors; that is, the vertical widths of shaded, colored, or hatched surfaces, strata, or bands communicate an impression of amount. They can be satisfactorily used to achieve greater emphasis than is possible with a line chart of the same data when amount is more important than ratio or change. They are not intended for exact reading, and should never be used when the layers or strata are highly irregular or where the lines plotted intersect. Gradual, regular movement or change can best be charted by this means (see Figure 7).

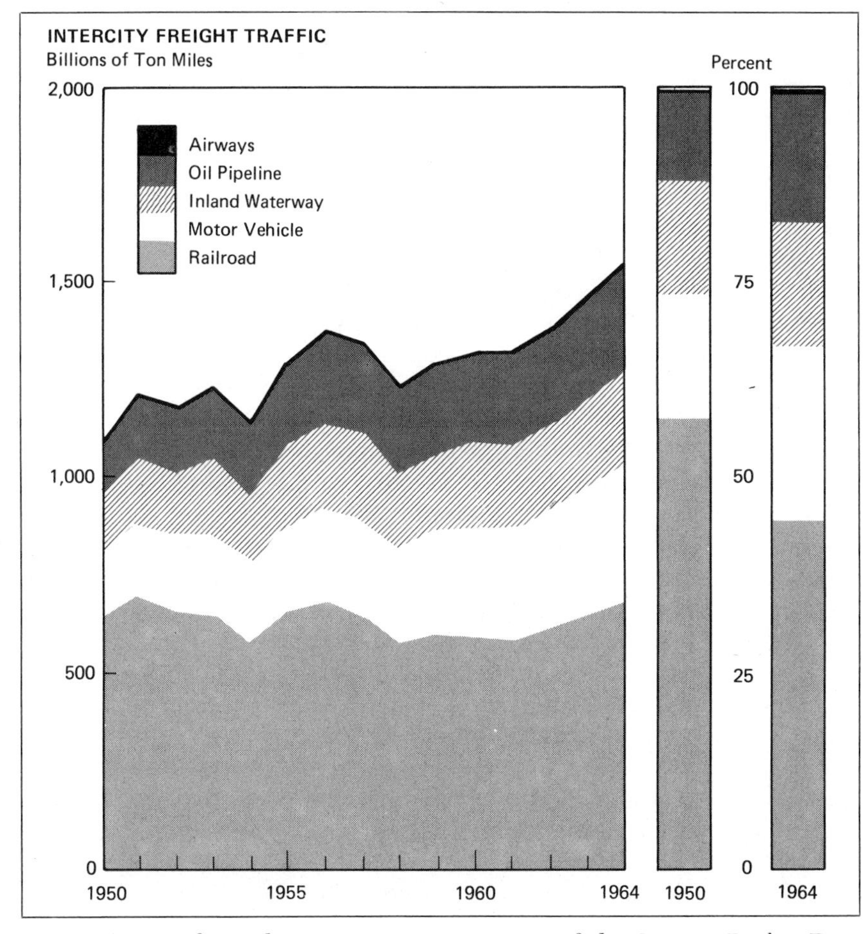


*Figure 7. A Surface Chart. Source: U.S. Bureau of the Census,* Pocket Data Book, USA, 1967 *(Fig. 24).*

*Circle or "Pie" Charts.* A circle, or pie, chart is simply a circle of convenient size whose circumference represents 100 per cent. The segments or slices show percentage distribution of the whole (see Figure 8). Since it is difficult to estimate the relative size of segments, labels and percentages must be placed on or alongside each segment. Although it is not a particularly effective graphic aid, the circle chart may be used for dramatic emphasis and interest as long as the subdivisions are not numerous. Figure 8 (bottom) illustrates an interesting application. One point to remember when you use a circle chart is that the segments are measured clockwise from a zero point at the top of the circle.

Occasionally you may see circles, squares, cubes, or spheres of different sizes used to compare amounts. The difficulty of comparing relative sizes, especially of cubes or spheres, makes these devices of no real use. We recommend that they be avoided. The line, bar, or surface chart will do better.

*Flowsheets and Organization Charts.* A flowsheet is a chart which makes use of symbolic or geometric figures and connecting lines to represent the steps and chronology of a process. An organization chart is like a flowsheet except that instead of representing a physical process, it represents administrative relationships in an organization.

The flowsheet (Figure 9) is an excellent device for exhibiting the steps or stages of a process, but its purpose is defeated if the reader finds it difficult to follow the connecting lines. Flowsheets should generally be planned to read from left to right, and the connecting lines should be arrow-tipped to indicate the direction of flow. The units themselves, representing the steps or stages, may be in the form of geometric figures or symbols. The latter are simple schematic representations of a device, such as a compressor, a cooling tower, or a solenoid valve. Standards for such symbols have been adopted in a number of engineering fields today, and you should make it a point to familiarize yourself with the symbols acceptable in your field. Publications concerning symbols may be obtained from the U.S.A. Standards Institute. These symbols may be used, by the way, in drawings as well as in flowsheets. Labels should always be put on geometric figures. Whether labels should be used with symbols depends on the intended reader.

Since a generous amount of white space is essential to easy reading of a flowsheet and since the flowsheet reads from left to right, the display will often need to be placed lengthwise on the sheet. This makes it necessary for the reader to turn the report sideways to read the chart, but this is better than crowding the figures into too narrow a space. If space requires it, a sheet of larger than page size may be used and folded in.

Flowsheets are usually enclosed in a ruled border, and the title and figure number centered at the bottom inside the border. See page 302 for example.

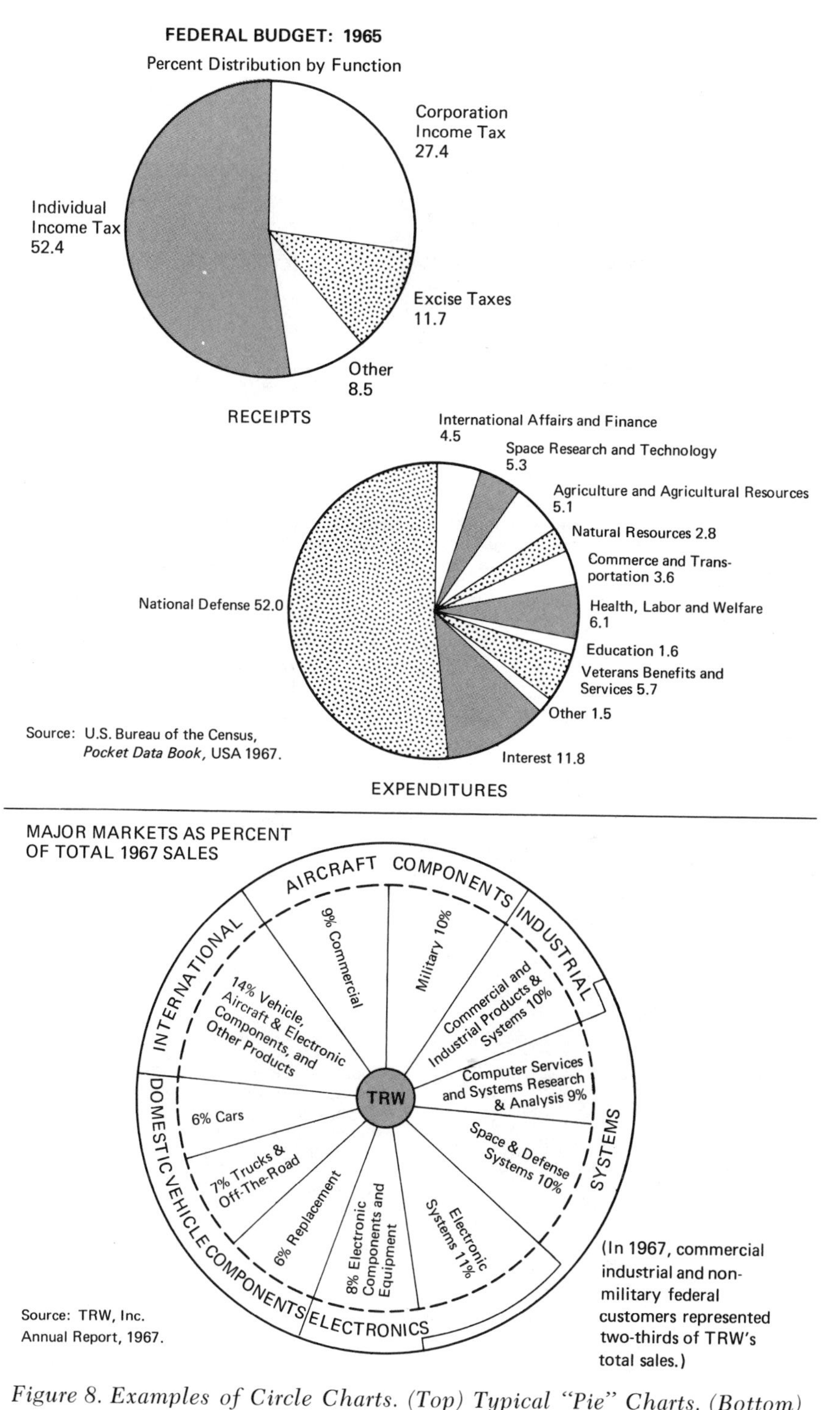


*Figure 8. Examples of Circle Charts. (Top) Typical "Pie" Charts. (Bottom) A "Double-Ring" Circle Chart Showing Major Divisions and Subdivisions.*

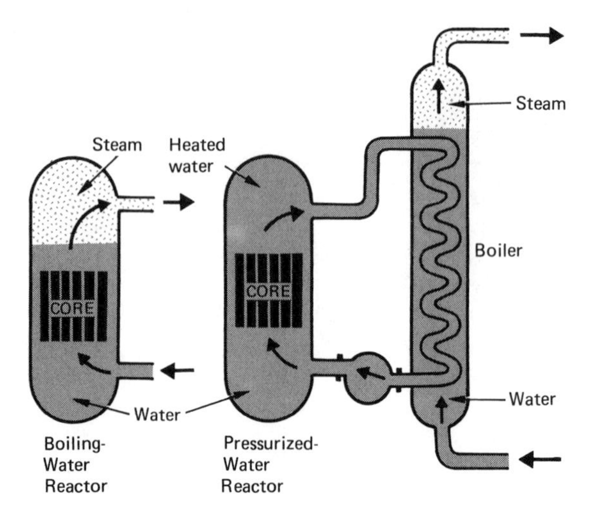


**Today's big reactors** work this way. Boiling-water type makes steam right in reactor; pressurized-water type, in boiler to which water carries heat.

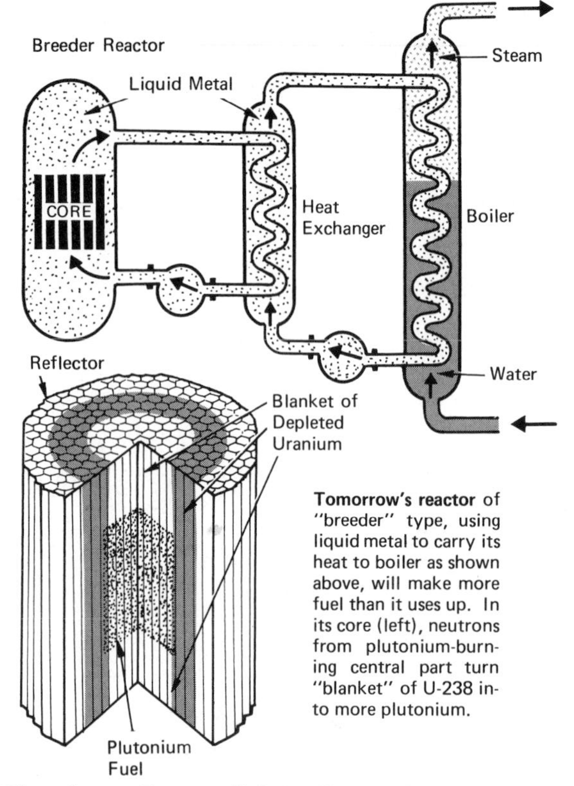


**Tomorrow's reactor** of "breeder" type, using liquid metal to carry its heat to boiler as shown above, will make more fuel than it uses up. In its core (left), neutrons from plutonium-burning central part turn "blanket" of U-238 into more plutonium.

*Figure 9. A Flowsheet: Types of Atom-Power Reactors. Source: "World's Biggest Atom-Power Plant" by Alden P. Armagnac,* Popular Science, *September 1969, pp. 95 ff.*

Organization charts are very similar to flowsheets (see Figure 10). Rectangular figures represent the units of an organization; connecting lines, as well as relative position on the sheet, indicate the relationship of units. Good layout requires that the figures be large enough so that a lettered or typed label can be plainly and legibly set down inside them, and they must be far enough apart so that the page will not be crowded. Organization charts usually read from the top down.

Colored flowsheets and organization charts are effective for popular presentation. Like the pictograph, such color charts require the services of trained artists and draftsmen.

*Map Charts.* The map chart is useful in depicting geographic or spatial distribution. It is made by recording suitable unit symbols on a conventional or simplified map or differentiated area of any sort (like electron distribution in a space charge). It is particularly important in a map chart that the symbols and lettering be clear and easy to read. Geographic maps suitable for use in making map charts are readily available from commercial suppliers. See Figure 11 for an example of a map chart.

## Drawings, Diagrams, and Photographs

Drawings and diagrams are especially valuable for showing principles and relationships that might be obscured in a photograph, but of course they are sometimes used instead of photographs simply because they are usually easier and less expensive to reproduce. A photograph, on the other hand, can supply far more concreteness and realism than drawings or diagrams (see Figure 12). We are using the terms "drawing" and "diagram" loosely to refer to anything from a simple electronic circuit diagram to an elaborate structural blueprint or a pictorial representation of a complex mechanical device.

Parts of drawings should be plainly labeled so as to make textual reference clear and meaningful. If the drawing is of a simple device with but few parts, the names of the parts may be spelled out on the drawing itself, with designating arrows. If the drawing is of a complex device, with a large number of parts, letter symbols or numbers with an accompanying key should be used. Figure number and title should be centered at the bottom, inside the border if one is used. If a source reference is necessary, it should appear in the lower right-hand corner.

Photographs should be taken or chosen with special attention to how prominently the elements important to your discussion stand

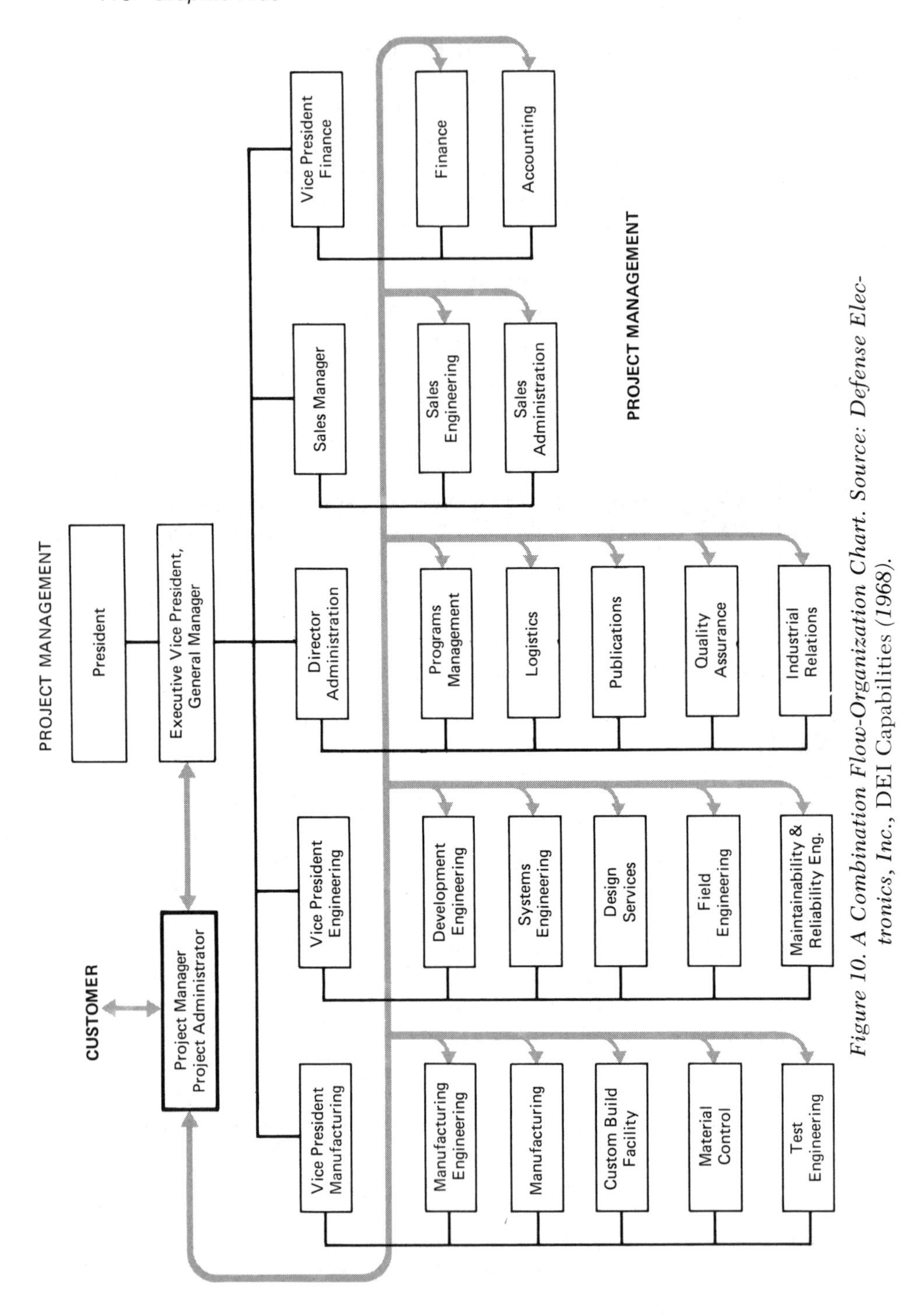


*Figure 10. A Combination Flow-Organization Chart. Source: Defense Electronics, Inc.*, DEI Capabilities (1968).

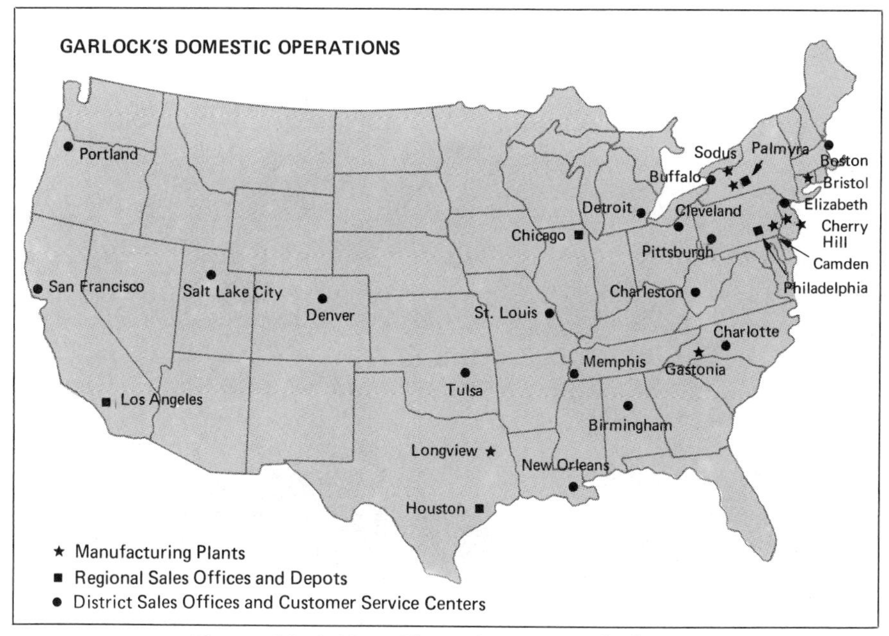


*Figure 11. A Map Chart. Source: Garlock, Inc.*

out. Very often this principle necessitates the use of an artificial background. A cluttered background distracts attention, not infrequently producing the impression that the photograph was originally intended as a puzzle, with a prize for anybody who could find gear B.

Glossy prints are better than flat prints because of their greater effectiveness in reproducing highlights and shadings. Each reproduced print should have an attractive margin of white space. If smaller than page size, prints may be satisfactorily mounted by use of rubber cement. Rubber cement has less tendency than paste or glue to wrinkle the page.

The figure number and title, as well as explanatory data, should be put directly on full-page photographs in black or white ink.

## Tables

The table is a convenient method of presenting a large body of precise quantitative data in an easily understood form. Tables are read from the top down in the first column and to the right. The first, or left, column normally lists the independent variable (time, item number, and so on) and the columns to the right list dependent variables (see Figure 13). The table should be designed so as to be self-explanatory, but textual comments on it should be made according to the same principles that apply to the use of a chart.

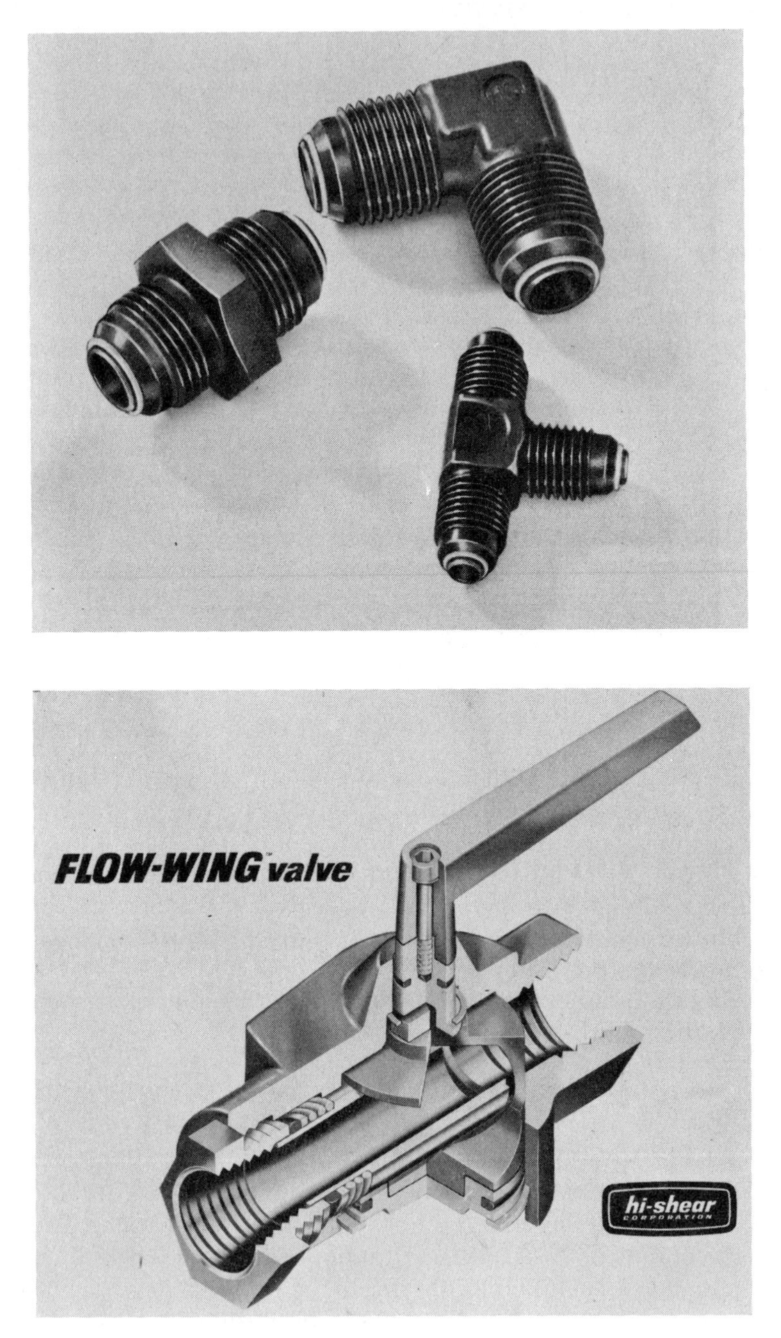


*Figure 12. A Photograph of a Fitting (Top) and a Drawing of a Valve (Bottom). Source: Hi-Shear Corporation.*

TABLE II

CENTRIFUGAL FORCE CALIBRATION DATA

| Element No. | Maximum g | Minimum g | Average g | Spread in g |
|---|---|---|---|---|
| 1 | 2.21 | 1.65 | 1.93 | 0.56 |
| | 2.20 | 1.65 | 1.93 | 0.50 |
| 2 | 2.48 | 2.25 | 2.36 | 0.23 |
| | 2.50 | 2.22 | 2.36 | 0.28 |
| 3 | 3.08 | 2.59 | 2.84 | 0.49 |
| | 3.12 | 2.58 | 2.85 | 0.54 |
| 4 | 3.07 | 2.60 | 2.84 | 0.47 |
| | 3.10 | 2.60 | 2.85 | 0.50 |
| | | | Average Spread | 0.45 |

*Figure 13. Illustration of Table Layout. Source: CAA Technical Development Report No. 48.*

To make a table easy to read, you should leave ample white space in and about it. If the table appears on a page on which there is also typed text, triple-space above and below the table. Leave a generous amount of space between columns and between the items in a column. The title and the table number should appear at the top. It is advisable to use Arabic numerals for the table number if Roman numerals have been used in the same report for numbering other kinds of graphic aids, or vice versa.

In separating parts of the table from one another—that is, one column from another, or one horizontal section from another—use single lines in most instances; but where you wish to give special emphasis to a division use a double line. Some people make it a practice to use a double line across the top of the table, under the title. Usually a single line should be drawn across the bottom of the table. The sides may be boxed or left open as seems most pleasing. But we will add this: in case of doubt as to whether a ruling should be used at any given point in the table, leave it out. More harm will probably be done by too many lines than by too few, provided ample white space has been left.

Align columns of numerals on the decimal point, unless units of different quantities (like 2000, representing lb, and 0.14 representing percent) appear in the same column, in which case the column should be aligned on the right margin.

A heading should be written for each column, with the initial letter of important words capitalized. The headings should be written horizontally if possible, but if that would use too much space they may be written vertically. Indicate units (like volts, Btu, cu ft) in the heading so that the units will not have to be noted in the column. If you have data in different systems of units, you should convert all to the same system before entering them in the table.

If the data in the table are not original, acknowledge their source in a footnote just below the bottom horizontal line of the table. Instead of using a superscript number in the table to refer to the footnote, use a letter symbol (Roman), an asterisk, or some other convenient symbol. Tabular footnotes, that is, notes that refer to specific items in the table itself, should be placed between the bottom line of the table and the source note, if one is needed. Tabular footnotes may be keyed with the asterisk or a Roman superscript letter, whichever is not being used with the source note.

## Conclusion

Graphic aids in their simpler forms are easy to prepare and easy to understand. In either their simple or more complex forms, they often convey information or provide dramatic emphasis with an effectiveness that would be difficult or impossible to achieve in writing. On the other hand, if they are relied on too much they may become a hindrance rather than a help. You should regard this chapter as merely a short introduction to the uses and techniques of one of the more valuable tools of science and technology.

### SUGGESTIONS FOR PRACTICE IN MAKING GRAPHIC AIDS

Although the professional man usually can make use of the services of draftsmen and technical illustrators in preparing graphs, charts, tables, drawings, etc. for his reports and proposals, it is not a bad idea to try your hand at making a few—for fun and to gain some idea of the problems involved.

1. If you are in the process of preparing a research report, explore the possibilities of making use of some of the graphic techniques described in this chapter for complementing your presentation. Doubtless your sources will contain many illustrations, but often they turn out to be difficult to reproduce. Sometimes it is possible to adapt and simplify an illustration from a source in such a way that it

is more useful than the original would be in supplementing a portion of your own presentation.

2. Drawing on what you have learned from this chapter, devise at least two graphic means of presenting the data tabulated below. How might color be effectively used?

*CAUSES OF FOREST FIRES AND EXPENDITURES FOR CONTROL**

| Item | 1950 | 1955 | 1960 | 1965 | 1967 |
|---|---|---|---|---|---|
| NUMBER AND CAUSE | | | | | |
| Total | 105,182 | 87,688 | 89,627 | 100,568 | 113,762 |
| Incendiary (arson) | 40,127 | 25,775 | 21,162 | 26,860 | 33,116 |
| Debris burning | 18,768 | 20,338 | 21,870 | 22,882 | 28,253 |
| Smoking | 18,287 | 15,882 | 16,030 | 16,517 | 16,844 |
| Lightning | 6,518 | 6,282 | 11,068 | 8,730 | 10,335 |
| Machine use | (NA) | (NA) | (NA) | 6,734 | 8,763 |
| Campfires | 3,802 | 3,924 | 3,614 | 3,404 | 3,405 |
| Miscellaneous and unknown | 17,680 | 15,487 | 15,883 | 15,441 | 13,046 |
| EXPENDITURES ($1,000) | | | | | |
| Total | 28,934 | 39,216 | 56,641 | 76,562 | 90,964 |
| Federal | 8,551 | 8,945 | 9,401 | 12,783 | 12,834 |
| State and county | 18,121 | 28,168 | 45,059 | 62,612 | 76,626 |
| Private agencies | 2,262 | 2,103 | 2,181 | 1,167 | 1,504 |

NA Not available.

*Table 302 from U.S. Bureau of the Census, *Pocket Data Book*, USA, 1969, U.S. Government Printing Office, Washington, D.C., 1969.

# SECTION SIX

# The Library Research Report

*The next best thing to knowledge is knowing where to acquire knowledge. That means, basically, knowing how to use a library. The professional man who has to write a report of several thousand words usually turns to the library. There, through the use of library materials, he will find sufficient subject matter for such a report.*

*For a student of technical writing, the preparation of a library research report has two distinct advantages:*

1. *An opportunity to use the library for research.*
2. *The experience gained in writing a report of this length. Nobody can learn to write without writing.*

*The two chapters in this section are devoted to a discussion of how to use the library and of how to write a library research report.*

# 22 Finding Published Information

## Introduction

The ability to find quickly and accurately what has been published on a subject is an essential skill for the student and the professional man. Students must frequently use the library to supplement information in their texts. Professional men use library resources to keep abreast of developments in their field and to get data for the solution of special problems. Almost all industrial concerns and laboratories nowadays maintain technical libraries for their personnel.

This chapter will introduce you to various sources of information about what has been published on a given subject. There are many more sources than could be presented in this chapter, but once you have begun to make a deliberate study of library resources, one thing will lead to another, and your ability to locate books and articles will steadily improve. The materials described in this chapter should be sufficient to provide for your needs in the preparation of the library research report, which is discussed in the next chapter.

Guides to the location of books will be considered first, then guides to the location of periodical articles, followed by a variety of other aids.

## Books

The most obvious source of information about books is the library card catalogue. In addition, there are special works of various sorts that should be noted.

*The Library Card Catalogue.* The card catalogue is an index to the books a library contains. It is a file, on 3 by 5 cards, of every book in the library. Some libraries have at least three cards for every book: an author card, a title card, and a subject card. In many libraries there are several subject cards for certain books, the number depending upon how many distinct subjects are treated in the book. In other libraries there may be no subject cards at all, only author and title cards, and sometimes only author cards.

It is important to remember that there may be several subject headings under which the book might be listed. For instance, if you are interested in the subject of beta rays you would naturally look for "Beta" in the alphabetical file. If you were to find nothing under that heading, other possibilities would be "Rays," "Electrons," and "Radioactivity." Cross-reference cards will often be included in the files to guide you to other headings in case book cards are not filed where you look first. Identical cards may appear in several places, of course.

A good deal can be learned about a book and its potential usefulness in a research project by careful examination of the card entry itself. Let's take a look at a typical card and see what information it contains and how that information may help you decide whether the book could be of use. See Figure 1.

321.041 ← 1
G621s
2→ **Goodman, Elliot Raymond, 1923–**
3→ The Soviet design for a world state. With a foreword by Philip E. Mosley. New York, Columbia University Press, 1960.

4→ xviii, 512 p. 24 cm. (Studies of the Russian Institute, Columbia University)

5→ Issued in microfilm form in 1957 as thesis, Columbia University. Bibliography: p. [489]–493. Bibliographical footnotes.

6→ 1. Internationalism. 2. Russia—For. rel. 3. International organization. 4. Communism. I. Title. (Series: Columbia University. Russian Institute. Studies)

JC361.G66 ←7 1960 321.041 60–7625

*Figure 1. A Library Catalogue Card*

1. The *call number* recorded on the upper left-hand corner of the index card is the book's classification (in this case according to the Dewey decimal system), which enables a library attendant to

find the book. You do not need to be concerned with the details of this system of book classification, but you do need to take pains to copy the call number accurately on a call slip when you request a book.

2. The *author line* gives the full name of the author of the book plus the date of his birth and, if he is dead, that of his death.
3. *Title and publishing data* are given next. This entry includes the full title of the book (plus subtitle and foreword if such appear), the place and date of publication, and the publisher. All this information, with exception of the subtitle, is needed for a bibliographical entry. It is also useful in determining the possible value of a book. You should note the date of publication especially, to learn whether the book's treatment of the subject is up to date. Experienced researchers tend also to attach some significance to the name of the publisher on the theory that well-established publishers of good reputation are more likely to publish books of merit than little-known publishers.
4. The next entry, called the *collation*, gives information about paging, illustrations, the size of the book (in centimeters), plus information, if pertinent, about the series the book appears in. The investigator of a technical subject may be particularly interested in whether a book contains illustrations. When a book does contain illustrations, a note like this may appear: "front., illus., plates, diagrs." If a book contains an introduction, its length will be given in lower-case Roman numerals.
5. Notes on bibliographies, contents, and the like may appear next. The researcher will be especially interested in notes about content, for from them he can often determine whether it will be worth while to check out the book. A notation about bibliography may suggest that he can find leads to other sources.
6. At this point *subject headings* or "tracings" are given, under which the book is also indexed. These too are sometimes useful in getting an idea of the subject matter coverage of the book and in locating other books.
7. *Classification data* appear at this point. Of little or no interest to the student researcher, they give the Library of Congress call number, the date of card code number, the order code number for this card, and so forth.

Altogether, then, quite a lot of information about a book is recorded on an index card, enough so that it is worth while to read it carefully to determine without checking out the book whether it will be of use.

*Other Guides to Books.* The guides to books on science and technology in general and to books on particular fields of science and technology are far too numerous to discuss here, but several sources of information about books besides the card catalogue deserve mention:

1. *Book Review Digest* (1905–). A subject-title-author index to book reviews published in selected magazines and newspapers; issued monthly and cumulated annually.
2. *Technical Book Review Index* (1935–). Besides guiding the researcher to reviews of new technical books, this index prints excerpts from the reviews.
3. *Scientific, Medical and Technical Books.* Published under the direction of the National Academy of Science and the National Research Council's Committee on Bibliography of American Scientific and Technical Books, these volumes cover the period from 1930 to the present. Entries describe a book's contents and its treatment of the subject.
4. *British Scientific and Technical Books* (1935–1952). Published by the Association of Special Libraries and Information Bureau (Aslib).
5. *International Catalogue of Scientific Literature.* This catalogue covers anthropology, astronomy, biology, botany, chemistry, mathematics, meteorology, paleontology, physics, and zoology.
6. *Guide to Technical Literature.* By A. D. Roberts, this publication provides bibliographies in nearly all branches of engineering and technology, but is of no value for publications after 1938.
7. *The United States Catalog of Books* (1900–1927). This publication indexes books published in this country; it has been replaced by the *Cumulative Book Index* which is issued monthly, quarterly, semiannually, and annually.
8. *Books in Print.* A subject guide to books in print, published annually in two volumes by R. R. Bowker Co.
9. *Cumulative Book Index.* A monthly publication, cumulated every six months and biennially, listing alphabetically by author, subject, and title all books published in English.

## Periodical Indexes

It is seldom that a research project can be completed without the use of articles appearing in professional journals, particularly because these publications contain the most recent information. Sometimes the researcher's first task is to find out what periodicals are published in a particular field. The chief bibliographies of periodicals are Bolton's *Catalogue of Scientific and Technical Periodicals*, the *Union List of Technical Periodicals*, Ulrich's *Periodicals Directory*, and the *World List of Scientific Periodicals*. The principal indexes to articles appearing in scientific and technical periodicals are listed below with brief comment.

1. *Agricultural Index* (1916–). A subject index to articles in selected magazines, bulletins, and books; issued monthly, cumulated quarterly, annually, and bienially.
2. *Applied Arts Index* (1957–). A subject index to articles on architecture, engineering, and applied science; issued monthly, cumulated yearly.
3. *Applied Science and Technology Index* (1958–; *Industrial Arts Index* before 1958). A subject index to articles appearing in more than 200 periodicals which deal with engineering, trade, science, and technology; issued monthly and cumulated quarterly and annually. This index to periodicals (sometimes books and pamphlets are indexed) is particularly useful to the student. Because its general subject headings are broken down into subdivision classifications, it often helps guide the student to a satisfactory topic for research—one sufficiently limited for practical investigation.

Shown in Figure 2 is part of a page from the *Applied Science and Technology Index*, which shows how information is handled. Let's

APPLIED SCIENCE & TECHNOLOGY INDEX

**WOOD**
Methoxyl groups in wood; revision of official standard T 2 m-43. bibliog diag Tappi 41: sup168A-70A D '58
Resistance of natural timbers to marine wood borers. C. H. Edmondson. bibliog diags Corrosion 14:33-6 N '58
*See also*
Aspen
Birch
Cellulose
Elm
Eucalyptus
Gumwood
Hard woods
Palms
Pine
Poplar
Printing on wood
Spruce
Wood pulp

**Analysis**

Chemical analyses of mountain hemlock. E. F. Kurth. Tappi 41:733-4 D '58

**Chemistry**

Effect of temperature upon the rate of hydrolysis of the sulphate acid ester in unstabilized cotton and wood cellulose nitrates. P. E. Gagnon and others. bibliog Tappi 41:515-17 S '58
...dence of branching in ethanol lignins. H. ... bibliog Tappi

**WOOD preservation**
Pressure preserved wood for permanent structures. C. M. Burpee. Am Soc C E Proc 84 [ST 7 no 1841]:1-10 N '58

**WOOD perservatives**
Deterioration of wood by marine boring organisms. H. Hochman. bibliog il Corrosion 15:61-4 Ja '59

**WOOD products**
*See also*
Cellulose
Lignin

**WOOD pulp**
Alkaline hydrogenation pulping. I. Sobolev and C. Schuerch. bibliog Tappi 41:545-51 O '58
Behavior of wood hemicelluloses during pulping. J. K. Hamilton and others. bibliog Tappi 41:803-16 D '58
Effect on paper wet strength of oxidation of the carbohydrate polymers in wood pulps. A. Meller. bibliog Tappi 41:679-83 N '58
Flexibility of wood-pulp fibers. O. L. Forgacs and S. G. Mason. bibliog il diags Tappi 41: 695-704 N '58
Groundwood and chemigroundwood from European poplar wood. W. Brecht. Paper Ind 40:449-50+ O '58
Hardwood versus softwood pulp. R. T. Trelfa. Tappi 41:sup185A-7A D '58
Kraft pulps from mixed hardwood ...

*Figure 2. Part of a Page from an Issue of* Applied Science & Technology Index. *Reproduced by Permission of the H. W. Wilson Company.*

take a single entry from it and see what it tells us. The fourth entry under "Wood Pulp" reads as follows:

> Flexibility of wood-pulp fibers. O. L. Forgacs and S. G. Mason. bibliog il diags Tappi 41:695–704 N '58.

This entry tells us that O. L. Forgacs and S. G. Mason have published an illustrated article containing diagrams and bibliography under the title "Flexibility of Wood-pulp Fibers," and that the article appeared on pages 695 to 704, in volume 41, of the November, 1958, issue of *Tappi.*

Because of the need for conserving space, entries are highly abbreviated. If you cannot identify the name of the magazine by the abbreviation, you will find an alphabetical list of full names of magazines and their abbreviations in the prefatory pages of each volume of the index. Note that volume and page numbers for a reference are given together, with a colon separating the volume from the pages. Some of the entries show a + after a page number. This mark means that the article is continued in the back pages of the magazine.

Other periodical indexes include:

4. *Art Index* (1929–). A cumulative author-subject index to a selected list of fine arts periodicals and bulletins; issued quarterly.
5. *Business Index* (1958–). Formerly a part of the *Industrial Arts Index*, this subject index is a guide to articles on business, trade, and finance; issued monthly, cumulated yearly.
6. *Engineering Index* (1906–). A subject-author index to articles appearing in professional technical periodicals in all engineering fields, government bureau publications, engineering college publications, and research organization publications. In addition to bibliographical information, this index also annotates publications. Published annually, it is produced after a review of more than 1200 publications. Publications listed may be somewhat advanced for undergraduate research projects, but students of engineering should certainly know about this index.
7. *Industrial Arts Index* (1913–1957). A particularly useful subject index to articles appearing in more than 200 technical and semitechnical journals covering the fields of engineering, business, trade, finance, and science. It was issued monthly, with annual cumulations.
8. *International Index to Periodicals* (1907–). An author-subject index to articles appearing in about 300 specialized professional journals dealing with science and the humanities; issued quarterly, and cumulated every three years.
9. *Poole's Index to Periodical Literature* (1802–1906). A subject

index to English and American periodicals of the nineteenth century.

10. *Public Affairs Information Service Bulletin* (1915–). A subject index to articles, pamphlets, and government documents on economics, government, and public affairs; issued weekly, cumulated annually.
11. *Readers' Guide to Periodical Literature* (1900–). An author-subject index to articles appearing in more than 200 magazines of general interest. It is issued monthly and cumulated annually or oftener. For the student researcher, this guide is valuable because it will lead him to articles written for a nontechnical audience. Such articles are good for getting a general knowledge of a topic on which research is being started.
12. *The New York Times Index* (1913–). This index, which is published monthly and cumulated annually, is quite useful for finding newspaper articles.
13. *Government-Wide Index to Federal Research and Development Reports.* A monthly guide to reports listed in *Nuclear Science Abstracts, Scientific and Technical Aerospace Reports, Technical Abstract Bulletin,* and *U.S. Government Research and Development Reports.* A useful subject and author index.
14. *NASA Index of Technical Publications.* Irregularly published listing of NASA publications.

## Abstracts

Publications of abstracts help the researcher by providing him with an opportunity of determining whether it will be worth his while to run down a particular periodical article for study. Since these abstracts are usually descriptive, rather than informational (see Chapter 4), they are guides to source material rather than sources themselves. We list below the more important abstract publications; a complete listing of abstracting and indexing services in science and technology for the world would include more than 1800 titles, more than 450 for the U.S. alone. (See *A Guide to the World's Abstracting and Indexing Services in Science and Technology*, Report No. 102, National Federation of Science Abstracting and Indexing Services, Washington, D.C., 1963.)

*Abstracts of Bacteriology* (1917–1925)
*Aeronautical Engineering Index* (1947–)
*Animal Breeding Abstracts* (1933–)
*Biological Abstracts* (1927–)

*Botanical Abstracts* (1918–1926)
*British Abstracts* (1926–1953)
*Ceramic Abstracts* (now included in *American Ceramic Society Journal*)
*Chemical Abstracts* (1907–)
*Dairy Science Abstracts* (1939–)
*Engineering Abstracts* (1900–)
*Excerpta Medica* (1947–)
*Field Crop Abstracts* (1948–)
*Forestry Abstracts* (1939–)
*Geological Abstracts* (1953–)
*Geophysical Abstracts* (1929–)
*Horticultural Abstracts* (1931–)
*Index Bibliographicus* (published in 1952 as a comprehensive guide to abstracts in periodicals)
*Index Medicus* (1879–1926)
*International Aerospace Abstracts* (published twice monthly by American Institute of Aeronautics and Astronautics)
*Mathematical Reviews* (1950–)
*Meteorological Abstracts and Bibliography* (1950–)
*Mineralogical Abstracts* (1920–)
*Nuclear Science Abstracts* (1948–)
*Science Abstracts* (1898–; Sec. A on Physics; Sec. B on Electrical Engineering)
U.S. Department of Agriculture's *Experiment Station Record* (1889–1946)
U.S. Department of Agriculture's *Index to Technical Bulletins Zoological Record* (1864–)

## Bibliographies of Bibliographies

Sometimes what you want most may be information about what bibliographies are available in your field of interest. The truth is that bibliographies are now so numerous that it has become necessary to publish bibliographies of bibliographies. Some of the best are as follows:

1. Besterman's *A World Bibliography of Bibliographies* (about 80,000 bibliographies are listed in this four-volume work)
2. *Bibliographic Index, A Cumulative Bibliography of Bibliographies* (1938–)
3. Dalton's *Sources of Engineering Information* (1948)
4. Collison's *Bibliographies* (1951)

5. Holmstrom's *Records and Research in Engineering and Industrial Science* (3d edition, 1956)

## Miscellaneous

A few items can conveniently be grouped here under the heading "Miscellaneous." This heading does not imply that they are unimportant.

1. *United States Government Publications: Monthly Catalog.* This guide (published since 1895) is invaluable in finding out what has been published by government agencies. Copies of government documents are obtainable from the Superintendent of Documents, United States Government Printing Office, in Washington, D. C., if your library does not possess them.
2. *U.S. Government Research Reports* (1946– ). This publication is a monthly issue of the Office of Technical Services, United States Department of Commerce. It is a subject index to nonclassified reports in science and industry.
3. H. S. Hirschberg and C. H. Melinat's *Subject Index of Government Publications* (1947) is a useful tool in the initial phase of a research project.
4. *Official Gazette* of the United States Patent Office provides information on patents issued. It is essential in much original research.
5. *Vertical File Service Catalog* (1935– ). This monthly and annual cumulative subject index is useful as a guide to pamphlets, brochures, folders, and leaflets.
6. New York Public Library's *New Technical Books* may provide useful leads to current literature; published since 1915.
7. U.S. Department of Agriculture's *Bibliography of Agriculture* (1942– ).
8. The Armed Services Technical Information Agency (ASTIA) holds hundreds of thousands of contract reports on file, covering the entire realm of science and technology. Reports received by the agency are catalogued and listed in its bi-weekly publication *Technical Abstracts Bulletin* (TAB). Copies of reports on file and bibliographies may be had by industries and research organizations which have established a need for information in registered fields of interest.

## Reference Works

Although encyclopedias and other special reference works are generally not essential basic sources of information for a student's research report, they are often extremely useful to the researcher in the course of his work on a subject. Encyclopedias may provide him with needed general information to enable him to read books and articles with greater understanding. Special engineering and scientific reference works, as well as technical dictionaries and biographical reference books, may be just what he needs to find out about difficult points which arise during the course of his reading.

Of particular use in determining what reference tools are available are Mudge's *Guide to Reference Books* (1936) and Constance Winchell's *Guide to Reference Books* (the latter being a continuation of Mudge's work). We list below, with brief comment, some of the principal—and most commonly available—works you ought to know about. You should realize, however, that there are numerous specialized encyclopedias and dictionaries for the various technical fields. There are encyclopedias for the fields of astronomy, physics, chemistry, chemical engineering, welding, petroleum, meteorology, horticulture, social sciences, and mathematics; and there are handbooks by the dozens for nearly all specialized fields. In addition to the dictionaries we mention, there are many others for such fields as aviation, mathematics, physics, chemistry, geology, agriculture, electronics, petroleum, meteorology, forestry, and medicine.

### *Encyclopedias*

1. *Encyclopedia Britannica.* A general encyclopedia, commonly regarded as the best in English. The 14th edition is in 24 volumes, supplemented annually with revisions and new information.
2. *Encyclopedia Americana.* This encyclopedia is regarded as especially useful on technical, business, and government topics. It contains maps, illustrations, and selective bibliographies.
3. *Encyclopedia of Science and Technology.* This 1960 publication of McGraw-Hill contains in its 15 volumes more than 7224 articles covering the entire span of today's scientific, technical, and engineering knowledge. An annual supplement keeps the work up to date.
4. *Engineering Encyclopedia.* This two-volume work is edited by Franklin D. Jones and published by the Industrial Press.
5. *Hutchinson's Technical and Scientific Encyclopedia.* In four volumes, this reference combines the functions of an encyclopedia and a technical dictionary.
6. *Kingzett's Chemical Encyclopedia.* Edited by Ralph K. Strong and published in 1952 by Van Nostrand.
7. *Motor Service's New Automotive Encyclopedia.* Edited by Wil-

liam K. Tobaldt and Jud Purvis, this work was published in 1954.

8. *Van Nostrand's Scientific Encyclopedia.* A one-volume reference work which is very useful for its short articles on topics in all branches of engineering and science.
9. *Encyclopedia of Chemical Technology.* This is a 15-volume work.
10. *Watkin's Cyclopedia of the Steel Industry.* This specialized work was published in 1947.
11. *Cyclopedia of American Agriculture.* Four volumes.
12. *Standard Cyclopedia of Horticulture.* A three-volume work.
13. *Encyclopedia of Engineering Signs and Symbols.* Published in 1965 by Odyssey Press.
14. *Encyclopedia of the Biological Sciences.* Published by Reinhold in 1961.

### *Dictionaries*

1. *Chamber's Technical Dictionary.* This book, revised and supplemented in 1948 by C. F. Tweney and L. E. C. Hughes, is a good all-around technical dictionary.
2. *Engineering Terminology.* The second edition of this work by V. J. Brown and D. G. Runner was brought out in 1939.
3. *Dictionary of Scientific and Technical Words.* This convenient-sized book published by Longmans contains very clear definitions of about 10,000 words.
4. *A Dictionary of the Sciences.* This paperback for the layman is published by Penguin.
5. *Dictionary of Electronic Terms.* Published by Allied Radio Corporation in 1955, this work is helpful to the special student.
6. *A Dictionary of Dairying.* This special dictionary was published in 1955 in London.
7. Crispin's *Dictionary of Technical Terms.* The subtitle points out that this book defines terms commonly used in aeronautics, architecture, woodworking, electrical and metalworking trades, printing, and chemistry.
8. Henderson's *Dictionary of Scientific Terms.* This book, revised and enlarged by John H. Kenneth in 1949, is chiefly devoted to the biological sciences.

### *Biographical References*

1. *Who's Who in America*
2. *Dictionary of American Biography*
3. *American Men of Science*

4. *Who's Who in Science*
5. *Who's Who in Engineering*

## Leads to Trade Literature

In addition to the sources of information which you may find in a library, there is a vast amount of information published by industrial concerns in the form of bulletins, catalogues, reports, special brochures, and the like which can be of great usefulness. Most of these publications may be had free of charge by simply writing for them. The problem is to find out about them and to know where to write. Three books in the library may be of special value in solving this problem:

1. *Thomas' Register of American Manufacturers.* This two-volume work gives both an alphabetical and a classified list of American industrial organizations. Although it is designed as a purchasing guide, you can use it to find out where to write for information on special subjects.
2. *Sweet's Engineering Catalogues.* Commonly referred to as *Sweet's File,* this compilation of catalogues can also guide you to organizations which may be able to furnish you with information unavailable in your library.
3. *The Gebbie Press House Magazine Directory.* This is a guide to magazines published by industrial organizations for circulation among employees, stockholders, and interested outsiders. Many of these house "organs" are valuable sources of information. One example is *The Lamp,* a magazine published by the Standard Oil Company in a handsome format. It contains many articles of interest to the petroleum technologist.

Most of the official periodical publications of the engineering societies contain a column or so naming and describing new industrial bulletins, pamphlets, and the like.

## Conclusion

It is almost impossible to overestimate the value of a thorough acquaintance with library resources, and yet it is not difficult to learn your way around a library once you recognize that doing so calls for deliberate and methodical study. The aids which have been noted in this chapter are by no means sufficient in themselves to solve all of your future library research problems, but with these aids as a nucleus and with a genuine interest in the subject you will have no trouble in developing the knowledge that you will need.

## SUGGESTIONS FOR WRITING

One of the best ways of achieving a sense of professional confidence, aside from knowing your subject well, we believe, is knowing where to find information when it is needed. It will be to your advantage to become acquainted with the resources of your library, particularly in your own professional field. The suggestions below are intended to help you gain a knowledge of those resources—and enhance your sense of professional confidence.

1. Investigate the resources of your library in relation to your professional major. Find out which of the following are available: abstracts, biographical references, bibliographies, book review publications, directories, encyclopedias, dictionaries, handbooks, newspaper indexes, periodical guides, and indexes to government reports. You may have to ask the help of the reference librarian, but you will likely find him flattered to be asked to help. Then write a brief report to a hypothetical student who is interested in specializing in your field. Give him the basic information about each of the resources you find, together with information about the particular usefulness of the items, as well as information about where the books can be found.

2. Find out what professional periodicals relevant to your field of professional interest your library subscribes to. Examine an issue or two of each to determine what specialized sphere of interests each periodical appears to deal with (that is, what sorts of articles the magazine publishes). Write a letter-report to a freshman who intends to major in this field, telling him about these periodicals: how they may be useful to him as a student or as a professional man.

3. Prepare a bibliography of your library's resources on a topic of interest to you (perhaps on the subject you will write a research report on). Include books, periodical articles, theses, dissertations, and government-sponsored research reports. Use the bibliographical form described in the following chapter (or the form prescribed by your instructor).

4. Find out where and in what publications you would find information about commercial products. Write a brief report explaining to your reader where he could find information on a particular kind of product (say, high-fidelity speakers, oil additives for his car, transistors, laminated wood products, and so forth).

# 23
# Writing the Library Research Report

## Introduction

A study of technical writing that did not include the preparation of at least one report of four or five thousand words in length would be unrealistic. As you know, reports of this length, and indeed of much greater length, are common in industry and business. And there are some problems in writing a long report that are different from the problems of writing a short one, particularly in organizing and handling data. Furthermore, in an academic study of technical writing it is only in a fairly long report that you are likely to find a realistic synthesis or combination of the writing problems you have previously been studying more or less in isolation.

These are good reasons for writing a long report. Usually the most feasible way of preparing a long report in a course in technical writing is to write a library research report.

The library research method itself may seem somewhat unrealistic, however, consisting as it does of study and discussion of what other people have said about a subject. This method does not require you to wrestle with a mass of raw data, as was usually required of the workers whose books, articles, and reports you would read in the library. This is a serious disadvantage. But in most respects the library research method is quite satisfactorily realistic. The problems of style, organization, format, handling of transitions, and so on, are not different from what you will encounter in a report on original work.

Moreover, the writing of library research reports is often required in science and industry—not to mention advanced courses in college. When a new technical project of any kind is begun, one of the first tasks to be accomplished is frequently a search for everything in print that may have some bearing on the subject. The results of such a search are usually written up in the form of a report or used as the basis for a portion of a report.

Two other factors connected with the preparation of a library research report deserve mention. One is that it provides an excellent opportunity for increasing your knowledge of how to use the library. A second is that it is an unusual opportunity to study in detail some technical subject that you would like to know more about. Students sometimes find the preparation of such a report a first step toward mastery of a special subject; and such specialized knowledge is often helpful in securing a position. Numerous students, within our own knowledge, have shown prospective employers a report of this kind as part of the evidence of their fitness for a job.

Considered as a process, the library research report requires the following steps: (1) selecting a subject, (2) making an initial, tentative plan of procedure, (3) finding published materials on the chosen subject, (4) reading and taking notes, (5) completing the plan, writing the rough draft, and documenting the text, and (6) revising the rough draft and preparing the completed report. These steps will be discussed in the order stated, with the exception of step (3) to which the preceding chapter was devoted.

## Selecting a Subject

In selecting a subject for your long report you should look for one that has the following qualifications: (1) the subject is interesting to you, (2) it is related to your major field, (3) it is a subject about which you already know enough so as to be able to read intelligently but not one about which you have nothing to learn, (4) it is restricted enough in scope so that it can be treated adequately within about 5000 words, and (5) it is a subject on which sufficient printed material is available to you.

Most of these qualifications are self-evident and need no particular comment. As for the last of them we would suggest this caution: don't assume that any library book or article or other document will be available for you until you have it in your own possession. You may find that a highly important article is off at the bindery, or lost, or charged out to an unco-operative faculty member. One other point in the list above—the problem of scope—will be discussed at length in a moment.

The five qualifications listed can in general be met by three different kinds of subjects: (a) subjects representing a project you are actively working on, like building a boat, designing a gas model airplane, or remodeling a room in a residence; (b) subjects concerned with the making of a practical decision, like choosing the best tape recorder within a given price range, or the best outboard motor, or the best type of foundation for a given residence at a given location; (c) subjects which will add to your store of practical knowledge, like cross-wind landing gear for aircraft, offshore drilling platforms in the Gulf of Mexico, or rammed-earth construction for small homes. Which one of these three types of subject matter is the best for you depends on your interests and your background.

You will find it helpful to make a list of possible choices under each of the three headings as one of the first steps in making your selection. As the next step, go to the library to find out whether sufficient material is available on the most attractive subject on your list. Right at this point, however, there is a strong possibility that you will run into the problem of limiting the scope of your subject. Chances are that your first formulation of the subject will prove too broad for a report of the length required. To make some suggestions about what to do, we'll consider an exaggerated example.

Suppose you are interested in the subject of television and decide that you will write a report on it. You consult the card catalogue and the appropriate indexes to periodical literature in the library, and discover scores of articles and books which deal with television. Obviously, you cannot read them all. Even if you could, you would find from an examination of titles that they deal with so many different aspects of the subject that unification of the material would be next to impossible. Further examination would show that much of the material is superseded by later developments and so is of no value. But the main thing would be the impossibility of covering all the material available. Two courses are open: you may reject the subject of television entirely or limit your investigation to some particular phase of it.

Assuming that you are unwilling to give up the subject altogether because of your interest in it, you will find several opportunities for limiting the scope of the subject. Let's consider just a few. You might begin with a time limitation and see what has been published on television developments during the past few years. It would quickly become evident that, for this subject, this way of limiting the subject would be wholly inadequate. So you might try subdividing according to subject matter rather than time. Here the classifications in the card catalogue and indexes would be of service. For instance, you might find a number of articles devoted to various parts of television apparatus, such as antennas, amplifiers, or cathode-ray tubes. Or special broadcasting problems might prove of interest, such as the feasibility of the use of satellite relay stations. In short, examination of the titles of publications on the general subject in which you are interested should suggest any number of ways of limiting the subject to manageable proportions. Ultimately, of course, you must examine the publications themselves to be absolutely certain that they offer adequate but not too abundant material for a report of the length you are expected to write. But most of the work of limiting the scope of a subject can be done by careful thought initially and by careful study of available sources of information.

In general, subjects of current interest are best, subjects that for the most part are treated in periodical articles. If whole books have been devoted to a subject, it is likely to be too broad for report treatment. This does not mean, of course, that books may not be used as sources of information. To return to our illustration above for a moment, you would be fairly certain of finding books on television in which chapters might be devoted to antennas, and these would be useful for a report on antennas. If you were to find, however, that an entire book or several books had been written on antennas, you would probably decide that the topic is too broad for adequate treatment in a five-thousand word report.

## Making an Initial Plan

Once a subject has been chosen and approved by your instructor, it is time to lay a few plans for general organization and coverage of the subject, so as to simplify and give direction to the task of reading and taking notes.

First, make a list of the things you want to find out about your subject. Add to this list those things you think your reader will want to know. Sometimes these items will be identical, sometimes not. This list will prove most useful to you when you read and take notes: you will have some idea of what to look for in the reading. Of course it will undoubtedly prove necessary to revise the list, perhaps a number of times. You may discover, for instance, that nothing has been published on some particular aspect of the subject which you thought ought to be discussed. You may discover important aspects of the subject discussed which had not occurred to you when you were making your tentative guidance outline. You may discover that certain aspects of a subject will have to be eliminated because of space limitations. Such a list or outline should not in any sense be regarded as final but merely as a general guide, something to give you a sense of direction. It should be subject to change at any time.

If your knowledge of a chosen subject is so slight to begin with that you do not feel able to compile such a list, you should do some general reading on your subject first in order to acquire the necessary acquaintance with it. Then you can make a tentative outline. The importance of making a list for guidance will be made clearer in the discussion of note taking.

An example of the relation of an initial list of topics for guidance to a final report outline is shown below. The subject is the magnetic fluid clutch.

*Initial Guidance List*

1. What functioning parts does it have?
2. How does it operate?
3. How much does it cost to make?
4. How efficient is it?
5. Is it difficult to maintain and repair?
6. When was it developed and by whom?
7. How does it compare with other types?

*Final Outline*

I. Introduction
 A. Definition of a clutch
 B. The need the magnetic clutch can fill
 C. Object of this report
 D. Scope and plan of this report
II. Principle of operation
III. Description of clutch
 A. Driving assembly
 B. Magnetic fluid
 C. Driven assembly
 D. Electric coil
IV. Advantages and disadvantages
 A. Inertia
 B. Simplicity of design
 C. Leakage
 D. Ease of control
 E. Number of parts
 F. Smoothness of operation
 G. Fluid trouble
 H. Centrifugal trouble
V. Applications
 A. Automotive
 B. Servo-mechanisms

## Reading and Taking Notes

Once a preliminary list of sources (bibliography) has been compiled and a guidance list has been set down, it is time to begin reading and taking notes. In deciding what to read first from a list of sources, you should choose a book or article which promises to give a pretty general and complete treatment of the subject and to be simply and clearly presented. How a book or article rates in these qualities can be guessed at by examining titles and places of publication. For instance, an article entitled "Color Television Explained" appearing in a popular magazine is certain to be easy to understand and nontechnical in its treatment. By reading simply-written articles covering your subject broadly, you will be better able to understand and use the information you find in books and specialized periodicals. It may happen that your judgment of a title and place of publication will turn out to be wrong, but in general you will simplify your job by following this procedure.

You can now begin reading and taking notes. This is a job that should be highly systematic from the start. The following paragraphs outline an efficient method.

Three basic requirements of any good system of reading and note taking are (1) the reading should be conducted according to a plan, not haphazardly; (2) the method of arranging the sequence of the notes should be highly flexible; and (3) the system should be economical of time.

The first of these three requirements has already been discussed. Its observance requires the preparation of an initial guidance list or outline so that pertinent materials can be selected and irrelevant materials ignored. We can go on, then, to discuss the second and third requirements.

Flexibility of arrangement of the notes is easily achieved by the use of cards. In theory, the method requires that only one note be written on a given card (4 by 6 cards are a convenient size). It is next to impossible to define "one note," of course, For our discussion, however, it will be sufficient to say that one note is any small unit of information that will not have to be broken up so that the parts can be placed at separate points in the report. When all note cards have been prepared and arranged in the proper order, it should be possible to write the report without ever turning forward or backward as you go through the pack of cards. Naturally, such perfection is scarcely to be expected in practice.

Several symbols or labels can be used to save time. The first of these is a heading, put at the top of the card (see Figure 1). This heading is useful in the process of sorting and arranging the cards. It can be taken from the tentative outline. Some people like to add a symbol from the outline (like II A).

Secondly, it is convenient to use a symbol to indicate the source from which the note was taken. It is imperative that you indicate the source of every note so that the text can be documented. One way to keep a record of sources would be to write complete bibliographical data on each note card. This method would be inefficient, however, whenever you had more than two or three note cards from the same source (imagine writing eight or ten times the data for an article with a long title from a journal with a long title, and written by two or three people).

A better way is the following. Write the bibliographical data on a blank card. You may find it helpful to use a card of a different size from the note card, or a different color, so that it is easy to distinguish the bibliographical cards from the note cards. On this bibliographical card, put a capital letter in the upper left-hand corner. Now, when you make a note from this source, instead of writing the complete bibliographical data on the bottom of the note card, all you need do is write the capital letter that will key the note card to the proper bibliographical card.

The last entry on the note card is the number of the page from which the information in the note was taken. This is shown in Figure 1.

One thing we have not yet mentioned is the nature of the notes themselves. What kind of note should one make—an outline, a series of words and phrases intended to recall complete discussions, full, almost word-for-word transcriptions, or what? Our advice is that your notes should first of all be entirely in your own words (except for quotations, about which we shall say more in a moment) and in summary form as far as possible. Secondly, we strongly recommend that notes be made full enough so that you will not be confused as to their meaning and significance later on when you come to use them. It is important to avoid using the same phrasing and sentence structure that the author of the article uses. To make the material your own, you should first read it carefully, making sure you understand it, and then put down what you want to use, briefly and in your own words. We would emphasize the importance of economy of words in note taking, to simplify the job of studying and using the notes later on. It is quite discouraging to read a note, perhaps some weeks after it was made, only to discover that it does not make sense to you and that you have to return to the source to find out what you had tried to get down in your notes.

Direct quotations are not essential to a research report, but there are several reasons why the writer of a research report may wish to quote the words of an author directly. First, he may want to lend force and emphasis to a section of his report by quoting a well-known authority. Second, he may feel that the statement of an original idea deserves to be presented in the originator's own words. Third, he may feel that an author has stated a point in such a way that inclusion of the original will enhance the interest of his own

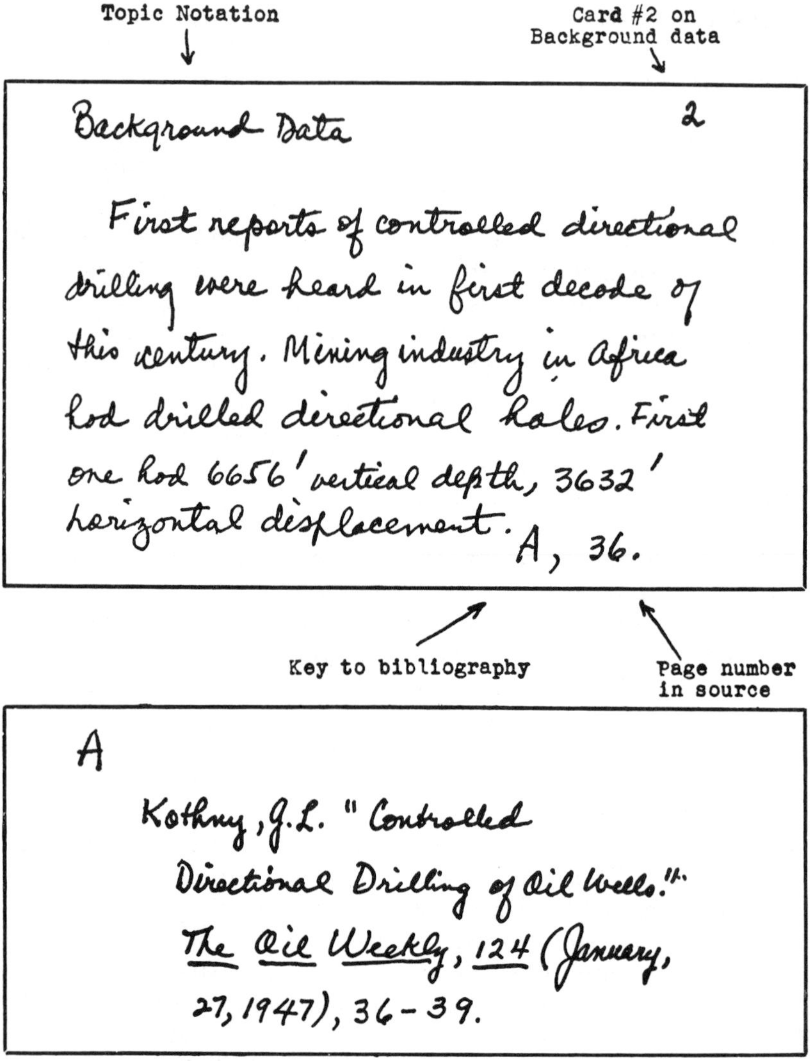

Topic Notation

Card #2 on Background data

Background Data 2

First reports of controlled directional drilling were heard in first decade of this century. Mining industry in Africa had drilled directional holes. First one had 6656' vertical depth, 3632' horizontal displacement. A, 36.

Key to bibliography

Page number in source

A

Kothny, G.L. "Controlled Directional Drilling of Oil Wells." *The Oil Weekly*, 124 (January, 27, 1947), 36-39.

*Figure 1. A Note Card and a Bibliography Card*

report. Fourth, he may wish to reproduce an author's opinion on a topic. Whatever the reason for a quotation, it is essential that the quotation be absolutely exact. Quotation marks should be placed around it in your notes and the page number of the source noted.

In concluding this discussion of note taking, we should like to urge you to remember that the system we have described is not a magic formula. You will have to use intelligence at every step, and you will inevitably have to do a lot of work with the cards, discarding repetitious notes and filling in gaps. On the other hand, the system is efficient. It comes closer than any we know to satisfying the three basic requirements stated at the beginning of our discussion. Its advantages are attested by the fact that it is widely used.

## Completing the Plan and Writing the First Draft

Once you have completed taking notes, it is time to prepare a final outline of your report. To do this properly, you will have to read through your notes carefully, perhaps a number of times. First of all, you will want to make yourself thoroughly familiar with the content of your notes to be able to write about your subject naturally and clearly. While you are mastering the content of your notes, you will be devoting some thought to the best order in which to present the topical divisions of your subject matter. It is at this stage that the usefulness of the note-taking system just described becomes most apparent. With topic headings on each card, you can now rearrange the cards in the order in which you think the contents should be presented. This rearranging may call for several experimental tries before you are satisfied with the result.

When the above preliminaries have been carried out, writing the rough draft amounts to little more than transcribing your notes to paper in a connected, coherent discussion. They will not be transcribed verbatim, of course. Although your notes, if properly taken, will be in your own words rather than in those of the authors of your sources, you will do well to rephrase and reword many of the passages in your notes as an additional safeguard against reflecting the style of writing used in your sources. You will do a good deal of this anyway (that is, without making a conscious effort) if you have mastered the content of the notes thoroughly. Furthermore, you will be adding transitional statements, developing and clarifying some points of fact by supplementary discussion, making comments (evaluations and conclusions) about the facts from your sources, and the like. In short, the report itself will contain a good deal of writing that is yours, personally, and not merely transferred from a source to your paper.

Most inexperienced writers attempt too much when they undertake the first draft of a report. They try to devote attention not only to the subject matter itself but also to style and correctness of expression. In writing the first draft, forget about style and correctness. Concentrate on the subject matter alone. Get down on paper what you want to say; there will be time later for smoothing out your sentences, correcting your spelling, punctuation, and choice of words. If you have a lot of inertia to overcome in getting started with your writing, do not conclude that you have no talent. Most writers are slow in getting started, even professional ones. Once a start is made, however bad, the task usually becomes easier. Awkward beginnings can be remedied later on.

## Documenting the Report

Documentation is the recording of published source materials for the research report. Sources are recorded in two forms: (1) a bibliography which appears at the end of the report, and (2) footnotes which appear throughout the report at appropriate places. A bibliography is included in the report because it is a convenience for the reader to have all the source materials listed in one place and because the presence of the bibliography makes it possible to use simpler footnote forms in the text than would otherwise be possible. Footnotes satisfy the ethical obligation of a statement of indebtedness and make it possible for the reader to check the authenticity of the text or find additional information.

*The Bibliography.* The bibliography is an alphabetized list, according to authors' last names, of all the written sources you have consulted—books, magazine articles, pamphlets, bulletins. Since you will already have a card for each of your sources with the necessary bibliographical information on it, making the formal bibliography is simply a matter of listing the items in proper order and recording the entries in correct form. For a book or independent publication, such as a booklet, you need to put down, in order: the author's name (surname first), the title (along with data pertinent to it), the place of publication, the publisher's name, and the date of publication. Here are some typical examples:

Buck, R. C. *Advanced Calculus.* New York: McGraw-Hill Book Company, Inc., 1965.

De Russo, P. M., R. J. Roy, and C. M. Close. *State Variables for Engineers.* New York: John Wiley & Sons, Inc., 1965.

Kuh, E. S., and R. A. Rohrer. *Theory of Linear Active Networks.* San Francisco: Holden-Day, Inc., 1967.

The second item above shows how to present multiple authorship.

Be careful about the form of bibliographies. Notice that a period follows each element: authorship, title, and publishing data. Each entry is single-spaced. The first line begins even with the left-hand margin; additional lines, if needed, are indented five spaces. There is double spacing between entries in a typescript.

The same general form is used for magazine article entries: authorship, title, and publishing data. The publishing data include the name of the magazine, the volume number of the magazine, the date of publication of the issue in which the article appears, and the page references. Study the following examples:

Chua, L. O., and R. A. Rohrer. "On the Dynamic Equations of a Class of Nonlinear RLC Networks." *IEEE Trans., CT-12* (1965), 475–488.

Livingstone, F. C. "Getting Science to Pay Off." *Science News, 96* (September 13, 1969), 224.

"SALT: A Season for Reason: Strategic Arms Limitation Talks." *Time, 94* (August 29, 1969), 16.

Notice that a period follows each main element of the entry, just as in an entry for a book. The title of the article is enclosed in quotation marks; the title of the periodical is underscored on the typewriter. The underscored number given just before the date of issue is the volume number of the magazine. It is underscored to distinguish it from the page number which follows the date. The volume number may also be written in Roman capitals to distinguish it in the entry. The third item in the magazine listing above shows what to do with an article for which no authorship is given. Enter it alphabetically according to the first important word in its title. Many magazines print articles by staff members without the signature of the author. As a matter of fact, often a number of people on the staff of a periodical have had a hand in the authorship of an article. This is especially true of news magazines like *Time* or *Oil and Gas Weekly*. If an article is labeled "Anonymous" in the magazine in which it appears, it should be so labeled in a bibliography, of course.

Oftentimes the question arises as to whether an article read but not made use of should be listed in the bibliography. As we have noted before, the researcher often finds several articles containing substantially the same information. If he has made notes on the first of these articles, he will not have taken notes on the others. Should he list them in his bibliography? The answer depends on whether the bibliography is to be regarded strictly as a list of sources actually used for facts within the report, or a list of articles on the subject. Our feeling is that only those published materials which have been used in writing the report should be listed. On the other hand, it certainly is defensible to list all sources dealing with a subject which have been read, whether or not all of them furnished data which are ultimately incorporated into the report. Perhaps your instructor will have an opinion on this matter by which you can be guided.

*The Use of Footnotes.* Now we come to the problem of writing notes at the bottom of pages to acknowledge indebtedness for facts and ideas presented in the discussion. Later we shall illustrate the form and content of various kinds of footnotes, but first let's clearly understand what needs to be footnoted. In other words, when do you need to write a footnote? The answer is simple: every fact, idea, and opinion which you have secured from your reading, quoted or paraphrased, must be acknowledged in the form of a footnote. Although footnoting may seem like an alarmingly difficult task at first thought, it really isn't so very difficult. Remember, your first draft has been largely a transcription of your notes, and your notes contain precise indications of the source and page number of each fact. You can, of course, put footnotes into your report during the process of writing the first draft. Or you may wish to wait until you have finished getting your discussion down on paper. It doesn't really matter when it is done. The important thing is to do it, and do it completely.

A number of questions naturally arise. Suppose one sentence contains information from two or more sources—does this call for only one footnote, or for more than one? On the other hand, suppose several pages of discussion in the report are based on a single source—does a footnote need to appear after each sentence, each paragraph, or at the end of the discussion?

Every unoriginal statement must be documented. That means two footnotes must be written if a single sentence contains data from two distinct sources. If a paragraph contains information from a dozen sources, a dozen footnotes appear at the bottom of the page. If several pages are based on one source, just one note is needed, at the end of the discussion. To put it another way, a footnote must appear at the end of each portion of discussion which is based on a particular source. The portion may be a phrase, a sentence, a paragraph, or a longer part of your composition. Let us repeat, your notes will have each fact identified as to source and page number. Except for the work, there will be no difficulty in documenting each fact.

"But suppose," you may say, "that in between two paragraphs of information taken from sources appears a paragraph which is original, like an evaluative comment, or a transitional paragraph. Do I need to footnote that?" The answer is no; you do not need to footnote yourself. "But," you may object, "how will the reader know that what I am saying is original and not taken from one of my sources? Perhaps he will think I've simply forgotten to put in a note." The answer to that is: the reader can usually tell from the nature of the comment that you are advancing—its content and style—that it is you speaking, not one of your sources. Just remember to document all the facts you have secured in your research, all the information you acquired *after* beginning your investigation, and you will have done a satisfactory job of documenting your report.

Although the foregoing discussion may make it appear that you will have an extremely large number of footnotes in your report, it doesn't usually take nearly as many as you might think. Let's consider an actual case. One of our students wrote a report entitled *Tantalum as an Engineering Material.* He organized it according to the following main headings: Introduction, Occurrence, Extraction of Tantalum, The Working of Tantalum, Tantalum Alloys, Uses for Tantalum, and Costs of Tantalum. His bibliography contained fifteen items. His report, which was twenty-one pages long, contained thirty-two footnotes.

Here's how they were distributed. None was necessary for the introduction because in this section he simply introduced the reader to tantalum as one of the rarer metals and explained what he proposed to discuss in the remainder of the report; he explained the purpose of the report, its plan of presentation, its limitations, and its point of view. Section II contained eight footnotes, five of them references to one source. It happened, you see, that most of his information on the occurrence of tantalum came from a United States Bureau of Mines article entitled "World Survey of Tantalum Ore." The other three articles referred to in this section had provided him with bits of information not contained in the above-mentioned item. He could have got by quite adequately with four footnotes instead of eight for this section in view of the fact that most of his data came from the one source. The third section of his report, on the extraction of tantalum, was based on information from one source, and he used two footnotes, one for each of the two paragraphs in this section. He could have used just one note. The fourth section, on the properties of the metal, contained seven footnotes, all but two of them references to one source. This one source contained the most complete discussion of the properties of tantalum; the other two articles referred to gave him a few facts not contained in the chief source. Here again he could have reduced the number of footnotes from seven to three or four. His fifth section, on the working of tantalum, contained eight footnotes, six of them from three sources. In general, separate subdivisions of this section were based on different sources so that he found it necessary to put a footnote at the end of each topical subdivision of the section, plus an additional one for a direct quotation. The next to last section, on uses for tantalum, required four footnotes—he had found material on four uses in four different articles. The last section, on costs, contained two notes.

The ratio of report length to number of footnotes described above is fairly typical, but no particular significance should be attached to the example. Another report of the same length might contain twice as many footnotes, or half as many. It all depends on how many sources are used and the extent to which each is used. You should not use more footnotes than are needed, but you should use enough to make clear the source of all information secured during the process of investigation.

The method of footnoting just discussed provides for complete documentation of the content of the report. Some people, however, feel that a somewhat less demanding and rigorous attitude toward the need for footnotes may be taken. They feel that facts which are common knowledge to workers in the field in which the research subject falls do not need to be documented, even though they may have been new to the writer at the time he began research. This would mean that footnotes would be necessary only for (1) direct quotations, (2) controversial matters, (3) ideas of critical importance in the content of the report, (4) citation of well-known, authoritative writers, (5) acknowledgment of an author's originality in developing the idea presented, and (6) comments on additional material the reader might like to examine.

Although footnotes are primarily used for references to sources in a research report, they do have other uses. Definitions of technical terms used in the text may be put in the form of footnotes if it is felt that some readers may need the definitions. If a term is one of crucial importance, its definition should appear in the text. It may happen that a number of terms need to be used which may or may not be familiar to a reader; the writer will not want to interrupt his discussion repeatedly to supply definitions, and footnotes offer a satisfactory solution. Footnotes may also be used for other statements which do not properly belong in the discussion. Suppose, for instance, that you find all your sources but one in agreement on some point in your discussion. A footnote could be used to report this one exception to general agreement. Finally, let us say that footnotes should be kept to a minimum; although necessary for acknowledging sources of information and occasionally for supplementary discussion, they do constitute an interruption to the reader and certainly add nothing at all to the readability of reports. Do not put a lot of them into a report in the hope that they will make it more impressive.

*The Form of Footnotes.* Knowing when and what to footnote does not completely solve the problem of documentation. We have yet to consider the form of the notes. We shall describe what we consider to be a satisfactory form for various kinds of footnotes, but we must frankly say that no single standard exists among professional scientists and engineers for documentary forms. In general, professional societies, research organizations, and editors of technical periodicals agree on what information should appear in a bibliography and in a footnote, but they by no means agree on the form for this information. Although a standard would in some ways be desirable, the absence of one does not constitute a serious problem. Our advice is to try to find out what form is preferred by the person or organization to whom you are going to submit the report. If no specific form is preferred, make use of the form described in this book. The important requirements of documentary forms are that they be accurate and complete enough so that the reader could locate the source should he wish to. The forms should be as brief as possible consistent with these requirements.

When a bibliography is presented at the end of a report, as we suggest for the library research report, footnotes may identify sources as briefly as possible. If no bibliography is appended, first footnote references to sources must be complete. In discussing the form and content of footnotes, we shall assume that a complete and formal bibliography is to appear at the end of the report.

A first reference to a book source should contain the author's surname, the title of the book (underlined), and the page reference, with commas after the author's name, the title, and a period at the end of the note. In typed material, the underlining indicates copy to be set in italics. If you have read articles by different writers with the same surname, you will have to preface the surname with initials. The same form is used for references to magazine articles except that the title of a magazine article is enclosed in quotation marks rather than underlined. The comma after a magazine title comes inside the last quotation marks. Thus two notes, one to a book and one to a magazine, would appear as follows on the first reference to either source:

[1]Jones, *Color Television*, pp. 19–21.
[2]Smith, "New Color Television System," pp. 12–14.

Compare these with the following bibliographical entries which would have to be made for each:

Jones, H. B. *Color Television*. New York: Rinehart & Company, 1951.
Smith, T. S. "New Color Television System." *Electronics Review, 31* (February, 1969), 10–20.

Subsequent references to the same sources may be presented more briefly. If either of these sources is referred to again later in the report, the footnote need contain only the name plus the new page reference, as:

[7]Jones, p. 18.
[8]Smith, p. 15.

Obviously, however, if you have used two or more publications by the same author, you will have to include the title in every footnote to avoid confusion.

If subsequent references to the same source are consecutive and on the same page, the abbreviation *ibid.* (for *ibidem* meaning "in the same place") may be used, along with the proper page reference if it differs from its immediate predecessor. This abbreviation does not represent much saving of time and effort if it is used in reference to a source with a single author; it does save time if there are several authors for an item. Compare:

> [5]Gaum, Graves, and Hoffman, p. 20.
> [6]*Ibid.*

Incidentally, the first of the above illustrations could be written "Gaum and others."

The above illustrations are for simple book and magazine article references, but the forms are suitable for references to practically any printed source of information. Remember that the title of any publication which is published as a separate unit is underscored, and that titles of items which appear within another publication (which has a covering title) are enclosed in quotation marks. Thus the title of an advertising leaflet of no more than a half-dozen pages would be underlined. The title of an article in an encyclopedia would be enclosed in quotation marks, but the title of the encyclopedia would be underlined.

Some additional details of mechanical handling of footnotes in a report are as follows: (1) number footnotes consecutively throughout your report with Arabic numbers raised slightly above the level of the line (they are called superscripts); (2) type or draw a heavy line part or all of the way across the page between the last line of the text on the page and the first of the footnotes; and allow at least two spaces above and below this line—you don't want the last line of text to appear to be underlined; (3) place the superscript numbers after the word, paragraph, section, or quotation in the text to which they refer, and do not put any mark of punctuation after them; (4) indent the first line of a footnote five spaces but begin additional lines flush with the left-hand margin; (5) single-space a footnote of over one line, but double-space between separate notes; (6) in a footnote the name of the author need not be repeated if it has already been given in the text (if you have written, "John Doe says in a recent article . . . ," you need not repeat John Doe's name in the footnote giving title and page). The sample page (Figure 2) from a student report illustrates most of the details.

Dr. St. Clair reasons that:

> These acoustic forces, arising from radiation pressure, act to cause a concentration of the suspended particles in the regions of maximum displacement and to produce attractive and repulsive forces between the particles.[19]

Variables to Consider

Three variables must be considered in connection with the precipitation of aerosols in large industrial volumes: the sound field intensity in which the aerosol is treated, the exposure time, and the frequency of the sound.[20]

Intensity. Although noticeable agglomeration is caused at 140 decibels, an intensity of about 150 decibels is most efficient for industrial application.[21] Sounds above 120 db, incidentally, are painful to the human ear.[22]

Effective conversion of the energy of a generator to sound depends upon the design of the generator and treating chamber. A properly designed installation may convert 40% to 60% of the compressed gas's energy to sound energy.[23]

---

[19]St. Clair, "Agglomeration of Smoke, Fog, or Dust Particles by Sonic Waves," p. 2439.

[20]Danser and Newman, "Industrial Sonic . . . ," p. 2440.

[21]Ibid., p. 2441.

[22]Jones, Sound, p. 245.

[23]Danser and Newman, p. 2440.

*Figure 2. Illustration of Page with Footnotes*

The foregoing discussion makes no attempt to be complete. Although the form described is satisfactory for most references, numerous special problems may arise. You should consult your instructor in working out a solution to them.

For example, some organizations (and journals) prefer to avoid putting footnotes at the bottom of appropriate pages; they prefer to gather references at the end of the article or report. When this custom is followed, reference numbers appear at appropriate places in the text (either as superscripts or as parenthetical numbers), but the documentation data appears in a consecutive listing at the end, usually headed "References." This method of gathering references at the end of an article or report has two possible advantages: the typist is freed of the necessity of remembering to allow enough space at the bottom of pages for footnotes, and the author can combine direct reference data with full bibliographical data into a single listing.

For knowledge of reference forms employed by publications in your professional field, consult the official publications of the professional societies. (The American Society of Mechanical Engineers, for instance, publishes a style manual; the American Institute of Electrical and Electronic Engineers distributes a booklet entitled *Information for Authors*.)

## Revising the Rough Draft and Preparing the Final Copy

After the rough draft has been completed, the next step is to revise the rough draft and prepare the report for submission. We shall assume that the rough draft has been documented. We suggest that you plan your work so that you will have plenty of time for the revision. It is an excellent idea to allow enough time so that you can lay your rough draft aside and forget it for several days, perhaps a week. The reason for this suggestion is that you will have difficulty spotting your mistakes if you undertake revision immediately after finishing the rough draft. (See the second example in Illustrative Material, Chapter 11, for an illustration of the kind of error too-hasty revision may produce.) You want to be able to read your rough draft objectively and critically, putting yourself as much as possible in the place of the person or persons who will read the finished report.

You can use the time between writing the rough draft and revising it to clean up other tasks incident to completion of the report: preparing the illustrations, preparing the cover and the title page, the letter of transmittal, the table of contents, the list of figures, and writing a first draft of the abstract.

As you start the final revision of the report, remember that you are making a revision, not a final copy. Making the final copy should be a purely mechanical operation, requiring no significant changes in the text. If you try to make revisions and final copy simultaneously, you will find yourself in such troubles as making a change on page ten that in turn requires a change on page five—which is already typed!

After you have completed the revision of the report, it is wise to go through it again several times, from cover to cover, deliberately checking each time for only one or two specific elements. Certainly the entire text should be checked once for grammar, with special attention to dangling phrases, pronoun reference, and subject-verb agreement; once for transitions; and once for spelling if you have trouble with spelling. We know a professional engineer who says he reads his reports through backward to check the spelling! He claims that by going backward he avoids getting absorbed in the meaning and finds it easier to catch mispelled words.

Just what elements should be checked depends to a considerable extent upon the material and your own strengths and weaknesses as a writer. The technical writing check list which we have included may be helpful. To this list we would strongly recommend that you add the following questions:

1. Are all the necessary functions performed by the introduction and the conclusion or summary?
2. Are transitions properly handled?
3. Are the principles of the special techniques of technical writing, like description of a process, observed as such techniques are required in the text?

Methodical use of a check list is good insurance. If you can answer "Yes" to all the questions in it, you can turn in your report, confident that you have done your best and reasonably assured that you have done well.

## Report Appraisal

The list at the end of this chapter is intended to assist you in planning, writing, and editing your own reports or in indicating to others the specific weaknesses of reports submitted to you for editing.

Before appraising a report, be sure to determine its exact purpose. What response is desired from the reader—or readers?

### SUGGESTIONS FOR WRITING

A choice of topic for a library research report should be made in accordance with principles discussed in this chapter, the resources of the library accessible to you, and the advice and approval of your instructor. The following list of possible topics for investigation is not intended to be limiting; we hope, rather, that it will be suggestive. We might also point out that many of the listed topics may need to be modified or limited so as to permit a full treatment of the subject in the length your instructor specifies. We have found, incidentally, that subjects of current and timely interest are often best since most of the available information on them is likely to be found in periodicals (subjects about which entire books have been written are usually too broad to be dealt with adequately in a paper of the length normally specified).

Nuclear Propulsion for Aircraft
Problems of Air Traffic Control
Odor Control in Air Conditioning
Artificial "Grass" for Athletic Fields
Anti-Air Pollution Devices for Automobiles
Gas Turbine Developments for Automobiles
New Developments in Electrically Powered Automobiles
Scientific Results of the Apollo Moon Exploration
Concrete Additives
Automobile Safety Devices and Standards
Quality Control and Reliability Techniques
The Mechanical Heart: Current Developments
Marijuana: Physiological Effects
Microphotography
Miniature Television Cameras
Microminiature Integrated Circuits
Industrial Water Pollution: Developments in Combatting
Hallucinogenic Drugs
Forest Fire Prevention and Control
New Techniques in Desalination
Practical Applications of Time-Sharing Computer Systems
The Use of Lasers in Medicine
Ultrasonics in Machining
Teaching Machines
The Apollo Space Program (some aspect of)
Some Aspects of the ABM Controversy
Food for Astronauts
Disc Brakes for Automobiles
New Techniques in Flow Metering
Cyrogenics in Hydrocarbon Processing
Oyster Farming
Alluvial Gold Deposits: How to Discover
Torque Control in Gas Turbines
Experimental Living in the Sea

Dolphin Research
High Fidelity Component Compatibility
Tape Recorder Care and Maintenance
Computer Applications in ________ (select a field)
Government Specifications for Manual Writing
Information Retrieval by Machine

## A TECHNICAL WRITING CHECK LIST

*Before You Begin, Have You . . .*

1. Defined the problem?
2. Compiled all the necessary information?
3. Checked the accuracy of all information to be presented?
4. Taken into account previous and related studies in the same field?
5. Learned all you can about who will read your presentation?
6. Determined why they will read it?
7. Tried to anticipate questions your readers will want answered?
8. Determined your readers' attitude toward the objective of the presentation?
9. Decided on the slant or angle you want to play up?
10. Checked the conformity of your approach with company policy and aims?

*In Making a Plan, Have You . . .*

1. Planned an introduction that will introduce the subject matter and the presentation itself?
2. Arranged the parts of the presentation so that one part leads naturally and clearly into the next?
3. Included enough background information?
4. Excluded unnecessary and irrelevant detail?
5. Planned a strong, forceful conclusion?
6. Clearly determined the conclusions and recommendations, if any, that should be presented?
7. Settled on a functional format: headings, subheadings, illustrations, etc.?

*In Writing, Have You . . .*

1. Expressed yourself in language that conveys exactly what you want to say?
2. Used language that is adapted to the principal readers?
3. Used the fewest possible words consistent with clearness, completeness, and courtesy?
4. Achieved the tone calculated to bring about the desired response?

5. Tried to produce a style that is not only accurate, clear, and convincing but also readable and interesting?
6. Presented all the pertinent facts and commented on their significance where necessary?
7. Made clear to the reader what action you recommend and why?
8. Correlated illustrations and art work closely with text?

*In Reviewing and Revising, Have You . . .*

1. Fulfilled your purpose in terms of the readers' needs and desires?
2. Proofread painstakingly for errors in grammar, punctuation, and spelling?
3. Weeded out wordy phrases, useless words, overworked expressions?
4. Broken up unnecessarily long sentences?
5. Checked to see if headings serve as useful labels of the subject matter treated?
6. Deleted words and phrases that might be antagonistic?
7. Honestly judged whether your choice of words will be clear to the reader?
8. Checked whether transitions are clear?
9. Double-checked to see that the introduction sets forth clearly the purpose, scope, and plan of the presentation?
10. Let someone else check your work?

*Finally, Have You . . .*

1. Finished the presentation on time?
2. Produced a piece of writing you can be proud of?

# *APPENDIXES*

# Appendix A

# A Selected Bibliography

## Grammar, Usage, and Style

Bryant, Margaret M. *Current American Usage.* New York: Funk & Wagnalls,1962.

Evans, Bergen and Cornelia. *A Dictionary of Contemporary American Usage.* New York: Random House, 1957.

Flesch, Rudolf. *The ABC of Style—A Guide to Plain English.* New York: Harper & Row, 1964.

Follett, Wilson. *Modern American Usage.* New York: Hill & Wang, 1966.

Fowler, H. W. *A Dictionary of Modern English Usage*, 2d ed. Rev. by Sir Ernest Gowers. New York: Oxford University Press, 1964.

Gause, John T. *The Complete University Word Hunter.* New York: Crowell, 1967.

Hawley, G. G., and Alice. *Hawley's Technical Speller.* New York: Reinhold, 1955.

McCartney, E. S. *Recurrent Maladies in Scholarly Writing.* Ann Arbor, Mich.: University of Michigan Press, 1953.

Nicholson, Margaret. *American English Usage.* London: Oxford University Press, 1957.

Partridge, Eric. *Concise Usage and Abusage.* New York: Philosophical Library, 1955.

Perrin, Porter G., K. W. Dykema, and W. R. Ebbitt. *Writer's Guide and Index to English*, 4th ed. Chicago: Scott, Foresman, 1965.

Strunk, William, and E. B. White. *The Elements of Style.* New York: Macmillan, 1959.

Turner, Rufus P. *Grammar Review for Technical Writers.* New York: Holt, Rinehart and Winston, 1964.

Vallins, G. H. *Good English.* New York: British Book Centre, 1952.

'Vigilans.' *Chamber of Horrors.* New York: British Book Centre, 1952.

## On Simplifying Style

Bernstein, T. M. *The Careful Writer.* New York: Atheneum, 1965.

Bernstein, T. M. *Watch Your Language.* Great Neck, N.Y.: Channel Press, 1958.

Flesch, Rudolf. *The Art of Plain Talk.* New York: Harper & Row, 1946.

Flesch, Rudolf. *The Art of Readable Writing.* New York: Harper & Row, 1949.

Flesch, Rudolf. *How to Test Readability.* New York: Harper & Row, 1951.

Gowers, Sir Ernest. *The Complete Plain Words.* London: Her Majesty's Stationery Office, 1960.

Gunning, Robert. *Technique of Clear Writing.* New York: McGraw-Hill, 1952.

Masterson, James R., and W. B. Phillips. *Federal Prose.* Chapel Hill, N.C.: The University of North Carolina Press, 1948.

## Technical Publications Problems

Baker, C. *Technical Publications.* New York: Wiley, 1955.

Clarke, Emerson. *A Guide to Technical Literature Production.* River Forest, Ill.: TW Publishers, 1961.

Clarke, Emerson. *How to Prepare Effective Engineering Proposals.* River Forest, Ill.: TW Publishers, 1962.

Doss, M. P. (ed.) *Information Processing Equipment.* New York: Reinhold, 1955.

Emerson, Lynn A. *How to Prepare Training Manuals.* Albany, N.Y.: University of State of New York, State Education Department, 1952.

Fry, Bernard, and J. J. Kortendick (eds.) *The Production and Use of Technical Reports.* Washington, D.C.: Catholic University of America, 1955.

Gill, Robert S. *The Author Publisher Printer Complex*, 2d ed. Baltimore: Williams & Wilkins, 1949.

Godfrey, J. W., and G. Parr. *The Technical Writer.* New York: Wiley, 1959.

Jordan, R. C., and M. J. Edwards. *Aids to Technical Writing*, Bulletin No. 21, Engineering Experiment Station. Minneapolis: The University of Minnesota, 1944.

Lasky, Joseph. *Proofreading and Copy Preparation: A Textbook for the Graphic Arts Industry.* New York: Mentor Press, 1954.

Mandel, Siegfried, and D. L. Caldwell. *Proposal and Inquiry Writing.* New York: Macmillan, 1962.

Melcher, Daniel, and Nancy Larrick. *Printing and Promotion Handbook*, 2d ed. New York: McGraw-Hill, 1956.

Reisman, S. J. (ed.) *A Style Manual for Technical Writers and Editors.* New York: Macmillan, 1962.

Singer, T. E. R. (ed.) *Information and Communication Practice in Industry.* New York: Reinhold, 1958.

Weil, B. H. (ed.) *Technical Editing.* New York: Reinhold, 1958.

Weil, B. H. (ed.) *The Technical Report.* New York: Reinhold, 1954.

## On Graphic Aids

Arkin, Herbert, and R. R. Colton. *Graphs—How to Make and Use Them.* New York: Harper & Row, 1936.

Cholet, Bertram. *Technical Illustration.* Brooklyn: Higgins Ink Co., 1953.

Dicerto, J. J. *Planning and Preparing Data-Flow Diagrams.* New York: Hayden, 1964.

French, T. E., and C. J. Vierck. *Graphic Science—Engineering Drawing, Descriptive Geometry, Graphical Solutions,* 2d ed. New York: McGraw-Hill, 1963.

Gibby, J. C. *Technical Illustration,* 2d ed. Chicago: American Technical Society, 1962.

Hoelscher, R. P., C. H. Springer, and J. S. Dobrovolny. *Graphics for Engineers.* New York: Wiley, 1968.

Modley, Rudolf, and D. Lowenstein. *Pictographs and Graphs—How to Make and Use Them.* New York: Harper & Row, 1952.

Rogers, A. C. *Graphic Charts Handbook.* Washington, D.C.: Public Affairs Press, 1961.

Schmid, C. F. *A Handbook of Graphic Presentation.* New York: Ronald, 1954.

Stevenson, G. A. *Graphic Arts Handbook and Products Manual.* Torrance, Calif.: Pen and Press, 1960.

Turnbull, A. T., and R. N. Baird. *The Graphics of Communication.* New York: Holt, Rinehart and Winston, 1964.

## On Technical and Scientific Reports

Ball, John, and C. B. Williams. *Report Writing.* New York: Ronald, 1955.

Blickle, Margaret D., and K. W. Houp. *Reports for Science and Industry.* New York: Henry Holt and Company, 1958.

Comer, D. B., and Ralph R. Spillman. *Modern Technical and Industrial Reports.* New York: Putnam, 1962.

Cooper, Bruce. *Writing Technical Reports.* Baltimore: Penguin Books, 1964.

Douglass, P. F. *Communication Through Reports.* Englewood Cliffs, N.J.: Prentice-Hall, 1957.

Gaum, C. G., H. F. Graves, and Lyne S. S. Hoffman. *Report Writing,* rev. ed. Englewood Cliffs, N.J.: Prentice-Hall, 1950.

Holscher, Harry H. *How to Organize and Write a Technical Report.* Totowa, N.J.: Littlefield, Adams, 1965.

Jones, W. Paul. *Writing Scientific Papers and Reports*, 5th ed. Dubuque, Ia.: Wm. C. Brown, 1965.

Kerekes, Frank, and Robley Winfrey. *Technical and Business Report Preparation,* 3d ed. Ames, Ia.: Iowa State College Press, 1962.

Linton, Calvin D. *How to Write Reports.* New York: Harper & Row, 1954.

Nelson, J. Raleigh. *Writing the Technical Report*, 3d ed. New York: McGraw-Hill, 1952.

Racker, Joseph. *Technical Writing Techniques for Engineers.* Englewood Cliffs, N.J.: Prentice-Hall, 1960.

Rautenstrauch, Walter. *Industrial Surveys and Reports.* New York: Wiley, 1940.

Rhodes, Fred H. *Technical Report Writing*, 2d ed. New York: McGraw-Hill, 1961.

Rose, Lisle, B. B. Bennett, and E. F. Heater. *Engineering Reports.* New York: Harper & Row, 1950.

Santmeyer, S. S. *Practical Report Writing.* Scranton, Pa.: International Textbook, 1950.

Sawyer, T. S. *Specification and Engineering Writer's Manual.* Chicago: Nelson Hall, 1960.

Sklare, A. B. *Creative Report Writing.* New York: McGraw-Hill, 1964.

Souther, James W. *Technical Report Writing.* New York: Wiley, 1957.

Sypherd, W. O., A. M. Fountain, and V. E. Gibbens. *Manual of Technical Writing.* Chicago: Scott, Foresman, 1957.

Tuttle, Robert E., and C. A. Brown. *Writing Useful Reports.* New York: Appleton-Century-Crofts, 1956.

Ulman, Joseph N., and Jay R. Gould. *Technical Reporting*, rev. ed. New York: Holt, Rinehart and Winston, 1959.

Van Hagan, Charles E. *Report Writer's Handbook.* Englewood Cliffs, N.J.: Prentice-Hall, 1961.

Waldo, Willis H. *Better Report Writing.* New York: Reinhold, 1957.

Wyld, Lionel D. *Preparing Effective Reports.* New York: Odyssey Press, 1967.

Zall, Paul M. *Elements of Technical Report Writing.* New York: Harper & Row, 1962.

## On Scientific and Technical Writing (Not restricted to reports)

Crouch, W. George, and R. L. Zetler. *Guide to Technical Writing.* New York: Ronald, 1948.

Davidson, H. A. *Guide to Medical Writing.* New York: Ronald, 1957.

Ehrlich, Eugene, and Daniel Murphy. *The Art of Technical Writing.* New York: Bantam Books, 1964.

Emberger, Meta Riley, and M. R. Hall. *Scientific Writing.* New York: Harcourt, 1955.

Fishbein, Morris, and J. F. Whelan. *Medical Writing, the Technic and the Art.* New York: McGraw-Hill, Blakiston Division, 1948.

Gensler, Walter J., and K. D. Gensler. *Writing Guide for Chemists.* New York: Harcourt, 1961.
Gilman, William. *The Language of Science.* New York: Harcourt, 1961.
Harwell, George C. *Technical Communication.* New York: Macmillan, 1960.
Hays, Robert. *Principles of Technical Writing.* Reading, Mass.: Addison-Wesley, 1965.
Hicks, T. G. *Successful Technical Writing.* New York: McGraw-Hill, 1959.
Houp, K. W., and T. E. Pearsall. *Reporting Technical Information.* Beverly Hills, Calif.: The Glencoe Press, 1968.
Kapp, Reginald O. *The Presentation of Technical Information.* New York: Macmillan, 1948.
Klein, David. *The Army Writer*, 4th ed. Harrisburg, Pa.: Military Service Publishing Co., 1954.
Kobe, K. A. *Chemical Engineering Reports.* New York: Interscience, 1957.
Marder, Daniel. *The Craft of Technical Writing.* New York: Macmillan, 1960.
Mills, John. *The Engineer in Society.* Princeton, N.J.: Van Nostrand, 1946.
Mitchell, John. *Handbook of Technical Communication.* Belmont, Calif.: Wadsworth, 1962.
Mitchell, John. *Writing for Technical and Professional Journals.* New York: Wiley, 1968.
Morris, Jackson E. *Principles of Scientific and Technical Writing.* New York: McGraw-Hill, 1966.
Oliver, L. M. *Technical Exposition.* New York: McGraw-Hill, 1940.
Rathbone, R. R. *Communicating Technical Information.* Reading, Mass.: Addison-Wesley, 1966.
Rathbone, R. R., and J. B. Stone. *A Writer's Guide for Engineers and Scientists.* Englewood Cliffs, N.J.: Prentice-Hall, 1962.
Rickard, T. A. *Guide to Technical Writing.* San Francisco: Mining and Scientific Press, 1908.
Rose, L. A. *Preparing Technical Material for Publication.* Urbana, Ill.: University of Illinois, 1951.
Sherman, T. A. *Modern Technical Writing*, 2d ed. Englewood Cliffs, N.J.: Prentice-Hall, 1966.
Tichy, H. J. *Effective Writing for Engineers, Managers, Scientists.* New York: Wiley, 1967.
Trelease, Sam F. *Scientific and Technical Papers.* Baltimore: Williams & Wilkins, 1958.
Ward, Ritchie R. *Practical Technical Writing.* New York: Knopf, 1968.
Weisman, H. M. *Basic Technical Writing*, 2d ed. Columbus, Ohio: Merrill, 1968.
Wicker, C. V., and W. P. Albrecht. *The American Technical Writer.* New York: American Book, 1960.

Woodford, F. Peter (ed.) *Scientific Writing for Graduate Students.* New York: The Rockefeller Univ. Press, 1968.
Zetler, R. L., and W. G. Crouch. *Successful Communication in Science and Industry.* New York: McGraw-Hill, 1961.

## Some Industrial Style Manuals and Reporting Guides

Beck, L. W., and Phyllis K. Shaefer. *The Preparation of Reports,* 3d ed. Wilmington, Del.: Hercules Powder Company, 1945.
Gaddy, L. *Editorial Guide.* Denver: The Martin Company, 1958.
*Guide for Preparation of Air Force Publications,* AF Manual 5–1. Washington, D.C.: Department of the Air Force, 1955.
Holscher, H. H. *How to Organize and Write a Technical Report.* Toledo, Ohio: Owens-Illinois, 1958.
*Instruction Manual for Preparing Research Reports, Minutes of Steering Committee Meetings, and Proposals,* Revised. Chicago: Armour Research Foundation, 1958.
*Instruction Manual Writing Guide.* Doraville, Georgia: Scientific-Atlanta, Inc. (n.d.).
Martin, Miles J. *Technical Writing and Speaking.* Schenectady, N.Y.: General Electric Research Laboratory, 1953.
Middleswart, F. F. *Instructions for the Preparation of Engineering Department Reports,* Revised. Wilmington, Del.: E. I. du Pont de Nemours & Co., Inc., 1953.
*Preparation of Technical Publications.* Indianapolis, Indiana: U.S. Naval Avionics Facility, 1966.
*Procedure Manual.* Richmond, Va.: Reynolds Metal Company (n.d.).
*Report and Letter Writing.* Philadelphia: Sun Oil Company, 1955.
*Report Standard.* Dearborn, Michigan: Ford Motor Company, 1967.
*Style Manual,* 2d ed. New York: American Institute of Physics, 1959.
*Style Manual for Technical Writers,* Schenectady, N.Y.: General Electric Company, Research Laboratory, 1963.
*Style Manual.* Rochester, N.Y.: Stromberg-Carlson, 1968.
*Style Manual,* 6th ed. Revised. New York: Union Carbide and Carbon Corporation, 1954.
*Style Manual.* Ayer, Mass.: U.S. Army Security Agency, Training Center and School, 1966.
*Technical Report Manual.* Engineering Division, Chrysler Corporation, 1955.
*Technical Report Writing Procedure.* East Hartford, Conn.: Pratt & Whitney Aircraft, 1955.
*Technical Writing Guide.* Philadelphia: Philco Technological Center, 1959.
Wallace, J. D., and J. Holding. *Guide to Writing and Style.* Columbus, Ohio: Battelle Memorial Institute, 1966.

### On Business Letters and Reports

Anderson, C. R., A. G. Saunders, and F. W. Weeks. *Business Reports.* New York: McGraw-Hill, 1957.

Babenroth, A. C., and C. C. Parkhurst. *Modern Business English,* 5th ed. Englewood Cliffs, N.J.: Prentice-Hall, 1955.

Brown, Leland. *Effective Business Report Writing,* 2d ed. Englewood Cliffs, N.J.: Prentice-Hall, 1963.

Harberger, S. A., Anne B. Whitmer, and Robert Price. *English for Engineers,* 4th ed. New York: McGraw-Hill, 1943.

Hay, Robert O., and R. V. Lesikar. *Business Report Writing.* Homewood, Ill.: Irwin, 1957.

McCloskey, John C. *Handbook of Business Correspondence,* 2d ed. Englewood Cliffs, N.J.: Prentice-Hall, 1951.

Saunders, A. G. *Effective Business English.* New York: Macmillan, 1949.

Schutte, W. M., and E. R. Steinberg. *Communication in Business and Industry.* New York: Holt, Rinehart and Winston, 1960.

Sheppard, M. *Plain Letters.* New York: Simon and Schuster, 1960.

Shurter, R. L. *Effective Letters in Business,* 2d ed. New York: McGraw-Hill, 1954.

### Technical Writing Casebooks

Brown, James. *Casebook for Technical Writers.* San Francisco: Wadsworth, 1961.

Brown, James. *Cases in Business Communication.* San Francisco: Wadsworth, 1962.

Dawson, Presley C. *Business Writing: A Situational Approach.* Belmont, Calif.: Dickenson, 1969.

Levine, Stuart. *Materials for Technical Writing.* Boston: Allyn and Bacon, 1963.

Schultz, H., and R. G. Webster. *Technical Report Writing: A Manual and Source Book.* New York: McKay, 1962.

### Handbooks of Usage for Engineers

Guthrie, L. O. *Factual Communication.* New York: Macmillan, 1948.

Howell, A. C. *Handbook of English in Engineering Usage.* New York: Wiley, 1940.

Thomas, J. D. *Composition for Technical Students,* 3d ed. New York: Scribner, 1965.

### Scientific and Technical Exposition with Instructional Comment

Blickle, Margaret M., and M. E. Passe. *Readings for Technical* Writers. New York: Ronald, 1963.

Estrin, Herman A. (ed.). *Technical and Professional Writing: A Practical Anthology.* New York: Harcourt, 1963.

Gould, Jay R., and Sterling P. Olmsted. *Exposition, Technical and Popular.* New York: Longmans, Green, 1947.

Jones, E. L., and Philip Durham. *Readings in Science and Engineering.* New York: Holt, Rinehart and Winston, 1961.
Miller, W. J., and L. E. A. Saidla. (eds.) *Engineers as Writers.* Princeton, N.J.: Van Nostrand, 1953.
Ryan, L. V. *A Science Reader.* New York: Holt, Rinehart and Winston, 1959.

## Bibliographies on Technical Writing and Editing

*A Review of Literature on Technical Writing.* Boston Chapter, Society of Technical Writers and Publishers, 1958.
Philler, Theresa A., Ruth K. Hersch, and Helen Carlson (eds.). *An Annotated Bibliography on Technical Writing, Editing, Graphics, and Publishing, 1950–1965.* Washington, D.C.: Society of Technical Writers and Publishers and Carnegie Library of Pittsburgh, Pa., 1966.
Shank, Russell. *Bibliography of Technical Writing,* 2d ed. Columbus, Ohio: Society of Technical Writers and Publishers, Inc., 1958.

# Appendix B
# The Galt Manuscripts

The manuscripts in this appendix are presented so that you can study the process of revision. The materials include two versions of an oral report and two of a written report on the same subject—artificial refrigeration. The first version in each case is one of several rough drafts written by Mr. Galt. The second version in each case is the final one. Along with each version we have made editorial comments to draw your attention to some of the more significant differences between the two versions. We have presented versions of both the oral and written treatment of the subject in order that you may compare the techniques of these two forms of presentation.

The first and final versions of the talk are presented first, placed on facing pages for ease of comparison. Then the early draft and final draft of the written report are presented, also on facing pages.

These materials were prepared by Mr. John L. Galt, of the General Electric Company, for presentation in a contest sponsored by the American Institute of Electrical Engineers. They are reprinted here by permission of Mr. Galt. It may be of interest to you to know that Mr. Galt won the contest.

## Version 1 of Galt Talk

### Cold Facts

If you had a million dollars in your pocket, what would you do with it? The possibilities are unlimited, aren't they? You might go into the real-estate business, to the extent of 100 ten thousand dollar houses. You might buy 500 two thousand dollar cars and clean up a tidy little profit for a vacation in South America. I once knew a fellow who said if he had a million dollars he'd buy a half-dozen trainloads of peanuts. Some sort of a scheme on the order of the recent phenomenal developments in the soybean industry. That's what we all thought—but it wasn't that at all. He just liked peanuts.

*Mr. Galt realizes the importance of a good "lead" and here tries to capture the interest and attention of his audience at the very outset. He tries to do so by employing a common point of interest—the pleasure people get from speculating about what they would do with a lot of money. In this version, he develops his lead point by suggesting several possibilities—and he makes one of them humorous.*

Well, there was once a time when, if you had had a million dollars, you could have bought a ton of ice. Ice, that stuff you're getting so tired of scraping off your sidewalk. Ice, which is now being manufactured at a cost of less than half a cent per pound, less than ten dollars a ton.

*In this second paragraph—still a part of the introduction—Galt leads into his subject matter proper—ice. The technique here is the startling fact—designed to fix attention.*

You see, it hasn't been so many years that we've had artificial refrigeration and ice-making. In fact, as late as the nineteenth century, the summer supply of ice to the Southern states came from the North. It wasn't artificial ice, it was ice that had been cut from frozen lakes during the winter and packed away in sawdust. There were no railroads in the South then, so the ice was transported southward by means of waterways and oxcarts. It always sold at a very high price, but it was during the New Orleans yellow-fever epidemic of 1853 that ice sold for the fantastic price of 500 dollars per pound, to be used for crude air conditioning in sickrooms.

*The startling fact is explained in this paragraph.*

Was there ever a more challenging problem facing the engineering ingenuity and resourcefulness of man in his struggle for independence from nature? He had long since developed reasonably satisfactory methods of keeping himself and his house warm in the winter, but, other than ventilation, few means had been developed for his relief from heat.

*This paragraph and the next complete the process of introducing the subject matter proper.*

The problem has been admirably met by the men from whose work has evolved the science now known as "refrigeration engineering."

## Final Version of Galt Talk

### Cold Facts: The Story of Artificial Refrigeration

How many times have you sat in idle and pleasant speculation upon that characteristically American proposition, "Boy, what I couldn't do if I had a million bucks!"

Well, there was once a time, not so very long ago, when, if you had had a million dollars, you could have bought *a ton of ice—two thousand pounds of frozen water.* It was during the New Orleans yellow-fever epidemic of 1850; natural ice, transported from the North by oxcart and river barge, was at a premium for use in air-conditioning sick-rooms. That ice was selling for the fantastic price of *five hundred dollars a pound—a million dollars a ton.*

Here was a problem challenging to the engineering ingenuity of man—the forcefully demonstrated need for *controlled cooling* in man's habitat. That problem has been admirably met by a handful of pioneers, from whose work has evolved the modern science of "refrigeration engineering." Let's take a look at some of the aspects of this science, now so commonly taken for granted.

First of all, we may define "refrigeration" as "a process whereby heat is transferred from a place where it is undesirable to a place where it is unobjectionable." This transfer involves two distinct problems: first, the *collection* of heat from the space to be cooled, and second, the *transportation* of the heat away from the space. We will consider the solution of these problems in connection with "vaporization refrigeration."

Vaporization refrigeration is based upon the following principles: One, the evaporation of a liquid is accompanied by the absorption of a large quantity of heat; the common example is the cooling of the human body by the evaporation of perspiration. Two, lowering the pressure on a liquid lowers the temperature at which it vaporizes; on a hunting trip in the mountains, water for your coffee boils at a lower temperature than it does in the valley below. Let's see how we apply these principles to our refrigeration process. [Uncover Chart I.]

*In this version, Mr. Galt keeps his original lead idea, but pares it down by eliminating the suggestions which followed it in the original. By doing so, he gains emphasis, force, directness. Furthermore, he has cut the joke which, however funny it might be considered by his audience, would interfere with their grasping the really important point stated in the second paragraph.*

*The second and third paragraphs of the original become one here to give greater directness and force.*

*Rephrased for better emphasis, this paragraph usefully says, "Let's take a look . . ." to tell the audience what it can expect to hear discussed. Observe the shifting of the transition to "vaporization refrigeration" from the opening sentence in the first version to the previous paragraph here. The definition of vaporization refrigeration is clarified by the reference to perspiration.*

*"First of all" is a transition between the introduction and the body of the talk. The second sentence's opening ("This transfer") is clearer than the original's "It."*

*More than twice as long as the original, the explanations given here of evaporation and transportation are certainly clearer.*

"Refrigeration" may be defined as a process whereby heat is transferred from a place where it is undesirable to one where it is unobjectionable. It involves two problems, the *collection* of the heat, and the *transportation* of the heat away from the cooled space.

*In both this draft and the final draft, Galt uses sound technique by starting the main body of his discussion with a definition of the central topic—refrigeration—and by announcing the two problems involved.*

We will consider "vaporization refrigeration," which is based on the fundamental principles that liquids, on vaporization, absorb large quantities of heat with no change in the temperature of the fluid, and reducing the pressure on the liquid reduces the temperature at which it vaporizes.

*The explanation of evaporation and transportation of heat is given in technical terms in this draft. Compare it to the expanded, clearer final version where the more leisurely explanation gives the listener time to absorb the facts. This is a good example of "pace" in exposition. Brevity is not always to be equated with clarity.*

Since the *collection* of the heat is accomplished by providing a receptacle, or heat sink, at a temperature lower than that of the space to be refrigerated, we will maintain our receptacle at a relatively low pressure, allow our refrigerant to vaporize in it, and term it an "evaporator."

We will accomplish the *transportation* of the heat away from the receptacle by withdrawing the vaporized refrigerant from the receptacle, and since the refrigerant is normally too costly to throw away, provide some means of discarding the heat from it, recovering it as a liquid, and returning it to the evaporator in a closed cycle.

The four fundamental units needed in a closed cycle of refrigeration are (1) the evaporator to serve as our heat sink, (2) some device for withdrawing the vaporized refrigerant from the evaporator at a rate which will maintain the desired low pressure in the evaporator, and discharging it at a higher pressure so that its increased boiling point exceeds the temperature of the surroundings, (3) a condenser in which the heat is extracted from the refrigerant at its increased boiling point, returning it to the liquid state, and (4) an expansion valve to reduce the pressure of the liquid refrigerant as it is again passed into the evaporator.

The second of these steps may be accomplished by a number of devices, but they may all be grouped under two general types of system, the "compression" system and the "absorption" system.

*Note how the emphasis is changed in expressing this in the final version.*

The compression system is the one more

The collection of heat is accomplished by providing a receptacle at a temperature lower than that of the space to be refrigerated, so that heat tends to flow from the surrounding space to the lower temperature level of the receptacle. The receptacle is generally termed a "heat sink," just as a low-level ground area in a wet region is sometimes called a "water sink." We will fill *our* heat sink with liquid refrigerant, and then lower the pressure in the sink so that the refrigerant evaporates rapidly at a low temperature, absorbing heat from the surrounding space. This particular type of heat sink we call an "evaporator."

*Note the attempt to clarify an idea by means of comparison—the "water sink."*

With the heat collected, we transport it away from the space by exhausting the gaseous refrigerant from the evaporator. This step might be very simple if we could discard the heat, refrigerant and all, to the atmosphere. But normally the refrigerant is too costly to waste, so economy dictates that we shall provide in the transportation step some means for discarding *only* the *heat*, while recovering the refrigerant as a liquid and returning it to the evaporator in a *closed cycle*. [Uncover Chart II.]

*The original devoted one sentence to evaporation, one to transportation. This version devotes four to the former, three to the latter.*
*A speaker must cover important ideas carefully: he must not present them too fast. An audience lost at this point will be lost for good.*

Closed-cycle refrigeration involves four fundamental units: First, the evaporator to serve as the heat sink. Second, some device for withdrawing vapors from the evaporator, then discharging them at an increased pressure and temperature level. Third, a condenser with which to extract the latent heat of the refrigerant and return it to the liquid state. And fourth, an expansion valve with which to reduce the pressure on the liquid refrigerant as it returns to the evaporator.

*The transition here is strengthened by starting this paragraph with "Closed-Cycle." The sentences listing fundamental units have been shortened so that the ideas can be more readily grasped.*

The device used in the second of these steps may take a number of forms, but these forms may be classified under the two general types of refrigeration systems—the "compression system" and the "absorption system."

*The subject of the parallel paragraph in the original was "The second of these steps." The change to "The device" allows a more sensible and natural lead into a discussion of the two systems.*

The *compression system* is the commoner of the two, employing a centrifugal or reciprocating compressor to take in the vaporized refrigerant, do work on it, and then discharge it at a high pressure and temperature.

The *absorption system*, however, is the more

commonly used, employing a centrifugal or reciprocating compressor to withdraw the vaporized refrigerant, do work on it, and discharge it at an increased pressure.

The absorption system, however, is probably the more interesting, both from the engineering and the popular viewpoint.

Did you ever wonder how you can light a fire in the bottom of your refrigerator and take ice out the top? Well, it isn't quite as simple as it might appear from the way our befuddled friend here is making ice cubes to cool his glass of water.

*References in these paragraphs are to illustrations not reproduced here.*

The principles of the process were first discovered by the noted scientist Michael Faraday in 1823. Following Faraday's work, Ferdinand Carré, a Frenchman, developed a simple intermittent refrigeration system, and then, in 1850, patented the first practical continuous refrigeration machine.

*Note the short paragraphs in this section. They show Mr. Galt's recognition of the need for a slower pace in oral presentation than in formal written reports.*

As in Carré's machine, the modern continuous ammonia absorption machine operates as follows. Liquid ammonia from the condenser is passed through an expansion valve into the evaporator, where the ammonia vaporizes. In these steps the process is exactly like the compression system. However, the gaseous ammonia leaving the evaporator, instead of being passed through a compressor, is absorbed in water. The resulting solution of ammonia in water is pumped to a tank called the generator, maintained at the higher pressure of the condenser. Heat is applied to the generator, driving the ammonia out of solution. The water is returned to the absorber, and the gaseous ammonia to the condenser, from where the cycle is repeated.

But in 1926 there occurred a modification of Carré's machine which resulted in the most revolutionary development to occur in absorption refrigeration. It was conceived by two young undergraduate students at the Royal Institute of Technology in Stockholm, Baltzar von Platen and Carl Munters.

*This and the next four paragraphs explain the contribution of Platen and Munters.*

These students proposed to convert the Carré machine into a constant-pressure apparatus by introducing hydrogen into the evaporator and absorber to equalize the pressure with that in the generator and condenser. This might seem offhand to nullify the principle of a machine

complicated and the more interesting, both from the engineering and from the popular viewpoint.

Do you, perhaps, have in your home the type of refrigerator in which you light a fire in the bottom and take ice cubes out the top? [Uncover Chart III.]

It seems a bit paradoxical, doesn't it? Our friend here, though, an engineer relaxing from a hard day at the office, understands the system perfectly. He's in a hurry for ice cubes, so he's giving the refrigerator a boost by applying his blowtorch.

This is the *absorption* type of refrigerator. The principles of this system were discovered by the noted scientist Michael Faraday, in 1823, during his work on the liquefaction of gases. A Frenchman, Ferdinand Carré, patented, in 1860, the first practical continuous absorption refrigerating machine. [Uncover Chart IV].

*To make sure his audience is with him, he begins with a reminder that this is the absorption type. Compare with the original.*

*Mention of Carré's intermittent system is omitted here as unnecessary. See earlier version.*

Carré's machine uses ammonia as the refrigerant. High-pressure liquid ammonia from the condenser is passed through an expansion valve into the evaporator, where it vaporizes, absorbing heat. In these two steps the Carré machine is identical with the compression machine. The vapors from the evaporator, however, instead of being picked up by the suction side of a compressor, are absorbed into water. The resulting solution of ammonia in water is transferred by a suction pump to a tank called the "generator," maintained at the same high pressure as the condenser. A flame is applied to the generator, bringing the solution to a boil. The ammonia gas boils through a pressure-reducing valve, and the gaseous ammonia continues to the condenser for liquefaction.

In 1926 there occurred a modification of the Carré machine which resulted in the most revolutionary development in absorption refrigeration to date. Two young Swedish students at the Royal Institute of Technology in Stockholm, Baltzar von Platen and Carl Munters, conceived an invention to eliminate the valves and the pump from the Carré machine. This they proposed to do by the simple expedient of adding hydrogen to the "low-pressure side" of the system, thus converting the Carré machine into a *constant-pressure apparatus.*

*Only minor changes here —somewhat more detailed for clarity.*

*In this version the material in this and the next four paragraphs is differently organized to make it clearer and more interesting. Note the use of more pronouns.*

which produces refrigeration by the evaporation of a liquid at a reduced pressure. However, the hydrogen molecules in the evaporator, unlike the gaseous ammonia molecules, exert no effect whatever tending to keep the remaining ammonia molecules in the liquid state. Hence, passing liquid ammonia from an atmosphere of gaseous ammonia to one of mixed hydrogen molecules and gaseous ammonia molecules reduces the *effective* pressure on the ammonia just as surely as if it had been passed through an expansion valve.

Then there is no need for the expansion valve.

Since the total pressure in the absorber is the same as that in the generator, there is no need for the pump that formerly served to transfer the solution from the low-pressure to the high-pressure side of the machine—the transfer may be accomplished by gravity flow.

There is no longer a higher pressure in the generator to force the water back to the absorber, so we will replace the valve with a simple vapor lift, similar to that used in the coffee percolator.

The importance of the Platen-Munters features is emphasized by the fact that an American corporation paid five million dollars for the patent rights.

This corporation has since marketed a highly developed form of the Platen-Munters machine, unique in its valveless, pumpless control of fluids within a hermetically sealed space. It contains three interrelated fluid circuits, rotating in unison to produce continuous refrigeration, powered by nothing more than a small source of heat. These three circuits consist of an ammonia loop, an ammonia-hydrogen loop, and an ammonia-water loop.

*In this and the next four paragraphs Galt explains the operation of a highly developed machine in which the Platen-Munters features are incorporated. Note that the style is spare and economical.*

The process is shown diagrammatically and much simplified in Fig. V. Beginning with the ammonia-hydrogen loop, the ammonia enters the evaporator through a liquid trap which confines the hydrogen to its own circuit. In the evaporator it vaporizes, producing refrigeration. The heavier gaseous ammonia molecules mix with the hydrogen molecules, and the resulting increase in the density of the mixture causes the

Now here is a refrigerating machine which operates on the principle of vaporizing a refrigerant under *low* pressure, then condensing it under high pressure. It might appear that, on *equalizing* the pressure throughout the apparatus, we have immediately nullified the principle of operation.

The secret lies in the use of *hydrogen,* explained as follows: It is the *gaseous* ammonia in the atmosphere of the evaporator which exerts a pressure on the *liquid* ammonia, making it difficult for the liquid to vaporize. (It was for this reason that Carré reduced as much as possible the pressure in the evaporator.) The *hydrogen* molecules *exert no such effect* on the liquid ammonia. And so, on adding hydrogen to the atmosphere of the evaporator, we have reduced the *effective pressure* on the ammonia just as surely as if we had reduced the *total* pressure in the evaporator. Furthermore, by maintaining the total evaporator pressure the same as the condenser pressure, we have eliminated the need for the expansion valve.

With the hydrogen also present in the absorber, the pressure in the absorber is the same as that in the generator, and there is no need for a pump to transfer the solution of ammonia in water—the transfer can be done by gravity flow. The pump is eliminated.

We no longer have higher pressure in the generator to transfer water to the absorber through the pressure-reducing valve, but we *do* have in the generator a boiling liquid. So we substitute for the pressure-reducing valve a simple "vapor lift," like that used in your coffee percolator.

*This paragraph is new.*

No valves, no motor, no pump. The only external source of energy, a direct flame on the generator. *Ice*—from *heat!* It's amazing, isn't it?

The importance of the Platen-Munters features is emphasized by the fact that an American corporation paid *five million dollars* for their United States patent rights!

Since that time this corporation has marketed a highly developed form of the Platen-Munters machine, unique in its valveless, pumpless control of fluids within a hermetically sealed space.

If you own a gas refrigerator, you may be

*Note the personal element in*

heavy gas to sink down the vertical tube to the absorber.

In the absorber, the ammonia dissolves in the countercurrent stream of water, while the practically insoluble hydrogen, lightened of its burden of heavy ammonia molecules, ascends to the evaporator to again perform its task of mixing with and reducing the partial pressure of the ammonia.

Taking up the water-ammonia loop, the strong solution of ammonia in water, called "strong aqua," flows by gravity to the generator, where the application of heat drives the ammonia out of solution. A vertical tube, the inside diameter of which is equal to that of the bubbles of gas formed, projects below the surface of the boiling liquid, so that as the gas bubbles ascend the tube they carry with them slugs of liquid. This "liquid lift" empties into the separator, where the ammonia vapor is separated from the "weak aqua." The weak aqua then returns to the absorber by gravity through a second liquid seal.

Finally, the ammonia loop, which has been traced as far as the separator, next involves the "condenser," which removes the latent heat from the vapor, converting it into cool liquid ammonia. From here it passes through the liquid seal that marks its re-entry into the evaporator, and the cycle has been completed.

The industrial modifications of this machine are many and interesting, but we have traced the evolution of the artificial icemaking machine far enough to appreciate the resourcefulness that made it possible.

The frontiers of science are unlimited; the engineer thrives on the meat of research and development; it is not frequent that he can point to a feat of reduction of the price of a commodity to one one-thousandth of 1 per cent of its onetime cost, but as long as there is an unsolved problem in man's advancement, there will be an engineer.

*Galt closes with a reference to the cost of ice, which served as a point in his lead.*

interested in seeing how it works. [Uncover Chart V.]

*this paragraph—which is not in the other version.*

*In five paragraphs, beginning here, the action of a modern gas refrigerator is explained (by reference to a chart) as before, but this version is clearer, more direct and concrete, and personal.*

We'll pick up the flow at the generator, where our engineer friend applied his blow-torch. In the generator there is a solution of ammonia in water. The application of heat causes the solution to boil, releasing gaseous ammonia. The ammonia bubbles ascend the vertical tube to the separator, carrying with them slugs of water, which empty into the separator and flow to the absorber. The ammonia continues to the condenser, where it is liquefied.

Liquid ammonia flows from the condenser to the evaporator, where, in an atmosphere of hydrogen, it evaporates rapidly at a low temperature, absorbing heat from the water in the surrounding ice trays and producing ice.

The heavy molecules of gaseous ammonia, mixing with the lighter molecules of hydrogen, increase the density of the mixture, causing it to sink down the tube to the absorber.

In the absorber, the gaseous stream meets a countercurrent stream of water from the separator. The ammonia dissolves in the water, but the practically insoluble hydrogen, now relieved of its load of ammonia molecules, ascends again to the evaporator.

*In this last paragraph of the body of the talk, cost—sure to be of interest to all listeners—is mentioned.*

The solution of ammonia in water flows by gravity from the absorber to the generator. The cycle is complete. The motivating power, a tiny flame. The cost to you, some fifty cents a week.

*In these closing paragraphs, a more effective conclusion is given by omitting mention of other facts that might be discussed but cannot be, and by reverting more specifically to the cost factor which has been his "dramatic" lead item. Altogether, there are two observations to be made in conclusion: (1) the final version is clearer and simpler, and (2) it is more personal and therefore more likely to attract attention and interest.*

These have been some specific facts in the story of artificial refrigeration—the story of the successful development of a process for the industrial marketing of ice at about *six dollars a ton,* less than *one one-thousandth of 1 per cent* of its onetime cost. It is a process characterized by convenience and economy, the fruits of engineering research in the science of refrigeration.

The next time you open your refrigerator door for a glass of ice cubes, recall a picture of armies of slaves returning from mountainous regions with snow to cool the wine of medieval kings.

Remember that the people of New Orleans once *gladly* paid *five hundred dollars a pound* for ice!

## One of the First Drafts of the Galt Report

### Cold Facts

One of the chemical engineer's primary concerns is with economic problems. Hence it is of interest to him when he reads of a process whereby the cost of a common commodity has been reduced from 500 dollars per pound to half a cent per pound.

*Although similar to the lead in version two, this one is less lively and is somewhat stilted. Here we are bluntly told that the subject is of interest. No mention of the subject—artificial ice making—is made here. Compare with the second version where he sensibly announces his subject at the outset.*

Until the nineteenth century, the summer supply of ice to the southern part of the United States consisted of cartloads, cut from the northern lakes and ponds during the winter, packed in sawdust and canvas, and conveyed laboriously southward. During the yellow-fever epidemic of the early 1800's in New Orleans, this ice sold for as high as $500 per pound to relieve the suffering of fever-ridden patients.

*Here he explains the startling statement which ends the first paragraph. Compare with the expanded version.*

## The Final Draft of the Galt Report

### Cold Facts: The Story of Artificial Refrigeration

#### *Summary*

The following pages present a story exemplifying the engineering resourcefulness of man in his never-ending struggle for independence from nature.

Thousands of years ago man found reasonably satisfactory methods of keeping himself and his dwelling warm, but until comparatively recently, few means had been developed, other than simple ventilation, for his protection against heat.

Dr. William Cullen, a Scottish physician of the eighteenth century, is credited with invention of the first artificial refrigeration machine, and the passage of time since that date has been marked by the steady increase in the efficiency, and decrease in the cost, of man's efforts to control the cooling of his habitat. Modern ice-making machines are miracles of ingenuity—monuments to the progress of humanity.

*For this final version, Mr. Galt has added a summary. In the light of what this text has to say on the subject of summaries and abstracts, what do you think of this one?*

#### *I. Introduction*

One of the engineer's primary concerns, in his never-ending development of processes and products for the advancement of mankind, is with economic problems. Occasionally there occurs, through the medium of engineering development, a cost reduction which is nothing short of phenomenal. History reveals just such a feat in the evolution of the modern industry of artificial ice making, a process that has succeeded in reducing the onetime cost of that commonest of commodities from a *million dollars a ton* to its present cost of less than *six dollars a ton.*

*Since his readers are not all chemical engineers, he has dropped the adjective here. Galt makes the lead here more dramatic by changing the terms of the cost comparison from pounds to tons. The longer sentences of this version give a smoother style.*

Until the nineteenth century, the summer supply of ice to the southern part of the United States came from the North, ice that had been cut from frozen lakes and ponds during the winter, stored in sawdust and canvas, and conveyed laboriously southward later in the year. There were no railroads south of the Mason-Dixon Line, so transportation was accomplished by river barges and by oxcarts. Ice, needless to say,

*The chief difference between this second paragraph and that in the earlier draft is in the addition of details. These added concrete details help the reader understand the situation referred to in the opening paragraph and they add interest. The admission that ice was nor-*

The history of artificial ice making began in 1755 at the University of Glasgow, where Dr. William Cullen produced ice with an experimental vacuum machine. The first American machine was built by Dr. John Gorrie of Apalachicola, Florida, also a physician, in 1844, and development of the process proceeded rapidly in the United States thereafter.

*The style here is rougher than in the revision. Still no mention of refrigeration engineering, his subject!*

Refrigeration is a process whereby heat is transferred from a place where it is undesirable to one where it is unobjectionable. In order to do so, the heat must first be collected, then transported away from the cooled space.

The collection of the heat is accomplished by providing a receptacle which is at a temperature lower than that of the cooled space. This receptacle is termed a "sink."

The transportation of heat from the sink and the discard thereof depends on the method of collection. If an ample supply of a naturally cooled material, such as cold water, is available and applicable to the problem, the transportation of the heat is accomplished by the flow of the water to, through, and from the receptacle, after which the heated water may be discarded. More generally, a "vapor compression" system is used, in which case the discard device consists of a mechanical means for raising the temperature level of the heat collected at the sink to a value greater than that of the surroundings, so the heat may be dissipated to the air or to some other convenient cooling system.

*What do you think of the phrase "the discard thereof" in both versions?*

was a luxury item in the South, and brought a high price even under normal circumstances. But during the New Orleans yellow-fever epidemic of 1853, ice sold for the fantastic price of *five hundred dollars per pound* to relieve the suffering of fever-ridden patients.

*mally a luxury item in the South would forestall the possible objection from some readers that the figures given above were misleading.*

The history of artificial ice making, man's answer to the need for controlled cooling of his habitat, had begun in 1755 at the University of Glasgow, where Dr. William Cullen produced ice with an experimental vacuum machine. The first American machine was built in 1844 by Dr. John Gorrie of Apalachicola, Florida, also a physician, but it was a number of years before the work of these and other pioneers blossomed into the modern science known as "refrigeration engineering."

*The addition of the phrase, "man's answer . . . ," helps make the transition from the lead into this historical part of his introduction. What do you think of his use of the word "habitat"? The ending of this paragraph has been changed to get in a reference to refrigeration engineering. Note also how the statement varies in fact from that in the preceding version.*

*II. Definition of Refrigeration*

"Refrigeration" is a process whereby heat is transferred from a place where it is undesirable to a place where it is unobjectionable. The process involves two problems, the *collection* of the heat, and its *transportation* away from the cooled space.

*The heading is a transitional device. Note how the emphasis is strengthened by changing the structure of the second sentence and how this sentence now forecasts the topic sentences of the two following paragraphs.*

The collection of heat is accomplished by providing a receptacle which is at a temperature lower than that of the space to be cooled. This receptacle is termed a "heat sink."

The transportation of heat from the sink and the discard thereof depend on the method of collection. If an ample supply of a naturally cooled material, such as cold water, is available and applicable to the problem, the transportation of the heat is accomplished by the flow of the water to, through, and from the receptacle, after which the heated water may be discarded (Fig. I). More generally, however, the "vaporization refrigeration" method is used, in which case the transportation, as well as the collection, of the heat becomes a more complicated process.

Vaporization refrigeration is based on two fundamental principles: (1) Liquids, on volatilization, absorb relatively large quantities of heat with no increase in the temperature of the fluid. (2) Lowering the pressure on the liquid lowers the temperature at which it vaporizes.

*Note how he has proceeded more cautiously in this version, giving principles before the details of the operation. Is this version clearer?*

When a volatile liquid is used as the working substance, or "refrigerant," the heat sink is termed the "evaporator." When, as is usually the case, the refrigerant does not boil at a temperature less than that of the cooled space, nor is it replaceable at a low cost, economy dictates the need for inclusion of equipment which makes possible the reclaiming of the vapors and the re-use thereof in a closed cycle.

*Note how this material is handled in the final draft.*

The four fundamental units which are needed in a closed system of mechanical refrigeration are: (1) the evaporator to serve as the heat sink; (2) a device for removing vapor from the evaporator to maintain the operating pressure, and then to discharge the vapor at a higher pressure so that its increased boiling point exceeds the temperature of the surroundings to which its

*The discussion of fundamental units is the same in both versions.*

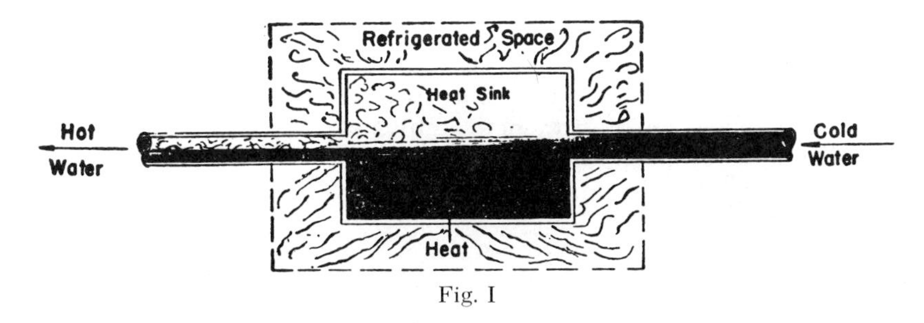


Fig. I

Using, then, as a refrigerant, a volatile liquid, the heat sink is maintained at a low pressure and is termed an "evaporator." Since the refrigerant is not usually replaceable at a low cost, economy dictates the need for inclusion, in the "transportation" step, facilities which make it possible to discard the heat to the atmosphere while reclaiming the vapors and re-using them in a closed cycle (Fig. II).

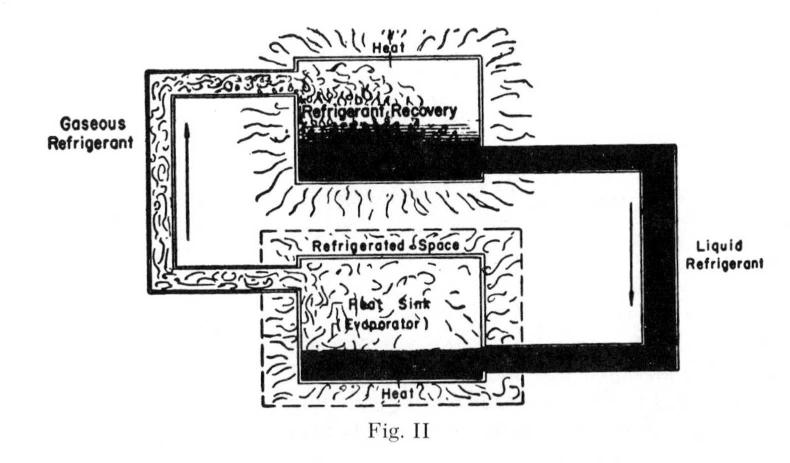


Fig. II

### *III. Units Required for Closed Cycle Refrigeration*

The four fundamental units which are needed in closed-cycle mechanical refrigeration are: (1) the evaporator to serve as the heat sink; (2) a device for removing vapor from the evaporator at a rate which will maintain the desired operating pressure therein, then discharging the vapor at a higher pressure so that its increased boiling point exceeds the temperature of the surround-

heat is to be discarded; (3) a condenser where the vapor gives up its sensible and latent heat and again becomes a liquid; and (4) an expansion valve to reduce the pressure of the refrigerant to that of the evaporator.

The device necessary to the accomplishment of the second step in this cycle may take a number of forms. The two most common ones are those shown in Fig. I, the "compression system" and the "absorption system."

*No figures were actually prepared for this version.*

The compression system is the one most commonly used, consisting merely of a reciprocating or centrifugal compressor which takes in low-pressure vapor, does work on it, and discharges it at a higher pressure.

The absorption system, however, is the most interesting of the two, using as a working substance a solution of two or more materials. The principles of the process were first demonstrated, unknowingly, by the noted chemist Michael Faraday in 1823. Faraday discovered that silver chloride would absorb ammonia, and, on heat-

*The discussion of Faraday remains about the same in both versions.*

*One of Mr. Galt's readers suggested that this discussion of Faraday's work needed clarifying. See figure*

ings to which its heat is to be discarded; (3) a condenser where the vapor may give up its latent heat and again become a liquid, and (4) an expansion valve to reduce the pressure of the liquid refrigerant as it returns to the evaporator (Fig. III).

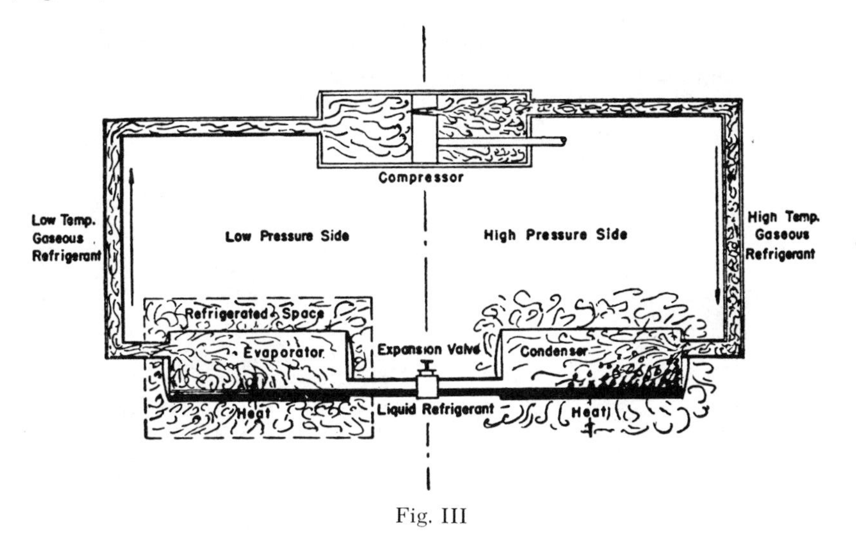


Fig. III

### *IV. Types of Refrigeration Systems*

The device necessary to the accomplishment of the second step in this cycle may take a number of forms. These forms may be grouped under two general systems of refrigeration, the "compression system" and the "absorption system."

### *A. Compression Refrigeration*

The compression system, illustrated in Fig. III, is the one most commonly used, employing a reciprocating or centrifugal compressor which takes in low-pressure vapor, does work on it, and discharges it at a higher pressure.

*Note the use of subheadings for emphasis.*

*The word "merely" in the first sentence of this paragraph in the first version has been deleted. What attitude did the term suggest in the earlier draft?*

### *B. Absorption Refrigeration*

The absorption system, however, is the more interesting of the two, using as a working substance a solution of two or more materials.

*Faraday's Work.* The principles of the process were first demonstrated, unknowingly, by the noted scientist Michael Faraday in 1823. Faraday discovered that silver chloride would ab-

*Why do you suppose he changed "chemist" to "scientist" in this version (in describing Faraday)?*

ing the mixture in one end of a bent glass tube, the ammonia would be evolved. He condensed the gas in the other end of the tube by immersing it in an ice-salt mixture, and found that on removing the tube from the bath, the ammonia boiled, producing a temperature far lower than that of the the ice-salt bath with which it had been condensed. Faraday had actually operated the first simple intermittent absorption system, using silver chloride as the absorbent and ammonia as the refrigerant.

*in the final draft. The sentence beginning "He condensed . . ." is split into two statements.*

The modern counterpart of Faraday's system is the Crosley Icy Ball, with its two balls connected by a length of pipe. In its simplest form, the first ball, containing a strong solution of ammonia in water, is heated, with the second ball being cooled by water. After ammonia has

*No change in the discussion of the Crosley Icy Ball.*

sorb ammonia; on heating the mixture in one end of a bent glass tube (Fig. IV), he caused the ammonia to be evolved. He condensed the gas in the other end of the tube by immersing it in

*The suggestion (from a reader of the early draft) that greater clarity was needed was met by the figure. Do you think the figure clarifies the discussion?*

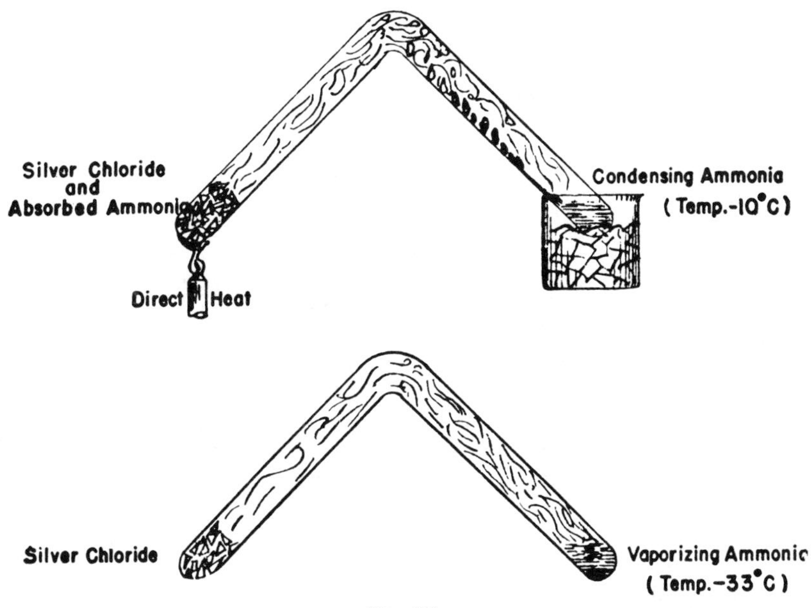


Fig. IV

an ice-salt mixture, and found that, on removing the tube from the bath, the ammonia boiled. The important aspect of the experiment was that the vaporization of the ammonia produced a temperature far lower than that of the cold bath with which it had been previously condensed. Faraday had actually operated the first simple intermittent absorption system, using silver chloride as the absorbent and ammonia as the refrigerant.

*Lest it be overlooked, Galt calls attention to the important significance of Faraday's experiment in this separate sentence.*

*Crosley Icy Ball.* The modern counterpart of Faraday's system is the Crosley Icy Ball, with its two balls connected by a length of pipe (Fig. V). In its simplest form, the first ball, containing a strong solution of ammonia in water, is heated, the second ball being cooled by water. After

been boiled out of the solution in ball No. 1 and condensed in ball No. 2, ball 1 is removed from the heat source and cooled. Refrigeration is then produced by ball 2, as the ammonia vaporizes and is reabsorbed in the weak solution in ball 1.

Following the work of Faraday, Ferdinand Carré developed the first practical continuous refrigerating machine in France in 1850. Carré's basic idea was to use the affinity of water for ammonia by absorbing in it the gas from the evaporator, then using a suction pump to transfer the liquid to another vessel where the application of heat caused the liberation of ammonia gas at a higher pressure.

*Note the difference in dates in the two versions.*

As in Carré's machine, the simple continuous ammonia absorption machine consists of four units: the evaporator, the absorber, the generator, and the condenser. In the ammonia-water system, high-pressure liquid ammonia from the condenser is allowed to expand through an expansion valve and the low-pressure liquid is then vaporized in the evaporator, absorbing its latent heat of vaporization from the surrounding refrigerated space. In these two steps the ammonia absorption system is exactly like the compression system. However, the gas from the evaporator, instead of being passed through a compressor, is absorbed in a weak solution of ammonia in water. The resulting strong solution

*The sentence beginning "As in Carré's machine . . ." does not make it clear whether the Carré machine is exactly the same as the ammonia absorption system or not.*

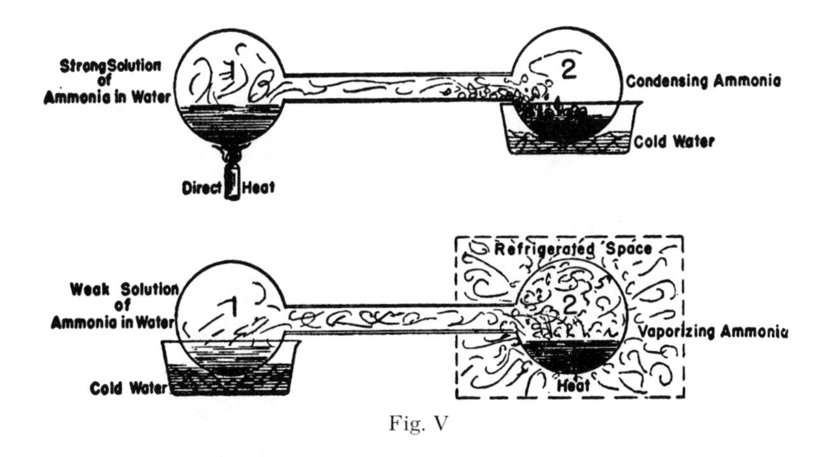


Fig. V

ammonia has been boiled out of the solution in ball No. 1 and condensed in ball No. 2, ball 1 is removed from the heat source and cooled. Refrigeration is then produced by ball 2, as the ammonia vaporizes and is reabsorbed in the weak solution in ball 1.

*The Carré Machine.* Following the work of Faraday, Ferdinand Carré developed and patented the first practical continuous refrigerating machine in France in 1860. Carré's idea was to use the affinity of water for ammonia by absorbing in water the gas from the evaporator, then using a suction pump to transfer the liquid to another vessel where the application of heat caused the liberation of ammonia gas at a higher pressure and temperature.

*This discussion of Carré remains about the same as that in the earlier version. He does add the fact that Carré patented his machine, and clears up a small matter of pronominal reference in the second sentence.*

Carré's machine is illustrated by the flow diagram in Fig. VI. In this ammonia-water system, high-pressure liquid ammonia from the condenser is allowed to expand through an expansion valve and the low-pressure liquid is then vaporized in the evaporator, absorbing its latent heat of vaporization from the surrounding refrigerated space. In these two steps the ammonia absorption system is exactly like the compression system. However, the gas from the evaporator, instead of being passed through a compressor, is absorbed in a weak solution of ammonia in water ("weak aqua"). The resulting strong solution ("strong aqua") is then pumped to the generator, which is maintained at high

*He leaves out the sentence in which he listed the four components of the system. Do you think he should have?*

is then pumped to the generator, which is maintained at high pressure. Here the strong solution is heated and the ammonia gas driven off. The weak solution which results flows back to the absorber, the highly compressed ammonia gas from the generator is condensed, and the cycle repeated.

Small ammonia absorption units have played an important part in the development of the absorption system. In 1900 Geppart proposed to convert the water-ammonia absorption machine into a constant-pressure device by adding some permanent gas to the system. His plan was to circulate the gas over the liquid ammonia in the evaporator, causing accelerated evaporation.

*This transition sentence (to the reference to Geppart) is not very effective—not very clear, in fact.*

But nearly twenty-five years later occurred the modification of Geppart's idea which resulted in the most revolutionary development yet to occur in absorption refrigeration. It was an invention to eliminate all valves and pumps in the Carré machine, conceived by two undergraduate students at the Royal Institute of Technology in Sweden, Baltzar von Platen and Carl Munters.

*This introductory paragraph about Platen and Munters stands in the final version.*

Its basis is Dalton's fundamental "law of additive pressures," which states that the total pressure of a gas mixture is the sum of the partial

*This paragraph remains unchanged in revision, except for a change of pro-*

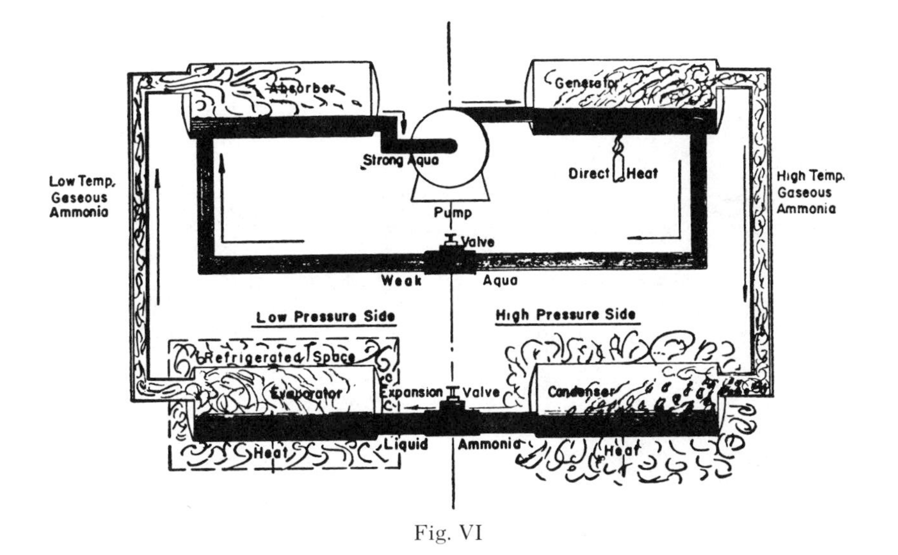


Fig. VI

pressure. Here the strong aqua is heated and the ammonia gas driven off. The weak aqua which results flows back to the absorber through a pressure-reducing valve, the highly compressed ammonia gas from the generator is condensed, and the cycle repeated.

*The phrase "through a pressure-reducing valve" has been added for the sake of greater completeness and accuracy.*

*Geppart's Modification.* In 1900 Geppart proposed to convert the water-ammonia absorption machine into a constant-pressure device by adding some permanent gas to the system. His plan was to circulate the gas over the liquid ammonia in the evaporator, causing accelerated evaporation.

*This version omits the ambiguous opening sentence of the earlier version and proceeds at once to Geppart.*

*Platen-Munters Invention.* But nearly twenty-five years later occurred the modification of Geppart's idea which resulted in the most revolutionary development yet to occur in absorption refrigeration. It was an invention to eliminate the pump and valves from the Carré machine, conceived by two undergraduate students at the Royal Institute of Technology in Sweden, Baltzar von Platen and Carl Munters.

The basis of this modification is Dalton's "law of additive pressures," which states that the total pressure of a gas mixture is the sum of the

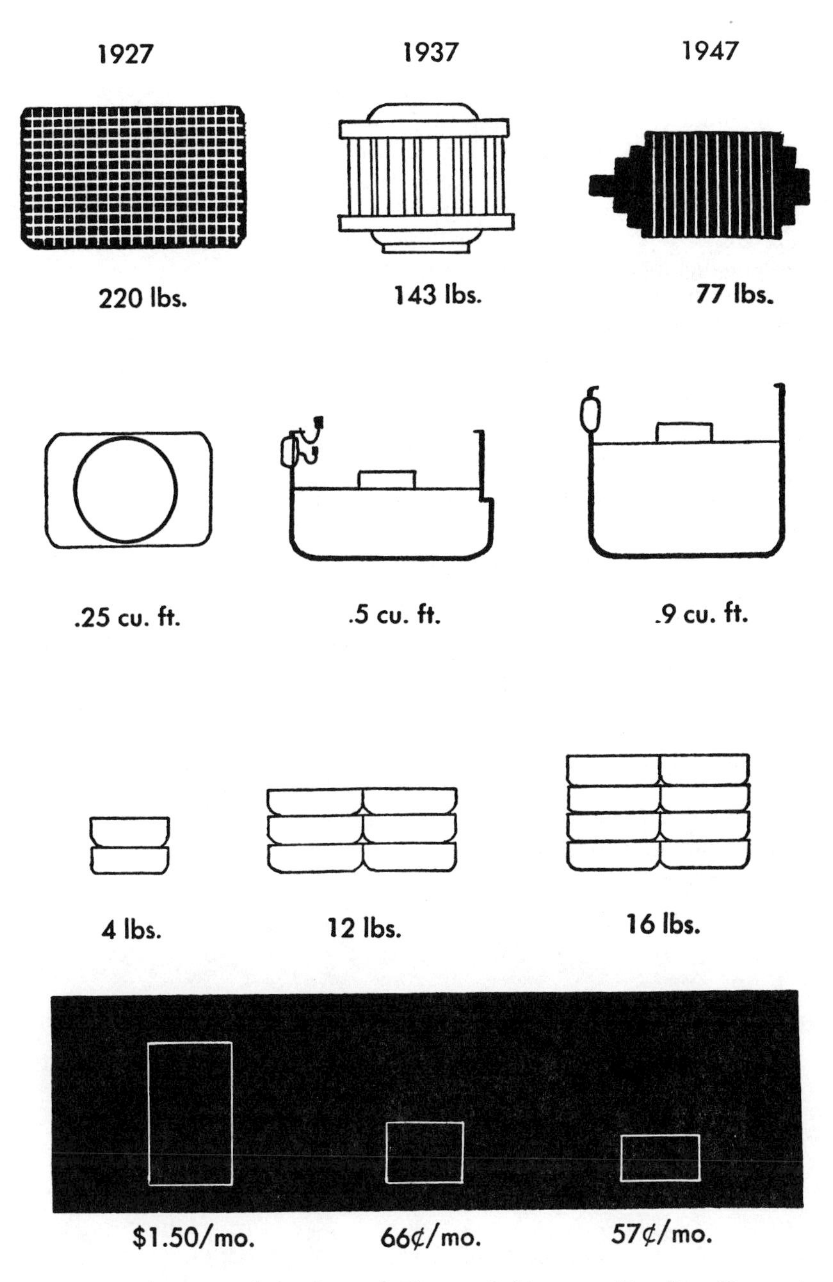


Fig. VII. Development of the General Electric Refrigerator. *Top line*: Decreasing weight of the refrigerating mechanism; *second line*: increasing volume of the freezing space; *third line*: increasing ice capacity; *bottom line*: decreasing cost of operation. Redrawn from the original photograph.

pressures of the individual components. The principle is applied by supplying an atmosphere of hydrogen in the "low-pressure" side of the apparatus, thus equalizing the pressure throughout the system. The desired drop in effective pressure on the ammonia as it leaves the condenser and enters the evaporator is accomplished by the change from an atmosphere of gaseous ammonia to one of a mixture of ammonia and hydrogen. The hydrogen molecules exert no effect tending to keep the ammonia molecules in the liquid phase; the concentration of ammonia molecules in the vapor phase is lessened by the presence of the hydrogen, and hence its pressure on the liquid ammonia is lessened. The result is that produced by the expansion valve in the Carré system—the ammonia evaporates rapidly at a relatively lower temperature and pressure, and the refrigerated space is cooled.

*nouns in the next to last sentence, and the omission of the last clause.*

Since, under these conditions, the total pressure in the absorber is also the same as that in the generator, there is no need for a pump to convey the strong solution from the absorber to the generator—the transfer can be accomplished by gravity flow. The return of weak solution to the absorber can no longer be done by the higher pressure in the generator, so a simple vapor lift, similar to that in a coffee percolator, is utilized.

*One of Mr. Galt's readers remarked that this portion of his discussion was not clear. A summary type of paragraph, as well as Fig. VIII, was added in the final draft to increase the clarity.*

partial pressures of the individual components. The principle is applied by supplying an atmosphere of hydrogen in the "low-pressure" side of the apparatus, thus equalizing the pressure throughout the system. The desired drop in effective pressure on the ammonia as it leaves the condenser and enters the evaporator is accomplished by the change from an atmosphere of gaseous ammonia to one of a mixture of ammonia and hydrogen. The hydrogen molecules exert no effect tending to keep the ammonia molecules in the liquid phase; the concentration of ammonia molecules in the vapor phase is lessened by the presence of the hydrogen, and hence their pressure on the liquid ammonia is lessened. The result is that produced by the expansion valve in the Carré system—the ammonia evaporates rapidly at a relatively lower temperature.

*Is anything gained—or lost—by omitting the last clause of this paragraph?*

Since, under these conditions, the total pressure in the absorber is also the same as that in the generator, there is no need for a pump to convey the strong aqua from the absorber to the generator—the transfer can be accomplished by gravity flow. The return of weak aqua to the absorber can no longer be done by higher pressure in the generator, so a simple vapor lift, similar to that found in a coffee percolator is utilized.

The result (Fig. VIII) is a continuous refriger-

*This one-sentence paragraph*

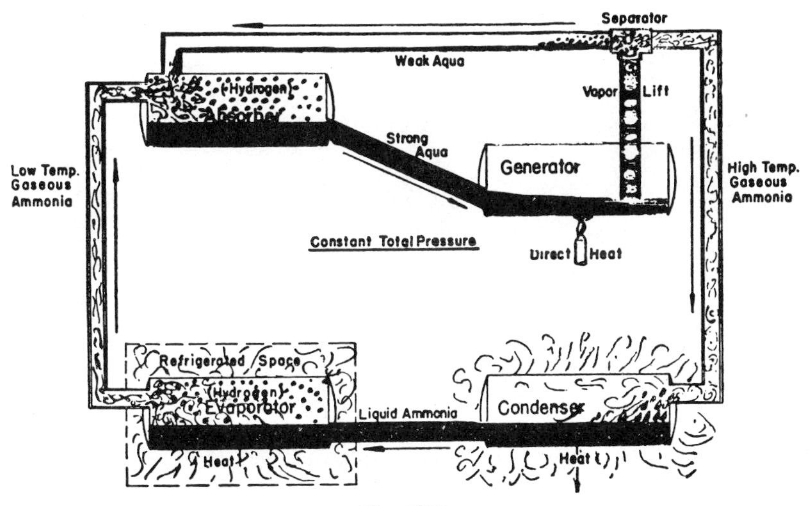


Fig. VIII

The importance of the Platen-Munters features is emphasized by the fact that an American corporation paid five million dollars for their U.S. patent rights.

Since that time, a highly improved form of their machine has been developed, a machine unique in its valveless, pumpless control of fluids within a hermetically sealed space. It contains three interrelated fluid circuits, rotating in unison to produce continuous refrigeration, powered by nothing more than a single small source of heat. As shown in Fig. III, the three circuits consist of the ammonia loop, the ammonia-hydrogen loop, and the ammonia-water loop.

The process is shown diagrammatically and much simplified in Fig. IV. Beginning with the ammonia-hydrogen loop, the ammonia gas enters the evaporator from the condenser through the liquid trap which confines the hydrogen to its own circuit. In the evaporator it takes up heat from the surrounding space and vaporizes,

ation machine using no valves, motors, or pumps, but motivated by a flame—"ice from heat."

The importance of the Platen-Munters features is emphasized by the fact that an American corporation paid five million dollars for their U.S. patent rights.

Since that time, a highly improved form of their machine has been developed, a machine unique in its valveless, pumpless control of fluids within a hermetically sealed space. It contains three interrelated fluid circuits, rotating in unison to produce continuous refrigeration, powered by nothing more than a single small source of heat. As shown in Fig. IX, the three circuits consist of the ammonia loop, the ammonia-hydrogen loop, and the ammonia-water loop.

*and the figure were added as a result of the criticism of a reader that this portion was not clear. Do you think the additions satisfactorily solve the problem?*

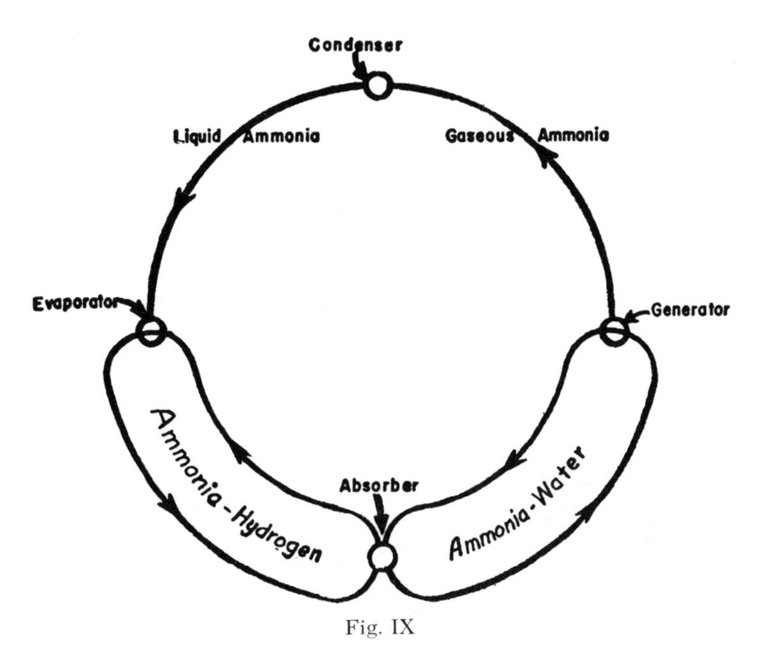


Fig. IX

The process is shown diagrammatically and much simplified in Fig. X. Beginning with the ammonia-hydrogen loop, the ammonia gas enters the evaporator from the condenser through the liquid trap which confines the hydrogen to its own circuit. In the evaporator it takes up heat from the surrounding space and vaporizes, its gaseous molecules mixing with those of the hydrogen. The addition of the heavier ammonia molecules increases the specific gravity of the vapor, and it sinks down the tube leading to the absorber.

In the absorber, the ammonia dissolves in the countercurrent stream of water, while the prac-

its gaseous molecules mixing with those of the hydrogen. The addition of the heavier ammonia molecules incréases the specific gravity of the vapor, and it sinks down the tube leading to the absorber.

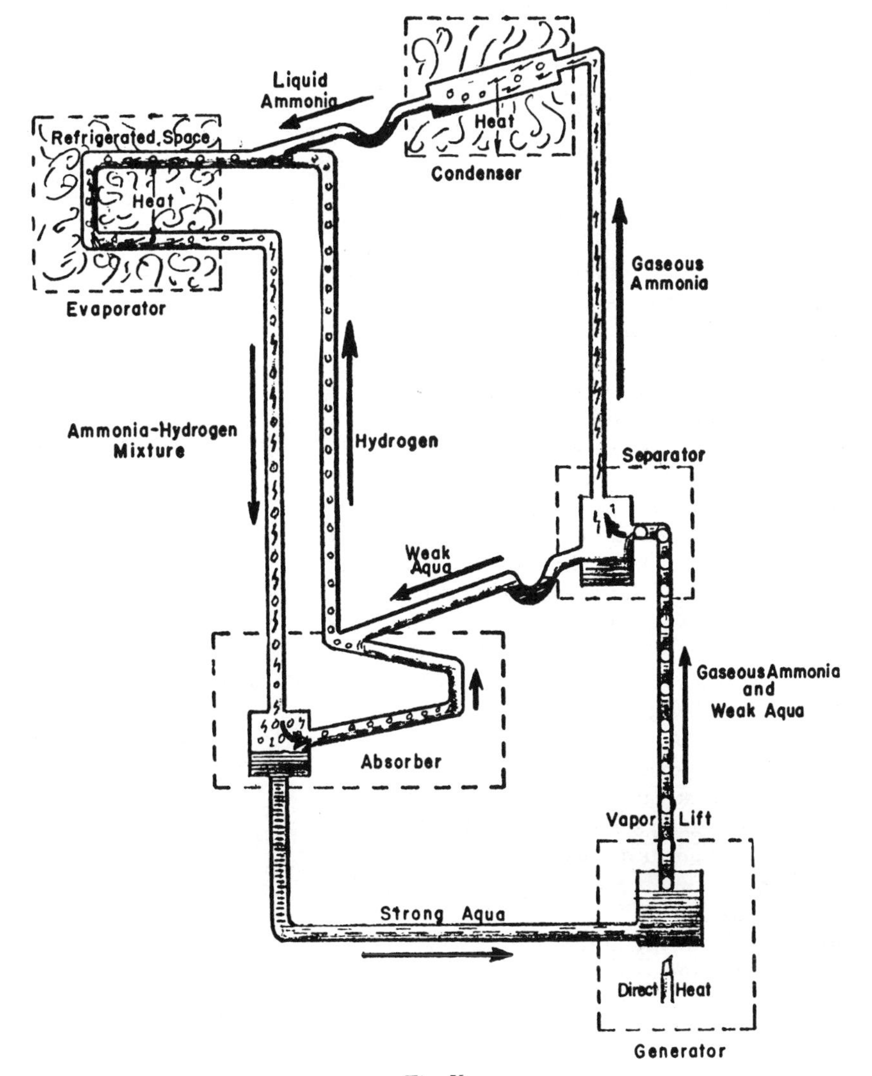


Fig. X

In the absorber, the ammonia dissolves in the countercurrent stream of weak aqua, while the tically insoluble hydrogen, lightened of its burden of heavy ammonia molecules, ascends to the evaporator to again perform its task of mixing with and decreasing the partial vapor pressure of the ammonia.

Taking up the ammonia-water loop, the strong solution of ammonia in water, called "strong aqua," flows by gravity to the generator, where the application of heat drives the ammonia out of solution. A vertical tube, the inside diameter of which is equal to that of the bubbles of gaseous ammonia generated, projects below the surface of the boiling liquid, so that as the bubbles ascend the tube they carry up with them slugs of the solution. This "liquid lift" empties into the separator, where the ammonia vapor is separated from the weak solution of ammonia in water, known as the "weak aqua." The weak aqua then returns by gravity to the absorber to pick up another load of ammonia.

Finally, the ammonia loop, which has been traced as far as the separator, next involves the "condenser," an air-cooled heat exchanger which removes the sensible and latent heat from the ammonia gas, converting it into a cool liquid. Here it passes through the liquid trap that marks its re-entry into the evaporator to serve its purpose in cooling the refrigerated space.

*He leaves the word "sensible" here out of the revision, an omission which changes the meaning. Sensible heat is measured by a thermometer; latent heat changes the state of a substance without changing its temperature.*

Not shown or discussed are the several refinements that go to produce the maximum of efficiency in this refrigerating process such as the heat exchangers that shunt the sensible heat of the several streams around to where it is most useful, the "analyzer" and "rectifier" that decrease the moisture content of the gaseous ammonia leaving the separator, and the "hydrogen reserve vessel" that balances change in atmospheric temperature with change in the total internal pressure on the system.

*Reference is made here to those refinements which are not shown or discussed. Compare with the final draft.*

Industrial applications of the absorption refrigeration system include appropriate large-scale refinements, such as that engineering triumph of chemical processing, the bubble

practically insoluble hydrogen, lightened of its burden of heavy ammonia molecules, ascends to the evaporator to again perform its task of mixing with and decreasing the partial vapor pressure of the ammonia.

Taking up the ammonia-water loop, the strong aqua in the absorber flows by gravity to the generator, where the application of heat drives the ammonia out of solution. A vertical tube, the inside diameter of which is equal to that of the bubbles of gaseous ammonia generated, projects below the surface of the boiling liquid, so that as the bubbles ascend the tube they carry up with them slugs of the liquid. This "liquid lift" empties into the separator where the ammonia vapor is separated from the weak aqua. The weak aqua then returns by gravity to the absorber to pick up another load of ammonia.

Finally, the ammonia loop, which has been traced as far as the separator, next involves the "condenser," an air-cooled heat exchanger which removes the latent heat from the ammonia gas, converting it into a cool liquid. Here it passes through the liquid trap that marks its reentry into the evaporator to serve its purpose of cooling the refrigerated space.

Omitted from the above discussion for the purpose of clarity, there are shown in Fig. XI the several refinements that serve to increase the efficiency of the machine, such as the heat exchangers that shunt the sensible heat of the several streams around to where it is most useful, the "analyzer" and "rectifier" that decrease the moisture content of the gaseous ammonia leaving the separator, and the "hydrogen reserve vessel" that balances change in atmospheric temperature with change in the total initial pressure on the system.

*Here the refinements, while not discussed, are illustrated in a new figure not included in the earlier version. This serves, of course, to make his discussion clearer.*

*V. Conclusion*

Industrial modifications of the refrigerating system are many, varied, and equally interesting. They include appropriate large-scale refinements, such as that engineering triumph

*This version has been expanded slightly and the last sentence in the next to last paragraph definitely marks*

tower, which replaces with unmatchable efficiency the analyzer and rectifier of the Platen-Munters machine.

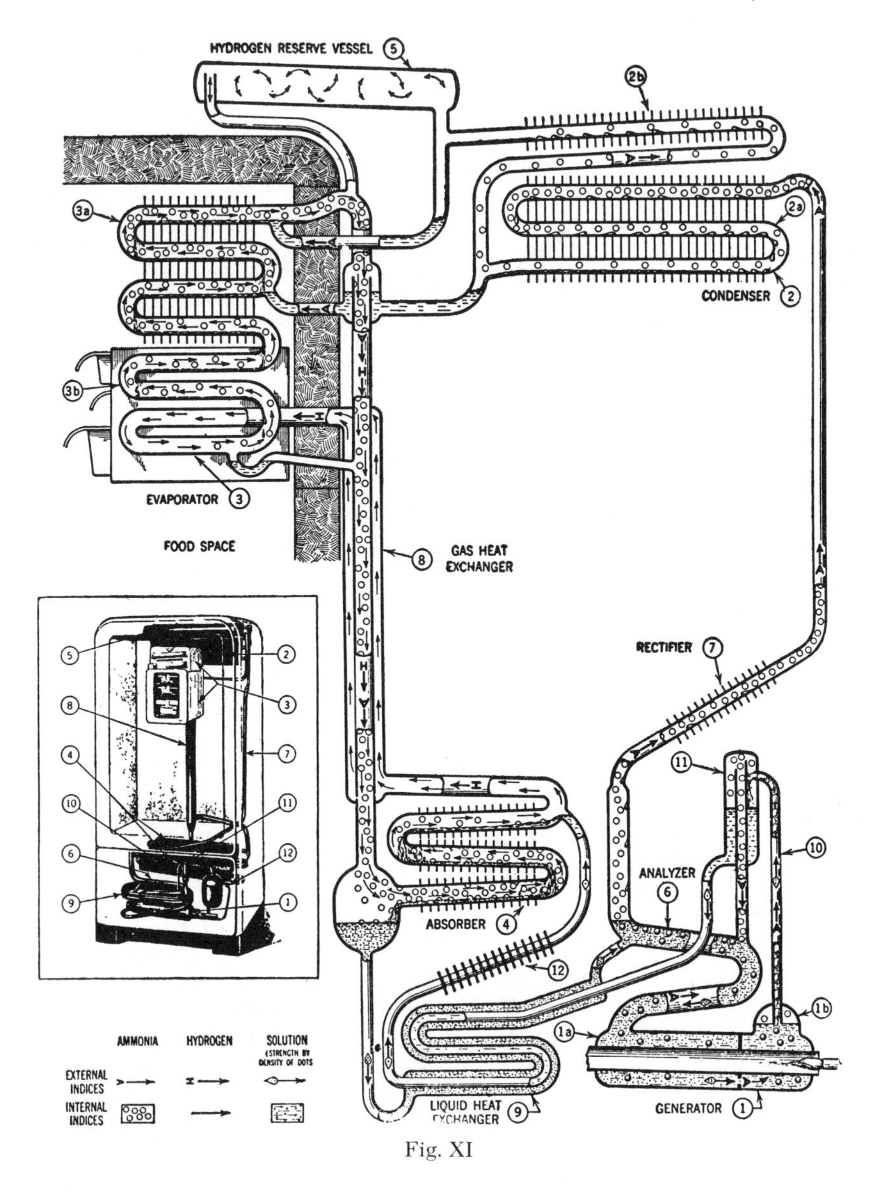


Fig. XI

of chemical processing, the bubble tower, which replaces with unmatchable efficiency the analyzer and rectifier of the Platen-Munters machine. Limitations on the scope of this paper prevent further elaboration, but it is hoped that the evolution of the artificial ice-making machine has already been traced far enough to as-

*this as the conclusion of the paper.*

Further advances undoubtedly lie in the future, but there is little possibility that they will ever overshadow the significance of the developments that have reduced the price of a valuable commodity to one thousandth of 1 per cent of its onetime cost, and earned undying fame for two young Swedish students.

*There is something of the same quality in this conclusion that was noted in the introduction—a certain bluntness and suggestion of haste. Note that considerable change has been made in the revision.*

sist the reader toward an understanding and appreciation of the resourcefulness that made it possible.

The frontiers of science are wide, its horizon unlimited. Possibly it is not frequent that the engineer can point with pardonable pride to such a feat as the reduction of the price of an everyday necessity to one one-thousandth of 1 per cent of its onetime cost. But he thrives on the meat of research and development, and as long as there is an unsolved problem standing in the way of the advance of science, there will be a place for the engineer.

*Note that in both versions, he has neatly tied his conclusion to the lead of his introduction—the startling reduction in the cost of ice.*

By way of summary, we may note several differences between the first and second versions. There are probably not so many differences between these two versions as one would normally expect because the earlier version is by no means the first draft of his report and because he has spent a good deal of time on the subject matter in preparing the oral version. Even so, we note the addition of figures for clarity, more careful transitions (including the use of headings), the polishing of style in several places, and the addition of some concrete details.

It is also significant to note the differences between the report and the oral version:

1. The tone of the oral version is more sprightly, more conversational. In the report, for instance, there is no mention of the pleasure of speculating about the spending of a million dollars—such a lead would not be suitable in tone and dignity for a report.
2. In the oral version, Galt often tells his audience what he is doing, as "Let's look at some of the aspects of this science . . ."
3. There is more detail in the written version.
4. There are more personal pronouns in the oral version, and more comparisons drawn from everyday experience to clarify points.
5. There are some differences in arrangement, or organization. See pages 482–487 of the report and pages 470–473 of the oral account.
6. There are more illustrations in the report.
7. The discussion of absorption refrigeration is given in lighter terms in the oral version; he even uses a humorous illustration of a man hurrying the refrigeration process by applying a blow-

torch to speed up formation of ice. In short, Galt realizes the effectiveness of the personal touch in giving a talk.

8. Background discussion of Dalton's law of additive pressures is omitted from the oral account.
9. Discussion of refinements is omitted from the oral version.
10. The oral version has a simpler, more dramatic conclusion.

# Appendix C
# Organization of Reports

The purpose of Appendix C is to give you the opportunity to examine the kinds of instructions on report writing that you may be issued on the job. Although instructions differ a great deal from one company to another, the following pages are fairly representative of content and point of view. They are extracted from *Report and Letter Writing*, prepared by the Research and Development Department of the Sun Oil Company.

## Organization of Reports

Much of what we are going to say about the organization of reports also applies to technical letters. But the report is a good deal more formal, and the sequence of sections is more firmly established. In a report these sections are given subtitles for the convenience of the reader and for ease of reference. Normally this is not done in a technical letter. In drafting long letters, however, it is helpful to set up section headings similar to those in reports. These may be deleted in the final revision if they do not add to clarity. Not all sections discussed need be included in a report or letter.

A report consists of several essentially independent parts, with each succeeding one supplying greater detail about the problem. You start in by merely stating the *subject* and *title*. Next you may give a summary of the report. Then you give much more information in the *body* of the report. Finally, detailed descriptions of apparatus, materials, or data are supplied in the *appendix* so that the experiments can be duplicated or the conclusions verified. To be sure, this involves some repetition, but permits the reader to stop whenever he has gotten enough detail to satisfy his needs.

Formal reports are bound with a heavy paper cover. Just inside the cover is the title page. This repeats much of the information on the cover and, in addition, tells who sent the report, who approved it, who did the actual work on the project, and who received copies of the report.

A technical letter, though less formal, should also conform to certain conventions. References to previous correspondence on the same or related subjects should be placed in the upper right hand corner of the first page. If the sender and receiver are company employees, reference to a letter from J. S. Blake to M. C. Dwyer, dated April 30, 19–, is usually given as JSB: MCD/4/30/–. In the case of a nonemployee, his last name should be given in full.

On all interoffice letters and memoranda it is customary in the Research and Development Department to omit the Salutation *Dear Sir* or *Dear Mr. Hicks*, as well as the complimentary closing *Cordially yours* or *Sincerely yours*. Letters to anyone outside the company should include these formalities, since their omission might be misunderstood.

## Opening Sections of a Report

### Title

Filing purposes require that the title consist of a subject name, plus descriptive words to cover the specific study reported. For example, "HF Alkylation – Effect of Reaction Temperature." This principle applies equally to letters. A title should be appropriate, informative, and unique. It should be no longer than necessary.

### Summary

The most important part of your report is the Summary. It is the section most often read; sometimes it is the only part read. There is a strong feeling that the Summary should be tailored to fit on a single page. In some instances this is not possible, for it must be long enough to do the job adequately. If your Summary must run to two or more pages, consider whether you should not precede it by a much shorter Abstract.

The Summary should briefly answer the questions:

1. What was the problem?
2. What are the facts and what do they mean?
3. What are the important conclusions?
4. What are the important recommendations?

There is no need to validate your data in the Summary; this is left to the Discussion of Results. Use a short insert table or chart, if you can tell your story more effectively, where it will take the place of a long descriptive passage. You may, if necessary, refer to a table or chart elsewhere in the report, but make sure it can readily be found.

### Conclusions and Recommendations

The writer's opinions about what was learned, based on the data presented, what steps should be taken as a result, and what additional studies are needed, should be given in the Conclusions and Recommendations. Both of these go beyond simple and readily apparent observations. In other words, saying that "the F-1 and F-2 Octane ratings are 95 and 82" is an observation. The statement "Because of the wide spread in F-1 and F-2 ratings, the fuel should have a good rich mixture performance" is a true conclusion in which the writer has applied his judgment to interpreting the data.

If there is no valid conclusion or recommendation, do not try to devise one. But since most work of value will support a sound conclusion, be sure one is not being overlooked. All the important conclusions and recommendations in the report should be listed in this section.

It is customary to open this portion with an introductory sentence or two which will set the stage for what is to follow. Your conclusions and recommendations may then be itemized, as (1), (2), (3), etc. Where this portion of your report is short, the section may be eliminated and the conclusions given as part of the Summary.

### Table of Contents

For short reports (up to 5-8 pages) a Table of Contents is usually not necessary. For longer reports it serves as a guide and tells the reader at a glance the scope of the report. Preparing it will give you an opportunity to check the logic of the organization of your report. It is, in essence, the outline from which you worked.

If a Table of Contents is worth preparing, it is worth doing well. Use the same care in choice of words that you have used in the selection of a title. Place the Table of Contents after the Conclusions and Recommendations; the reader who goes no farther than this has no need of the table. Separate lists of tables and charts giving page numbers may be included after the Table of Contents.

### Introduction

The first section in the body of the report is the Introduction. It includes a statement of the problem, the value of the work to the company, and the background and reason for the work. Here you can list references to previous letters or reports, tell why the particular field of investigation was chosen, the relation of the problem to other fields of endeavor, the scope and limits of the report, and some mention of future work contemplated. The Introduction is not a rehash of the Summary. It sets the stage for what is to follow.

If only a sentence of introductory material is needed, this section may be omitted and the introductory thought covered elsewhere in the body of the report.

## Miscellaneous Sections

In addition to the foregoing sections found in most reports, there are others that, with the Introduction, make up the body of the report. Your range of these section headings is practically unlimited, provided that the terms chosen are applicable to the problem. These may be either generic terms such as:

Preparation of Samples (or Charge Stocks)
Test Methods
Operating Procedures
Product Specifications
Presentation of Data
Discussion of Results
Apparatus (or Equipment)

or better still they may be specific titles dealing with the study, as:

Hydrotreating
Concentration by Vacuum Steam Distillation
Chemistry of Gum Formation
Effect of Sulfur Trioxide on Color
Possible Markets for Product
Analysis of a Two-Component Acid Mixture
Method of Aging

Some generic headings occur in reports often enough to make worth-while a few comments on them.

### Work Done

A section on Work Done can be used to cover the actual work performed and the equipment and materials when these are incidental. The title preferably is more specific, such as *Experimental Runs Made in 20 Liter Cases*. Omit details not pertinent as well as ones understood by the reader. On the other hand, one fault of technical writers is assuming that the reader is familiar with a test procedure because it has become commonplace to the writer.

Data may be presented in this same section along with the experimental work except when the volume of data is large, in which case it may best be put in a section by itself. Screen the data to eliminate any not pertinent to the results obtained, but don't delete data merely because they appear to be inconsistent. Short insert tables or graphs are effective. For example:

The values for competitive regular motor oils were:

Mean Piston Rating . . . . . . . . . . . . . . . . . . . . . . . . . . . 9.6
Mean Viscosity Increase . . . . . . . . . . . . . . . . . . . . . . . 14%
Mean Normal Pentane Insolubles . . . . . . . . . . . . . . . 0.90%

Voluminous data should be relegated to an appendix, using references in the text as necessary.

### Discussion of Results

This is the span between the factual data and the writer's conclusions. In it, you are leading the reader through the reasoning necessary to understand the conclusions and see that they are sound. Logic sometimes requires that you advance the opposing arguments in order to show that these are outweighed by the advantages, or that they are disproved by the assembled facts. Don't assume that the reader agrees with your conclusions, but establish their validity through the process of logical reasoning. Try to strike a balance between annoying the reader by too thorough a discussion, and making it necessary for him to verify your conclusions by analyzing the data himself.

This is probably the section in which it is most difficult to refrain from using long, involved sentences. Even though your reasoning may be complex, keep your style simple and straightforward.

Here is the place to include a statement of the limitations of the data and how far the information may be used in other connections.

Where the conclusions are fairly obvious and not much discussion is needed, this section may be left out and the points covered under another section such as *Work Done*.

### Bibliography

Many reports contain a list of references to prior articles, reports, letters, and patents directly related to the field of investigation. Usually these are referred to in the text, and in this case should be itemized and numbered in the order of reference. The text reference need only show the item number in parenthesis.

The bibliography is placed after the last page of the report (the signature page), and ahead of the appendices, if any.

### Appendix

In order to shorten reports and make them more readable, detailed descriptions of materials or apparatus used, operating procedures, detailed experimental results, and the like, are often included as appendices to the body of the report. There may be several in a report. Each appendix should be written as a separate unit. If detailed data are set up as an appendix, condensed tables or charts should be included in the main body of the report to support the conclusions reached. In other words, your report should be complete and self-sufficient in itself. Appendices are added to provide detail of value to future workers. Be sure to make reference in the body of the report to appended material.

# Appendix D

# A Study of the Use of Manuals

The report appearing on the following pages gives the most illuminating insight into the practical problems of technical writing that we have ever seen in print. From the annoyance of the schematic circuit diagram that has too many folds in it, to the analysis of a reader's probable vocabulary, here are the tasks faced by the technical writer. The particular subject being discussed is the preparation and use of manuals, but many of the questions that arise—like how to avoid "snowing" the reader in an analysis of circuit theory—are in principle common to a wide variety of technical documents.

The writing technique chiefly represented here is interpretation, and in form the report can be thought of as a recommendation report.

Limitations of space have made it necessary to delete several pages of the report. Comparison of the text shown here to the Table of Contents of the report will indicate what has been removed.

Research Report

Report 782

# Human Factors in The Design and Utilization of Electronics Maintenance Information

*J. H. Stroessler, J. M. Clarke, P. A. Martin (Soc, USN), And F. T. Grimm (Soc, USN)*

U.S. Navy Electronics Laboratory, San Diego, California

A Bureau of Ships Laboratory

## THE PROBLEM

Determine who uses technical manuals for Navy electronic equipments, how they are used and for what purposes, the extent to which present content is adequate for present users, and the functional relationship between technical-manual and other types of maintenance information.

## RESULTS

The following are the major findings of an NEL team which interviewed Navy and civilian maintenance personnel at a representative selection of Navy ships, shore establishments, and commands:

1. Technical manuals should meet the needs of users ranging from Navy maintenance personnel with minimum training to experienced civilian engineers.
2. Technical manuals are used in all phases of maintenance, in the development of local preventive maintenance procedures, in performance measurement, and in peaking of equipment. They are also used in installation, in designing new equipment and combining equipments to form systems, in planning by supply departments, and in instruction and the development of training materials.
3. The schematics and the alignment, adjustment, and calibration information are the most widely used parts of technical manuals. Also much used are theory of operation, block diagrams, parts lists, and installation information and drawings.
4. Improvement and simplification of technical manuals can speed up the performance of Navy maintenance tasks and lessen reliance on civilian engineers. The present complexity of technical manuals can be reduced by publishing separate volumes each covering a major maintenance task.
5. Preventive maintenance and corrective maintenance each require a separate volume. POMSEE (BuShips program for improved preventive maintenance) checklists would fulfill the requirements for the preventive maintenance part. Improved organization and indexing are required to facilitate location of information—the most time-consuming task imposed by current manuals. Present trouble-shooting guides are inadequate and little used; improved aids are needed.
6. A separate volume is required to cover planning, design, and installation.
7. Supplemental information provided by the Electronics Information Bulletin, the Electronics Maintenance Books, etc., is inadequate. Navy personnel are handicapped as compared to civilian engineers who receive a steady stream of timely information from their parent companies.
8. POMSEE materials are effective for training purposes.

## RECOMMENDATIONS

1. Consider ways in which Navy schools, particularly "A" schools, might develop greater student familiarity with the contents and organization of technical manuals and provide more training in their use.

2. Consider ways in which technical manuals might be given more systematic use in on-board maintenance training programs.

3. Establish a representative committee to draw up functional specifications for a prototype installation manual. Prepare a manual according to these specifications, and evaluate it under actual use conditions.

4. Develop an experimental manual combining operation and preventive maintenance and utilizing POMSEE materials. Evaluate the manual functionally.

5. Develop an experimental manual for corrective maintenance incorporating the best features of existing Navy and civilian technical manuals and a modification of POMSEE performance standards as a trouble-shooting guide. Give this manual field trials and evaluate its effectiveness.

6. Collect and evaluate against the findings of this survey all available experimental attempts to rearrange and clarify corrective maintenance information for greater comprehensibility and utility. From the most promising of these techniques develop and evaluate an experimental corrective maintenance manual.

7. Develop a "little black book" consisting of supplementary maintenance data on the equipments in one particular system, to be sent directly to maintenance personnel responsible for this system. The form of this booklet should be looseleaf, additional sheets to be supplied as rapidly as new information becomes available. Blank pages on which maintenance personnel can keep their own notes should be included. Evaluate in the field.

8. Investigate further the problems of information dissemination, and set up a system for more complete exchange of information among shipyard, ship, school, and Bureau of Ships personnel.

9. Devise a system for more adequate and efficient gathering and recording of maintenance information in the operating forces.

10. Study the feasibility and appropriateness of publishing Navy maintenance periodicals which are more specialized and personalized than the existing ones.

11. Devise methods for improving procedures used in making corrections to technical manuals and the integration of field changes into manuals.

12. Investigate the conditions surrounding the actual writing of Navy technical manuals, and consider means for obtaining a uniformly better product.

## CONTENTS

Tables



## INTRODUCTION

### *The Problem*

This study was designed and prosecuted under the following broad problem assignment:

> Develop organized electronics maintenance philosophy and principles to be applied by electronic designers for the purpose of improving electronic maintenance in the Fleet.

Under this general objective, the present study appears as the following individual task:

> Study and consult on human factors aspects of the design and utilization of electronic maintenance information—continuing.
>
> Determine who uses Instruction Manuals (IM), how they are used and for what purposes, the extent to which present IM content is adequate for present users, and the functional relationship between IM and other types of maintenance information.

This task is unique in that it represents the Navy's first study of maintenance publications* from the "human factors" point of view. In such an approach, knowledge of the techniques for investigating areas involving human behavior is combined with a background of maintenance information and insight into the Navy's problems concerning training and utilization of electronics maintenance personnel.

The problem was broken down into more specific components to arrive at questions which might be effectively answered by a field study. These are:

1. What are the tasks for which technical manuals and related materials are used?
2. What are the characteristics of the personnel performing these tasks?
3. How effective are the technical manuals and related materials as tools for the performance of each major task when used by particular categories of personnel?
4. To what extent are various parts of the technical manuals and related materials used in connection with each major task by each category of personnel?
5. How appropriate are the technical manuals and related materials to the needs and capabilities of the various user groups?
6. How are the technical manuals and other maintenance publications related in subject matter coverage, format, use, etc.?

## Related Factors

As the problem statement implies, technical manuals constitute only one of the factors which influence the effectiveness of electronic maintenance. These factors are so highly inter-related that none of them can profitably be considered in isolation. Throughout this study, therefore, the place of manuals in the total maintenance pattern has been considered. A summary of the most significant of the related factors is presented here:

1. *Complexity of Equipment and Systems.* This factor creates inescapable variations in technical manuals. Simple equipments require only small easily used manuals because there are relatively few data and instructions to be

*The term "technical manual" will be used in this study to designate a publication traditionally known in the Navy as an "instruction book." See MIL-M-16616(SHIPS).

presented. Thus the organization can be straightforward. In contrast, the interactions between circuits and parts of a complex equipment, which make fault location difficult, create needs for extensive data which can be presented only in a considerable number of pages. Data needed for particular tasks may, therefore, be widely dispersed.

2. *Maintainability.* The maintainability features of equipments, such as built-in test instruments and aids to ready access and repair, are important factors in determining the content and organization of technical manuals. Further, the nature of adjustment features, safety features, and controls influences the complexity and comprehensibility of alignment and calibration procedures all of which must be described in the manual.

3. *Environment.* Lighting, work space, stowage, and other aspects of the maintenance man's surroundings establish requirements for the physical characteristics of manuals, such as size, legibility, etc.

4. *Instruments and Tools.* The relevance of this factor derives from the fact that no advantage results from specifying tests and repair procedures unless the tools and instruments required for making them are provided. Conversely, Navy policy in regard to test instruments and tools determines what the technical manuals should recommend.

5. *Administration.* Administration heavily influences the use to which maintenance materials are put and the degree to which maintenance personnel are effective in their use. Some of the larger aspects of administration —particularly those affecting the formulation and implementation of policy in technical matters and those which influence the handling of the personnel turnover problem—also have an important bearing upon maintenance information.

6. *Training.* This factor determines whether maintenance personnel have the basic knowledge and the skills required to use technical manuals and related materials effectively.

7. *Information Distribution.* It is evident that there is a direct relationship between the effectiveness of maintenance information and the extent to which maintenance personnel receive publications and the processes by which they may obtain these publications.

### Scope and Limitations of the Study

The information for the study was collected by means of a field survey conducted by a Task Group consisting of two psychologists and two Chief Sonarmen. The extensive maintenance experience of the latter provided insight into many practical aspects of the problem and supplied detailed technical knowledge.

In general, the Task Group attempted, wherever possible, to diagnose the reasons why technical manuals and related materials are not fully effective, to determine what may be done to meet Navy needs more completely through maintenance publications, and to recommend further research where the data were sufficient to support only tentative conclusions.

The Task Group tried to examine a *representative cross section* of Navy maintenance as it involves the use of technical manuals and related materials. The choice of maintenance locations and choice of subjects were limited both by availability and cost of travel in time and money. As a result, 138 subjects were interviewed on the Pacific Coast as against 67 subjects on the Atlantic Seaboard and Great Lakes Naval Training Station.

Availability also limited opportunities to study the use of the technical manuals recommended for special investigation in an earlier statement of the problem assignment. It was impractical to delay the work while waiting for ships having equipments to which these manuals pertained. Nevertheless, a good deal of material on these manuals was obtained, and is presented in Appendix A.

It was possible to study the effects of the 1956 Bureau of Ships Specifications for Technical Manuals only insofar as features required by these specifications are contained in existing manuals. Some of the findings in the study emphasize the need for certain features required by the new specifications. Relationships between the findings and the new specifications are summarized in Appendix B.

The results of the survey are summarized in the report proper and are given in full in Appendix C. Actual interview questions and responses have not been included herein, but the data have been reproduced separately as Appendix D. Appendix D is available to users of this report upon request to the Commanding Officer and Director, Navy Electronics Laboratory, San Diego 52, California.

## PROCEDURES

### *Method of the Study*

The terms of the problem called for information derived directly from maintenance situations afloat and ashore, rather than data from laboratory experiments, or maintenance and personnel records. It was readily apparent that a field survey was the most appropriate method for gathering such information.

Interviews capable of bringing out each subject's experience with technical manuals were selected as being the device most adaptable to the requirements of the problem. Other investigators have found this technique to be highly effective in the prosecution of similar research. Its particular advantages for this field study of maintenance publications are as follows:

1. It permitted a Navy-wide sampling of maintenance situations, and all major categories of maintenance personnel.
2. It permitted sampling of each subject's entire maintenance experience.
3. It facilitated the procurement of diagnostic information on which to base conclusions as to how technical manuals might be improved.
4. It provided insight into many of the critical relationships between technical manuals and other maintenance factors such as training, administration, information distribution, technical policy, etc.

It should not be overlooked, however, that data obtained in face-to-face interviews are vulnerable to possible error and misinterpretation. The scope of the information obtained may differ somewhat from subject to subject. A man may fail to remember, or he may be unable to describe, facts of considerable importance. He may even conceal, distort, or falsify his experience and reactions. The Task Group employed devices, described later, to minimize the introduction of errors from these sources.

### *The Instrument*

In developing the questions used in this field study, two basic criteria were kept in mind: (1) the interviews must be so structured that a large number of subjects would describe the same areas of experience, thus enabling the Task Group to tabulate and compare responses; (2) the interviews should stimulate the subjects to remember as much of their pertinent experience as possible and provide conclusions from it.

Background knowledge from several sources was drawn upon in determining the main points to be emphasized and explored. Research reports and general literature covering the field were reviewed, and to data thus obtained was added information from maintenance conference notes and findings based on an analysis of a large number of technical manuals. Much practical knowledge was furnished by the two Chief Sonarmen assigned to the project. Both of these Chiefs had many years of maintenance experience behind them. Service on various projects at NEL had sharpened their ability to recall, select, organize, and reason about their fund of first-hand information.

A preliminary set of questions was tried out in the field, and on the basis of experience thus gained a final set of 21 questions, some of which were divided into a number of sub-questions, was developed. A number of these questions were designed to bring to light experiences and ideas which might have been missed without a rather full discussion. The interviewer supplemented or restated questions to bring out more information when the subject's first response indicated that the question had been misunderstood, or had elicited a superficial or less-than-candid reply. As a further safeguard against incomplete or distorted answers, several questions were asked concerning single important areas. The interviewer was thus able to detect inconsistencies in answers, and to approach sensitive and/or difficult areas from more than one angle. These overlapping questions also gave the subject an opportunity to consider his experience in more than one light.

### *Population*

The interview sample was selected to represent as complete a cross section of Navy electronics maintenance situations and personnel as was possible within the time and money limits established for the problem. A total of 216 people were interviewed consisting of 135 Navy personnel and 81 civilians.

### *Interview Procedures*

The Task Group operated as two-man teams under widely varying circumstances. Available space aboard ship, duty assignments, the wishes of officers and supervisors who made the subjects available, as well as differences in the subjects themselves all had to be considered. The interviewers made every effort to adapt themselves to circumstances, varying their methods as required to extract the most information from each situation.

In most cases the prepared questions were used in their entirety. Occasionally, when circumstances made it necessary to interview several men at a time, the questions served as a guide to be followed by the interviewers as systematically as possible. In talking with civilian engineers and technicians, the interviewers selected those questions most suitable for drawing upon the subject's specialized experience and knowledge. This latter procedure was followed also in the case of officers responsible for maintenance supervision.

The interviewers found that most men preferred to talk at length, giving many examples and providing considerable information on the relationships between technical manuals and other maintenance factors. The average time for an interview was 1.5 hours.

Customarily, one member of each team took notes while the other carried on the interview. In this way it was possible to make a complete record of the interview. The notes were discussed by the two interviewers as soon as possible following the interview, and points of confusion and differences of interpretation were resolved. In the case of enlisted subjects, a Sonar Chief Task Group member carried on the interview. When officers and civilians were interviewed, the tasks were reversed and the civilian scientist conducted the interview. In order to bring out complete and candid responses, all subjects were promised anonymity for themselves and their ships or shore activities.

## SUMMARY OF RESULTS

The complete results and a full discussion of their significance appear as Appendix C on which this summary is based.

### *The Effectiveness of Current Navy Technical Manuals*

The great variations among technical manuals make it impractical to evaluate their over-all effectiveness in precise terms, though it may be safely assumed that if manuals in general were more effective, electronic equipments would be maintained in more acceptable operating condition. No single manual appears to be effective in all particulars. Yet many manuals have one or more effective specific characteristics (see Appendix A). Despite these variations, it is possible to state the characteristics which maintenance personnel find helpful and desirable.

*SIZE*

The smaller manuals are easier to use than the larger ones such as the manual for AN/SPS-8.

*INCLUSIVENESS*

Experienced personnel—especially men who have been on active duty in combat zones—feel that the omission of any information contained in the most inclusive of the current manuals would be hazardous. They foresee the possibility of having to perform any and all maintenance functions with no aid or advice from outside the ship itself. At the same time a majority of personnel want to "stay in the book," finding it inconvenient to turn to supplementary sources. (For means of reconciling the conflict between size and inclusiveness, see Conclusion.)

*TEXT*

A majority of maintenance personnel interviewed reported the technical vocabulary and sentence structure used in the current manuals to be within their abilities.

A significant percentage of personnel interviewed wanted clearer, more precise alignment instructions.

A significant percentage of personnel interviewed desired complete, though not necessarily detailed, descriptions of all circuits. They desired more detailed explanations of unusual and/or complex circuits, and less detailed explanations of conventional circuits.

A great majority felt that more standardization, both of terms and symbols, would be beneficial.

A majority felt that a glossary of manufacturer's terms and new or recondite engineering terms would be helpful.

*SCHEMATICS*

The following features and characteristics of schematics were considered desirable by a significant proportion of personnel interviewed:

1. Clear, uncrowded, legible
2. A separate set of schematics, including one over-all schematic
3. Signal path emphasized
4. "In-line" schematics
5. Colored schematics
6. Road mapping
7. More waveforms at significant points
8. More realistic waveforms
9. Clear statement of conditions under which oscilloscope readings were made
10. Maximum and minimum voltages shown where a tolerance exists; critical voltages indicated
11. Inputs and outputs to and from each portion of circuit shown in schematics; also inputs to equipment as a whole and outputs from it if used as a system component

*ALIGNMENT*

A significant proportion of men interviewed believed that simpler, more concise alignment procedures were necessary for relatively inexperienced men.

*SYSTEMS DATA*

A significant proportion of men interviewed believed that more systems data are required as follows:

1. Manuals covering all—even the most minor—system components
2. Individual system layout, including interconnections
3. Input and output data for all system components

*ORGANIZATION*

A significant proportion of personnel interviewed, experienced men predominating, were of the opinion that the present technical manuals:

1. Should require less hunting in order to find data needed
2. Should require less "flipping back and forth" of pages to gather the information needed for a particular task
3. Need better indexes

*REPAIR INFORMATION*

A significant proportion of personnel interviewed believed that:

1. Explicit directions for disassembly and assembly were very helpful, especially in the case of mechanical units
2. Exploded drawings were the most practical way of communicating mechanical disassembly and assembly information

*INSTALLATION INFORMATION*

A majority of personnel engaged in planning, design, and installation believed that:

1. It would be helpful to have all the information needed for their task concentrated in one section or publication instead of having it scattered through the manual as is the case at present.
2. More appropriate mechanical drawings would be helpful, a large percentage of those in current manuals being suitable for construction rather than installation.

*ACCURACY OF TECHNICAL MANUALS*

A majority of personnel interviewed reported finding errors in technical manuals. It was reliably estimated that at least 50 per cent of the manuals in current use contained errors. The following causes were given for the majority of these inaccuracies:

1. The difficulty of learning about and procuring data on changes
2. The difficulty of learning which field changes have been made, and which should be ordered
3. The time consumed in making pen-and-ink changes, and the illegibility resulting from an accumulation of these changes
4. Undependable distribution of manual changes which are necessitated by field changes

## CONCLUSIONS AND RECOMMENDATIONS

### *Conclusions*

*GENERAL*

1. Time-pressure rather than the inability to maintain equipments is the most important of the several reasons why maintenance personnel rely on

civilian engineers. It must be borne in mind that a ship may have as few as 15 technicians for 1000 equipments. When the maintenance load is heavy, slowness in accomplishing particular repairs may cause tasks to accumulate and may even result in the progressive degradation of equipments. It is believed that improvement of manuals can speed up the performance of maintenance tasks, thus lessening reliance on civilian engineers and improving the over-all operating condition of equipments.

2. Technical manuals should meet the needs of users ranging from third class Navy technicians to civilian electronics engineers. Personnel throughout this entire range may have sole responsibility for maintenance in a particular location and can rely on no one else to interpret the manuals for them.

3. The need to reduce the size and complexity of manuals can be met by publishing separate volumes, or sets of small volumes, covering each of the major maintenance functions (see specific conclusions below). The conflicting need to have available all information that may be needed in emergency situations can be met by giving to each ship and shore activity manuals on all phases of maintenance for the equipments aboard. Individuals engaged in a particular phase of maintenance could use the manuals on other phases for reference only, and thus be able to "stay within the book."

*PREVENTIVE MAINTENANCE*

1. Operation and preventive maintenance are so closely related that they should be presented in a separate volume or set which can be kept in a place handy to the pertinent equipment.

2. POMSEE checklists appear to be suitable for the preventive maintenance part of this combination.

*CORRECTIVE MAINTENANCE*

1. Information required for corrective maintenance should be presented in a separate volume or set.

2. The major functions to be provided for are: trouble-shooting, alignment (including adjustment and calibration), disassembly, replacement, and assembly.

3. Trouble-shooting is the most complex and time consuming of these functions. Information should be provided which enables the technician to: (a) gain a functional understanding of the circuits and their interrelationships; (b) trace circuits and signal flows; (c) make measurements which, when compared against the standards provided, enable him to localize malfunctions; (d) locate parts and components; (e) identify parts and components.

4. Most of the difficult, time-consuming searching for information and "flipping back and forth" occur during trouble-shooting. Putting corrective maintenance information into a separate volume will in itself eliminate much hunting and flipping—if the information is complete. Improved indexing will also reduce these handicaps. More convenient arrangement and less cross-referencing are also needed. The most usable and effective arrangement has yet to be determined.

5. Present trouble charts appear to have little acknowledged usefulness. Evidence accumulated in this survey corroborates the Hewlett-Packard find-

ing that a trouble chart should "check out the equipment completely." It appears that POMSEE standards, with minor modifications, would do this. The illustrations and instructions on testing in the POMSEE standards would provide valuable help to the inexperienced technician.

6. The Parts List should be part of the corrective maintenance volume or set. Stock Number Identification Tables (SNITS lists) should be in loose-leaf form to be inserted as a replacement for the present Stock Number Lists in the manuals.

*SYSTEM INFORMATION*

Two kinds of system information are lacking. The first is principles-of-operation type data explaining the interactions of system components and giving the basic performance standards. The second is information on the actual installation as completed by a Naval Shipyard, and system performance standards determined after installation. The development and integration of these types of information need to be studied and made more systematic. One possible solution is a system technical manual into which pages containing data supplied by Naval Shipyards could be fitted.

*INSTALLATION INFORMATION*

1. Information required for planning, design, and installation should be presented in a separate volume or set. This should probably contain the wiring diagrams, which are used relatively little in corrective maintenance.

2. This volume or set should be made available early. It should be the first of the instructional materials on new equipments to appear.

*ACCURACY OF TECHNICAL MANUALS*

1. If practicable, all corrections to technical manuals should be in the form of printed inserts. They should be sent directly to the ships and shore activities having the subject equipments.

2. Technical manuals will be subject to inaccuracies until the announcement, issuance, recording, and identification of field changes are made more systematic.

3. Technical manual changes necessitated by field changes should reach the appropriate ships unfailingly and promptly.

*SUPPLEMENTARY INFORMATION*

1. The Electronic Information Bulletin appears to be less effective than the bulletins previously published for sonar, radar, and communications.

2. The all-inclusive nature of the Electronics Maintenance Books has reduced their use to that of a last resort source of information. The volumes are so large that they are forbidding in appearance and physically difficult to use. Pamphlets on specific equipments or pages to be inserted in maintenance manuals might make the information contained in the EMB's more usable.

3. There are no adequate media for communicating technical explanations (e.g., the limited value of tube testers) and more general explanations (e.g., the nature and purposes of the POMSEE program) directly to maintenance personnel. Because of this lack, personnel often fail to understand the directives and technical materials which they receive. These appear to be the

reasons why personnel of long service regret that *Electron* magazine is no longer published.

4. Standard technical reference books should be made more readily available to maintenance personnel. These references should be cited in technical manuals.

5. In general, the long lag between the development of maintenance information and the time it reaches Navy maintenance personnel constitutes a severe handicap. It would be beneficial if this information reached Navy personnel as promptly as it reaches civilian contract engineers, who receive a steady feedback from their parent companies. The field engineer's "little black book" containing company-supplied information supplemented by his own notes in many cases is a tool which might fulfill a number of the Navy technician's maintenance needs.

*TECHNICAL MANUALS AND SUPPLEMENTS IN TRAINING*

1. Technical manuals are primarily maintenance tools rather than instructional aids. Although used in schools and other training situations they should not include materials and devices that are of use only in teaching. The student should learn to use the manual as he would any other maintenance tool.

2. POMSEE materials appear to be very effective for training, especially instruction in test instruments and test procedures, lack of which is severely felt throughout the Navy.

3. Over-use of the lecture method of teaching appears to cut down the time available for practical training in the use of instruction books and test instruments.

## *Recommendations*

1. Consider ways in which Navy schools, particularly "A" schools, might develop greater student familiarity with the contents and organization of technical manuals and provide more training in their use.

2. Consider ways in which technical manuals might be given more systematic use in on-board maintenance training programs.

3. Establish a representative committee to draw up functional specifications for a prototype installation manual. Prepare a manual according to these specifications, and evaluate it under actual use conditions.

4. Develop an experimental manual combining operation and preventive maintenance and utilizing POMSEE materials. Evaluate the manual functionally.

5. Develop an experimental manual for corrective maintenance incorporating the best features of existing Navy and civilian technical manuals and a modification of POMSEE performance standards as a trouble-shooting guide. Give this manual field trials and evaluate its effectiveness.

6. Collect and evaluate against the findings of this survey all available experimental attempts to rearrange and clarify corrective maintenance information for greater comprehensibility and utility. From the most promising of

these techniques develop and evaluate an experimental corrective maintenance manual.

7. Develop a "little black book" consisting of supplementary maintenance data on the equipments in one particular system, to be sent directly to maintenance personnel responsible for this system. The form of this booklet should be looseleaf, additional sheets to be supplied as rapidly as new information becomes available. Blank pages on which maintenance personnel can keep their own notes should be included. Evaluate in the field.

8. Investigate further the problems of information dissemination, and set up a system for more complete exchange of information among shipyard, ship, school, and Bureau of Ships personnel.

9. Devise a system for more adequate and efficient gathering and recording of maintenance information in the operating forces.

10. Study the feasibility and appropriateness of publishing Navy maintenance periodicals which are more specialized and personalized than the existing ones.

11. Devise methods for improving procedures used in making corrections to technical manuals and the integration of field changes into manuals.

12. Investigate the conditions surrounding the actual writing of Navy technical manuals, and consider means for obtaining a uniformly better product.

## APPENDIX A: COMMENTS ON INDIVIDUAL TECHNICAL MANUALS

The following are significant observations made by personnel interviewed on the technical manuals suggested for particular study by the problem assignment.

### *AN/SPS-8*

#### *HELPFUL CHARACTERISTICS*

The cable layout is outstanding as to function, origin, and termination of each line.

#### *ADVERSE CHARACTERISTICS*

1. Volume I is too big and bulky.
2. Tracing a single circuit involves use of too many schematics.
3. The schematics are folded too many times.
4. It is hard to maintain continuity in going from one schematic to the next.
5. The extra schematics in the folder are too large to be practical.
6. The theory of operation "snows" both Navy electronic technicians and the technical representatives of commercial companies.
7. The manual lacks a description of the purpose of the equipment.
8. The checks called for are too complicated.
9. Information on particular components and circuits is scattered through too many sections (antenna information cited as an example).

10. There is no antenna alignment procedure; no gear tolerances are given.
11. Mechanical parts are not identified well enough.

### *SRT Series*

(Note: This manual was very difficult to obtain. The only copy the Task Group could get for study was in possession of a systems design group at NEL. This was an "Advance Form Instruction Book." Comments should be interpreted with the latter fact especially in mind since maintenance personnel occasionally must work with just such incompletely developed books.)

*ADVERSE CHARACTERISTICS*

1. Many photographs and schematics are omitted.
2. The manual is very confusing; there are inconsistencies between schematics.
3. Schematics contain errors.
4. Corrective Maintenance sections are too vague.
5. The circuitry is very difficult to follow.*

### *SPA-4*

*ADVERSE CHARACTERISTICS*

1. Too many corrections to be made in pen-and-ink.
2. Many difficulties arise due to differences in terms and their meanings between manuals for BUORD and BUSHIPS portions of the system.

### *SPA-8A*

*HELPFUL CHARACTERISTICS*

1. Pages of waveforms are useful; time base relationship diagrams and inclusion of distortion components in idealized waveforms are particularly helpful. Waveforms would be even more useful if more information on timing within the cycle were included. It would also be helpful if the origin of the signal resulting in each particular waveform were indicated. Waveforms should be in sequence. Wealth of details helpful—it has *all* the useful information.
2. Theory of Operation section is outstanding for understandability.
3. The folder of separate schematics is helpful.
4. Listing of inputs is very useful.

*ADVERSE CHARACTERISTICS*

1. Repeater alignment procedure is so detailed that the continuity of steps is lost. Technical representatives have a simpler, better alignment procedure.
2. Adjustment sequence needs improving.

*Expert personnel find this equipment difficult to maintain due, in part, to crowding. Certain difficulties in using the manual probably originate in the poor maintainability features of the equipment itself, which are reflected in the manual in the form of hard-to-trace circuit diagrams, vagueness, etc.

3. Lacks enough data to enable personnel to substitute a different type of oscilloscope for the one recommended.

### *UQN-1B*

*HELPFUL CHARACTERISTICS*

1. Clearly written and helpful as a whole.
2. Simplified schematics very good.

*ADVERSE CHARACTERISTICS*

1. Has many inaccuracies in schematics.
2. Some simplified schematics in Theory of Operation section which originally appeared in the manual for UQN-1 are inaccurate in reference to UQN-1B.
3. Corrections are difficult to obtain.

### *Miscellaneous Manuals*

The following list contains desirable features of various current technical manuals as pointed out by personnel interviewed.

1. TBS: Uses color well, especially in showing individual signal paths on block diagrams. Uses color effectively on schematics, but not so extensively as in Sangamo books.
2. VF: Overlays very effective in showing mechanical construction. A series of photographs and exploded views would probably accomplish the same results at less cost. Use of color to differentiate mechanical parts involved in some operations is very effective. Servicing block diagram is very helpful.
3. AN/CP6: Very good layouts on test equipments and test hookups.
4. AN/SPS-8A: Cabling diagrams are very good, but manual is too bulky.
5. AN/SQS-4: Good layout. Test set-up for checking preamplifier stages, etc., with oscilloscope are helpful. Pictures of waveforms on schematics would be an improvement.
6. AN/SQS-12: Display of variable pulse length in color is helpful. In general a good manual, but lacks waveforms.
7. BLR: Has two extremely helpful tables. One lists test equipments, their characteristics, and use data in reference to particular units and sections of units. The other is a guide by unit numbers to corrective maintenance data in the manual itself.

## APPENDIX C: RESULTS AND DISCUSSION

### *Effectiveness of Technical Manuals*

*USE AND FUNCTIONS*

The various sections in current technical manuals are so different in content and purpose that they must be evaluated individually in terms of their functions and users. In doing this, there are three main questions to be answered: (1) Does a given section provide material that is useful for the function it is intended to serve? (2) Can the necessary information be readily found and assembled (in the sense that each piece joins with and supports others) for application to particular maintenance problems? (3) Is the information presented—through text and visual aids—in a manner that makes it readily understood and usable? The first question may be answered in a very general way, by examining the extent to which each section is used by major groups. Table 11 presents such a breakdown.

One of the most important differences between civilian and military maintenance personnel lies in their use of the Installation section. Among the civilians interviewed, this section was the one most frequently reported as receiving extensive use. Most of these users are in the planning, design, and installation groups in Naval Shipyards. Civilian personnel differ from the military also in that their responses indicate little or no use of the Operation, Operators Maintenance, and Preventive Maintenance sections.

*INFORMATION LOCATION*

Fifty-seven per cent of 132 subjects interviewed concerning corrective maintenance information stated that necessary data were often hard to locate, and that even when they could be found, this was at the cost of considerable "flipping back-and-forth" among the pages. Highly trained and experienced men, both civilians and Navy personnel, reported these experiences more frequently than did inexperienced men. The probable causes of this difference are: (1) inexperienced men try to solve fewer difficult trouble-shooting problems; (2) inexperienced men are inclined to blame their difficulties on themselves rather than on the technical manuals. The following direct quotation from an interview subject typifies the attitude of inexperienced personnel: "It's not for a third-class to question the book."

TABLE 11. *Use* of technical manual sections by 107 military personnel and 44 civilians.*

| Section | Frequently Used | | Seldom Used | |
|---|---|---|---|---|
| | Mil. | Civ. | Mil. | Civ. |
| 1. General description | 3 | 0 | 8 | 0 |
| 2. Theory of operation | 76 | 14 | 0 | 4 |
| 3. Installation | 2 | 23 | 76 | 4 |
| 4. Operation | 7 | 0 | 6 | 1 |
| 5. Operators maintenance | 3 | 0 | 8 | 0 |
| 6. Preventive maintenance | 13 | 0 | 3 | 0 |
| 7. Corrective maintenance | 101 | 16 | 0 | 0 |
| 8. Parts list | 82 | 10 | 10 | 0 |

Further reasons for experiencing difficulty in locating corrective maintenance information were also indicated. First, the information needed for trouble-shooting is, in many cases, so scattered that the technician frequently

*Infrequent mention of a section may reasonably be interpreted as lack of use.

makes several frustrated attempts before finding what he needs. Second, there appears to be no standard organizational procedure that tells the reader where to look for the information he requires. A member of the Task Group in collaboration with a highly trained engineer who had been head of a technical manual writing group made a limited test of the time required to find needed information. The technical manual selected was typical of recent publications, NAVSHIPS 92501 (A), for the AN/FRT-27. The output from the oscillator buffer stage in the transmitter was selected at random as the information to be sought. The first step was to try to find the voltage reading on the simplified schematic, where it did not appear. The second step was to try to find it on the detailed schematic; it was not there either. Nor could it be found in the Theory of Operation section, which was examined as the third step. The fourth step was to look into the trouble chart where the voltage reading was finally found after a search through three pages of the chart. Total time required to find this information was about 20 minutes. Graduate engineers have reported spending days in off-and-on search before locating needed data.

The third main cause of difficulty in locating information was said to be the indexing system, which is quite different from that used in standard civilian references. The index in a Navy technical manual, when it has one, consists of the Table of Contents arranged alphabetically with clusters of heterogeneous topics under main headings. Many current technical manuals are not indexed at all. Of 38 subjects who discussed indexing, all agreed that a better index would be of help to them.

*INFORMATION ARRANGEMENT*

Closely related to the need for less hunting of information and less "flipping back-and-forth" are certain problems of information arrangement. The trouble-shooter finds it necessary not only to locate information, but to assemble it, mentally, for use. This is difficult with the present arrangement of technical manuals. Information useful in locating circuits and stages is found in block diagrams. Information required to gain a functional understanding of circuits is for the most part in Principles of Operation, but these data must be supplemented with the graphic representations and component values found on schematics. Part of the information required for identifying parts and components is obtained from photographs, and the rest from schematics. Data from tables as well as data from schematics are required in order to trace signals by comparing a sequence of measurements with values given in the manual. The significance of these measurements may be derived from the Principles of Operation, from a trouble-chart, or even from the General Description.

One experimental arrangement having the object of integrating trouble-shooting data was shown to personnel interviewed. Eighty-eight per cent thought it helpful. This arrangement was developed by the RCA Service Company for use in their Inter-Level Electronics Training program. In this presentation a circuit is laid out on a single fold-out showing a block diagram, a simplified schematic, and an actual schematic in left-to-right order. The block diagram includes a short statement of the purpose of each stage, and a resume of the stage's function. On the same page with the simplified schematic, there are condensed analyses of the circuit and possible component failures. All components are included in the complete schematic with a brief description of each stage.

## *INFORMATION PRESENTATION*

### *Block Diagrams*

Functional block diagrams are useful to most maintenance men interviewed during this survey. They are most frequently used in gaining an over-all understanding of the relationship between circuits and components and the manner in which each major unit or stage contributes to the transformation of inputs into outputs. It appears necessary to visualize these relationships, and a good block diagram appears to be the best, if not the only, device which can aid in developing an adequate mental picture at the proper level of generality. It is important to bear in mind that trouble-shooting (except for "Easter-egg hunting") means going from the general to the particular, from the whole to a part, eliminating as one goes until the faulty segment is reached. As one 20-year chief put it, "No equipment is complicated once you get to the place where the fault is."

Block diagrams represent the general, gross aspect of the equipment. Thus they often serve as a starting point in trouble-shooting, helping the maintenance man to identify the area and/or circuit where the fault is located, and enabling him to begin reasoning as to what may have caused the unwanted effects. A block diagram may also serve as a bridge between the equipment itself and the schematics. Maintenance men—particularly the inexperienced—will frequently go from a part they have identified as faulty to the block diagram and thence to the schematic. Or they may reverse the process, going from the circuit segment or component they have been studying on a schematic, to a block diagram and thence to the equipment.

These aspects of the trouble-shooting process emphasize the need for close integration between the Theory of Operation section—where block diagrams are generally located—and the Corrective Maintenance section. They also make it apparent that locating useful block diagrams in the General Description section may cause the maintenance man needless flipping back-and-forth between widely separated aids.

### *Simplified Schematics*

Simplified schematics are also found very useful, particularly at the stage when the maintenance man is trying to understand how a circuit works. They aid in tracing the *course* of a circuit simply because they are not encumbered with all the data required for signal tracing.

"Circuit tracing" and "signal tracing" are often used interchangeably. It is useful, however, to think of circuit tracing as the following of a path, or course, without regard to the characteristics of the energy which follows this course. Signal tracing may be defined as the effort to follow the flow of energy along a course, determining the segments in which it is normal and those in which it is not normal.

*Photographs and Drawings*

Seventy per cent of the 124 men interviewed on the subject considered the photographs and pictorial drawings in manuals helpful. The main uses of these visual aids are: (1) in gaining a concrete understanding of the way an equipment functions, i.e., circuit and component interrelationships, and their contribution to outputs; (2) in identifying parts; (3) in locating parts. Pictures should be clear with sharp definition to make individual parts stand out. Callouts and captions should be large enough, and spaced far enough apart, to be easily read. It appears best—when a choice must be made—to sacrifice completeness of coverage to legibility. Good, clear drawings are as helpful as photographs, in the opinion of maintenance personnel interviewed.

Pictorial block diagrams of complex equipments and systems are especially effective in giving an over-all understanding. These may be made up of photographs, or drawings, or a combination of both. Their usual location —the General Description section—is not the best, however, because these diagrams are most often used in conjunction with the Theory of Operation section. It is held by most maintenance personnel that visual aids and the text to which they are related should be as close together as possible.

*"Sams Approach"*

The above discussion leads logically to a consideration of the frequently-mentioned "Sams approach" since the Sams Company maintenance manuals make use of a large number of photographs with callouts. Further characteristics of this company's manuals are (1) schematics having standardized format, symbols, and location of components; (2) waveforms and voltages on the schematics themselves; and (3) necessarily limited text.

Although many Navy maintenance personnel who have used Sams Company books have found them effective, most of these men doubt whether they would prove satisfactory for Navy equipments without considerable adaptation. There are several reasons behind this doubt. First, most Navy equipments are much more complex than the civilian equipments for which Sams books are written; thus, Sams type books for Navy equipments would necessarily have to be more complex and bulky. Further, it has yet to be proved that the standardized schematics used by the Sams Company are easier to read out than the best of the schematics in Navy technical manuals. Finally, there are reasons to believe that the Sams books do not contain *all* the information required in Navy maintenance situations.

## *SECTION ANALYSES*

*Corrective Maintenance Section*

The extensive use made of this section (see table 11) makes it profitable to consider each of its separate sub-sections individually.

*Schematics.* All available evidence shows that schematics are the heart of the trouble-shooting process.* They are the maps showing current and signal flow—charting their courses between input and output and indicating the modifying points through which they pass. Schematics also provide a large part of the data required to understand a circuit and measure its performance.

In general, dependence upon schematics increases with specialization and experience. Experienced Navy personnel (chiefs and first class) tend to rely on schematics more heavily than do the less experienced rates. Some shipyard and tender repair specialists use very little of the manuals beside the schematics. The reasons seem to be: (1) specialists require less complete information on a given equipment than other technicians because a majority of their problems are repetitions of—or similar to—problems they have solved before; (2) highly experienced men generally have a reasonably complete understanding of the way each equipment under their care functions. Consequently, good schematics provide both these categories of personnel with most of the information they require for corrective maintenance, except when faced with a completely unfamiliar equipment or a very unusual malfunction. In such cases, experienced men may use all parts of the Corrective Maintenance section, and parts of other sections as well.

Most experienced maintenance personnel feel that wherever the waveform has significance it should be shown on the schematic. The oscilloscope on which the waveform was obtained should be indicated, and time base and peak voltage data should be given.

The complaints most commonly made against schematics by personnel interviewed were:

1. Lines were spaced too close together.
2. Too many lines crossed one another.
3. The signal path was not emphasized graphically.
4. The course of a current flow was highly intricate.
5. Too many symbols were crowded onto a page.
6. Symbol numbers were too small to read easily.

Conversely, a sizeable majority of the respondents found schematics relatively easy to read when the following conditions obtained:

1. Color was used to trace a signal path.
2. The "in-line" convention was used to simplify the pathway.
3. Signal paths were emphasized with heavy lines.
4. Schematics were relatively uncrowded.
5. Symbols and numbers were readily visible.
6. "Roadmapping" was used to facilitate part location.
7. Realistic waveforms were shown instead of idealized ones.

According to 86 per cent of the comments, the indication of inputs from, and outputs to, related equipments is desirable. Ninety per cent of mainte-

*The "color test" is the shipboard technician's favorite way of showing how much more the schematics are used than other portions of the manual. The edges of the schematics soon take on dark hues ranging from a greasy gray to black-brown. The rest of the pages have edges ranging from pale gray to virgin white.

nance personnel would like to have a voltage tolerance range given on the schematic, and a special marking to indicate each point at which the voltage level is critical and no tolerance is permitted. Most maintenance personnel reported finding related voltage and resistance (V/R) tables very useful, particularly if they were of the pullout type and were not hidden when the manual was closed.

A separate book or package of schematics was regarded as useful and desirable by a majority of respondents, particularly those who are experienced. Such a unit of material would be easier to carry about than the complete manual. In addition, when maintenance men work with several schematics and the text as well, fold-out drawings in the manual mask one another, whereas separate schematics taken from a package can be laid out in easily visible sequence.

*Alignment Procedures.* Alignment procedures in technical manuals are frequently criticized as being too vague and/or too elaborate to be easily understood and followed by inexperienced technicians.

The most significant data on this point comes from a Mobile Electronics Technical Unit (METU) that works with technicians aboard minesweepers and minelayers. Many of the technicians aboard such ships have had no shipboard experience under men of greater competence than themselves, since the senior technician is usually no higher than ET-3. Consequently, they frequently have difficulty in maintaining their equipments, and depend heavily on advice and instruction from the local METU. It was found by METU personnel that the greater part of the assistance required had to do with alignment, the procedures given in the technical manuals being too cumbersome for the shipboard technicians to follow. The fact that the EIB (Electronic Information Bulletin) frequently publishes simplified alignment procedures is further evidence that the procedures given in the technical manuals are often unnecessarily and inefficiently complex, at least for maintenance in operational locations.

The evidence indicates that alignment procedures should be made a separate sub-section, in chart form. This chart should be simple and straightforward, and should contain the standards necessary for reference.

*Trouble-shooting Aids.* Trouble-shooting aids take four main forms: (1) servicing block diagrams; (2) diagrammatic trouble charts; (3) tabular trouble charts (specified by MIL-M-16616); (4) charts of typical troubles.

a. Servicing block diagrams when available are used considerably by experienced personnel. The servicing block diagram was reported by these men to be an excellent transitional device for localizing a fault to a stage or chassis area sufficiently restricted to facilitate identification of the appropriate schematic. A servicing block diagram does not, however, tell the maintenance man what to measure, how to measure it, or indicate the most appropriate series of trouble-shooting steps.

b. Part of these functions are provided for by the *diagrammatic* trouble charts found in some manuals (for example, NAVSHIPS 91522 (A), Volume One, for the AN/SPS-8 radar). A diagrammatic trouble chart illustrates the most promising series of steps for localizing the causes of recognizable symptoms. The presentation is graphically similar to that of a flow chart. Experienced technicians and engineers report that this graphic representation is helpful in holding the series of steps in mind. They also help the maintenance man working on an unfamiliar equipment to identify (on a "go-no-go" basis) the areas where faults causing observed symptoms probably lie.

Even when used together, a servicing block diagram and a diagrammatic trouble chart do not make up a complete or satisfactory set of trouble-shooting aids. A servicing block diagram hits the high spots, so to speak, and it takes considerable experience to know how to fill in the detailed procedures between the checks described in the diagram. Data on test instruments and test procedures are lacking, and reference data in the form of standard measurements are scanty. Moreover, effective use of a servicing block diagram depends on a good grasp of the logic of the circuits. Consequently, still another type of aid is required.

c. Tabular trouble charts (see Mil Spec MIL-M-16616) do not appear to meet this need satisfactorily. Though most inexperienced men and some experienced men use them "to get a start" or "to get ideas" they are widely and severely criticized—especially by the more experienced personnel. The principal faults found are as follows: (1) It is often difficult to find on the chart the particular symptom or failure actually encountered. (2) Very frequently the chart fails to carry the maintenance man far enough to find the malfunction; the steps end too soon, leaving him to proceed without assistance from the manual. (3) It is easy to lose the sequence of steps; that is, to forget what one has already done and become puzzled as to what steps to take next. The reasons for these difficulties are that the charts give the steps as discrete items which are not easy to relate to one another (in contrast to servicing block diagrams where the relationships are visualized); and that the abstract statements on the chart are difficult to hold in mind.

A promising solution to this problem was suggested during this survey by an Electronics Officer in the Pacific Fleet and an experienced submarine sonarman in the Atlantic Fleet. Both of these men believed that the POMSEE Maintenance Standards, Part I, would make an excellent trouble-shooting guide. These preventive maintenance publications have excellent illustrations and simple, clear instructions for making tests. These are well integrated with an easily followed step-by-step procedure with which maintenance personnel may check out the equipment completely, circuit by circuit. According to the civilian maintenance technicians interviewed by the Hewlett-Packard Corp., instructions and data for making such a complete checkout are indispensable to an effective trouble-shooting guide. Interviews conducted during the present survey substantiate this finding. It therefore seems probable that POMSEE Maintenance Standards, Part I, can be adapted to trouble-shooting. They would have the additional advantage of providing realistic measurements for reference.

d. "Typical trouble" charts are not used by experienced men. They consider this form of aid a waste of time because the troubles listed are so seldom encountered. Inexperienced men use "typical trouble" charts because they seem (superficially) to be a quick and approved way of locating faults. Instructors and senior maintenance personnel deplore the type of trouble-shooting which these charts encourage. They feel that inexperienced men waste time in hit or miss hunting for listed faults (known as "Easter-egging" or "Christmas-treeing") and may, by following such a procedure, delay or avoid altogether the use of a systematic approach which will achieve results more quickly and certainly.

*Test Equipments.* Information concerning test instruments and tests forms an important part of the guidance required for effective trouble-shooting. A few current manuals—for example the Galvin Electric Company's manual for Radio Beacon AN/CPN (NAVSHIPS 900.38)—have complete and well illustrated directions for making the necessary tests. Such instances are so rare that a large proportion of personnel interviewed pointed out lack of test information as a general weakness of manuals. The data maintenance personnel need include alternative test instruments, settings, test hook-ups, and photographs of waveforms taken from oscilloscopes available to Navy personnel. Neither technical manuals for test instruments nor supplements such as *Handbook of Test Methods and Practices* (NAVSHIPS 91928 (A) ), appear to meet the need completely because of the reluctance of maintenance personnel to "go outside the manual" for information. It is, in fact, time consuming and burdensome to consult a number of references under the working conditions that usually obtain when an equipment fails.

*Repair.* Repair is a separate section in the newer manuals, but in this report is considered as the second half of the corrective maintenance process—trouble-shooting being the first. Actually, repair takes only about one-fifth of the time maintenance men typically spend in correcting a malfunction. The greater part of repair work is mechanical, since repair of electrical parts consists mainly of removal, replacement, and re-alignment. Mechanical repairs are often difficult and complex. Moreover, Navy electronic maintenance personnel are usually better trained in electronics than in mechanics. For these reasons, the need for adequate mechanical information is great.

A majority of personnel interviewed found the current manuals generally deficient in information required for mechanical repair. This whole area may require more attention than it is currently receiving. Sonarmen report that a very high percentage of the repairs made by them are mechanical and that the required information is frequently unavailable.

The Antenna Section of Shop 67 in one Naval Shipyard reported that lack of data concerning mechanical tolerances for antenna mounts prevented ships from making adjustments which they would have been capable of making had the data been available. The result was unnecessary expenditure of time and effort by shipyard personnel, who were themselves handicapped by lack of such data.

Mechanical stages of complex equipments frequently are explained and shown graphically in less completeness than electrical stages. Some manuals for teletype equipments provide striking illustrations of this lack. For example, Teletype Corporation Instructional Manual No. 67 (Mod 15 Printer Set for Navy Shipboard Use) had so many gaps in the information and such poor schematics that an ET school instructor reported needing six months to develop lesson plans on this one equipment.

More mechanical information of the following kinds was particularly desired by personnel interviewed during this study: (1) complete identification of mechanical parts; (2) more exploded drawings showing graphically the precise steps to be followed in disassembly and assembly; (3) critical data, such as mechanical tolerances.

Exploded drawings appear to be equally useful for mechanical repairs as elaborate overlays, and are certainly much less expensive. A significant number of experienced men also feel that the mechanical portions of circuits (which are usually represented by blocks or generalized symbols on schematics) should be drawn in more detail—though not necessarily on the schematic itself.

*Theory of Operation Section*

Next to the Corrective Maintenance section, the most used part of technical manuals is the Theory of Operation (or Principles of Operation) section. Practically all maintenance personnel—civilian and military—study this section when they are performing corrective maintenance on an equipment that is relatively new to them, or when they are trying to solve a particularly baffling maintenance problem on any equipment. It should be borne in mind that, to the inexperienced maintenance man, all equipments are unfamiliar and most maintenance problems are baffling. The reason for studying this section is to gain an understanding of how the equipment is supposed to work.

*Understandability and Readability.* The understandability of the language used in the text is as important in the Theory of Operation section as any other place in the manual. Ninety per cent of the instructors discussing the problem stated that the Theory of Operation section is the most difficult for their students. Consequently, the matter of difficulty level or readability of the writing (actually, level of comprehensibility) is of great significance.

The most realistic way to describe the appropriate level of comprehensibility is in terms of reader experience. The Theory of Operation section (and other textual material) should be written for personnel with a mean experience of about one year and ranging roughly from six months to two years. Text written at this level will meet the needs of men who have been converted from other rates and normally rated men whose actual experience with maintenance is limited but whose learning ability is high. It allows time for the average man to acquire the "real" (sensory) referents necessary for comprehension of practical—as contrasted to academic—electronics. To make the comprehensibility level higher would mean cutting off a large proportion of low-rated maintenance personnel who have responsible work to perform. To set it lower would mean that technical manuals would have to be written primarily for educational purposes since, on the average, the maintenance man with less than a year's experience does more learning than actual maintenance.

The average Navy maintenance man has at least 12 years of civilian schooling. "Readability" in the technical sense popularized by Rudolf Flesch is, therefore, not a pressing consideration, although improvement would doubtlessly result if technical manual writers held average sentence length down to 25 or 30 words and used a minimum of polysyllabic words (like "minimum" and "polysyllabic"). As for the terminology used, most "A" school graduates have acquired sufficient technical vocabulary to read the words.* Most of them consult more experienced men and reference books to obtain the meanings of words they do not know.

Slightly more than half of the personnel discussing this matter felt that the Theory of Operation section was written clearly with easily understood words and sentences. The remainder believed that understandability of the section needed improvement because it was usually too technical and assumed too much reader background knowledge. This was frequently expressed in the statement, "The Theory of Operation section is written from an 'engineer's point of view'."

*Explanations.* The explanation of concepts in the Theory of Operation section was frequently mentioned by the personnel interviewed as a characteristic in need of improvement. Here is the point at which the maintenance man's training in basic theory meshes with the more specific information in the manual. Some maintenance people feel that the less experienced men would be benefited if Theory of Operation sections were written in such a way as to recall the applicable principles to mind.

In contrast to this, other maintenance personnel feel that some Theory of Operation sections already contain too much engineering logic. This was expressed by the statement that the manual is "not written from the maintenance point of view."

Although opinion is divided as to whether or not the Theory of Operation sections currently are written from the "maintenance point of view," certain conclusions seem tenable on the basis of comments made: (1) The maintenance man should be required to make as few mathematical computations as possible. (2) The text should not attempt to justify the circuitry or explain its particular virtues. (3) The text should enable the maintenance man to bring his basic knowledge to bear in deriving a functional understanding of cause-and-effect relationships. The maintenance man is apparently more interested in what takes place along the course of a circuit than in why it happens; that is, he is more concerned with the direct physical causes of effects than with the scientific principles underlying the occurrences. In short, as one man put it, "maintenance personnel want maintenance information, not design engineering information."

About 40 per cent of the maintenance personnel commenting on the matter feel that explanations in the Theory (Principles) of Operation section are not complete (detailed) enough. They believe that references are not cited frequently enough, and that while simple, obvious stages are often given in

*In general, "A" schools give basic electronic training, "B" schools give more advanced training, and "C" schools give specialized training on individual equipments.

elaborate detail, unusual and complex circuits are often given cursory treatment. Elaborating the obvious wastes the time of experienced men, and misleads the inexperienced by making them think that simple stages are difficult—thus causing them to spend time hunting for mysteries that are not there. All circuits should be described rather than being merely labelled, even though it is unnecessary to explain them in detail. Statements such as "From this point on, the sweep circuit functions as an ordinary DC amplifier circuit" leave too much for the maintenance man to interpret for himself.

It has been suggested by some of the experienced civilians interviewed that the underlying cause of this imbalance in the explanations is the lack of understanding on the part of some technical writers. It would be natural for a writer confronted with an already-developed equipment, finding that the engineers who had developed it lacked time to explain everything to him, to elaborate the stages he understood best and treat the less understood parts obscurely or in a very general manner.

*Standardization.* The technical vocabulary of "A" school graduates is generally adequate so long as only standard terms are used. Maintenance personnel are, however, frequently puzzled by manufacturers' unique designations and the variations of engineering terms which have not become standardized. Over 80 per cent of the comments made on this subject indicated a strong feeling that the terms and symbols used should be standardized, including the standardization of inputs and outputs, and the "A" and "B" sides of tube symbols.

It is evident that standardization should be carried as far as possible. However, standardization is an ideal impossible of complete realization because engineering language is growing constantly. Several words having the same meaning may be in current use for a time before one becomes the standard term. The inclusion of a glossary containing any unique or unusual terms seems to be the most logical solution for the problem. Seventy-five per cent of the maintenance men commenting on the matter believed that such a glossary would be helpful. Obviously, the glossary ought to cover all the text, but, as a few men indicated, it might be best placed at either the beginning or end of the Theory of Operation section.

# Appendix E

# Organizing the Research Report

In this Appendix we are presenting the most interesting extension of the logic of outlining we have ever seen. We feel that it can be of great value in planning long reports, particularly reports such as proposals on which several, or many, people collaborate.

We are much indebted to Mr. F. Bruce Sanford for this material.

# Organizing the Research Report to Reveal the Units of Research

*by*

*F. Bruce Sanford*

United States Department of the Interior
Fish and Wildlife Service
Bureau of Commercial Fisheries

Circular 272

## Abstract

As a research project becomes increasingly complex, the traditional outline used to report the research becomes less satisfactory. The reason is that the traditional outline tends to dismember the basic units of the research and to regroup the parts in such a manner that the whole is obscured. Suggested here is a model that will help the researcher organize his report in such a way that the basic units are kept intact and their identity is revealed regardless of how complex the research may be.

## Introduction

A scientific investigation is composed of units. If the reader is to understand the report of the research, he must know what the units are and how they fit together.

Unfortunately, except for articles reporting on only a single unit, the traditional pattern used in the organization of research reports (I. Materials and Methods; II. Results and Discussion) conceals both the identity of the units and their interrelations. As a result, the communication efficiency of the scientific literature is low.

Since the traditional model is inadequate, why do we persist in using it? There are three reasons. First, because the traditional model is logical for reporting a single unit of research, we assume that it is equally logical for reporting a larger number. Second, because the subject matter of a research report is complex, we ascribe much of the reader's difficulty in assimilating the report to the complexity of the subject. Third, because we recognize that part of any difficulty the reader may have in understanding our report may be due to our lack of ability to express ourselves well, we attribute any remaining unexplained obscurity to poorly written sentences. So, we overlook the basic fault—poor organization.

What we need then is an organizational model that will reveal the units of research and their interrelations regardless of the number of units being reported. The aim of this article is to suggest such a model.

The outline of a research report is determined by the design of the research—or, at least, it should be. The design of the research, in turn, is determined by the purpose of the research. Accordingly, to gain insight into our problem of organizing reports logically, let us first look at the relation of purpose to research design and then at the relation of research design to the outline of the research report.

## I. Relation of Research Purpose to Research Design

When designing our research, we must consider the complexity of its purpose. If the purpose is complex, we divide it into a branch chain of subpurposes (as shown in Figure 3). We can thereby simplify the purpose and reduce the scope of our research to fit our resources. Let us suppose, for example, that the ultimate purpose of our research is to increase the demand for fish-reduction products. On thinking about our problem, we see that we can accomplish our purpose if we can increase the demand for fish-reduction lipids and fish-reduction proteins (Figure 1). Upon examining Figure 1, we see

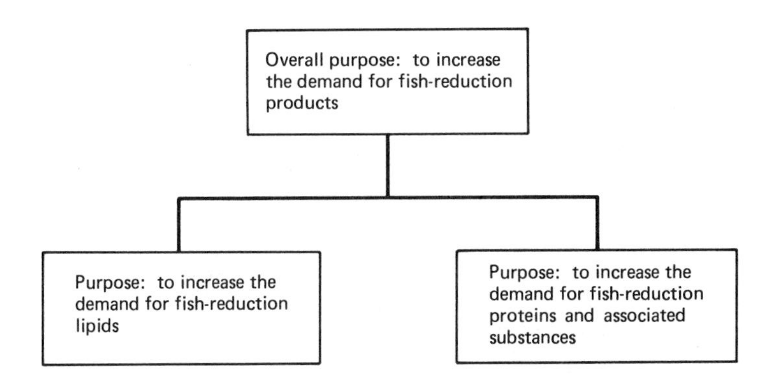


Figure 1. Division of Our Overall Purposes

that it is still too complex. We therefore continue subdividing, perhaps as shown in Figure 2, until we come to purposes that are sufficiently specific and sufficiently restricted in scope that we can handle them with our particular research capabilities. Figure 3 illustrates this process of division symbolically.

Now that we have our hierarchy of purposes in mind, we can design our research to fit our time, our money, and our facilities.

If, for example, our research budget is small, we might confine our research to Purpose IA1. In that event the design of our research would take the form shown in Figure 4. Note that the purpose determines the procedure and that the procedure determines the results. Note further that we draw a conclusion (or arrive at a recommendation) from the results and that this conclusion reflects back to the purpose.

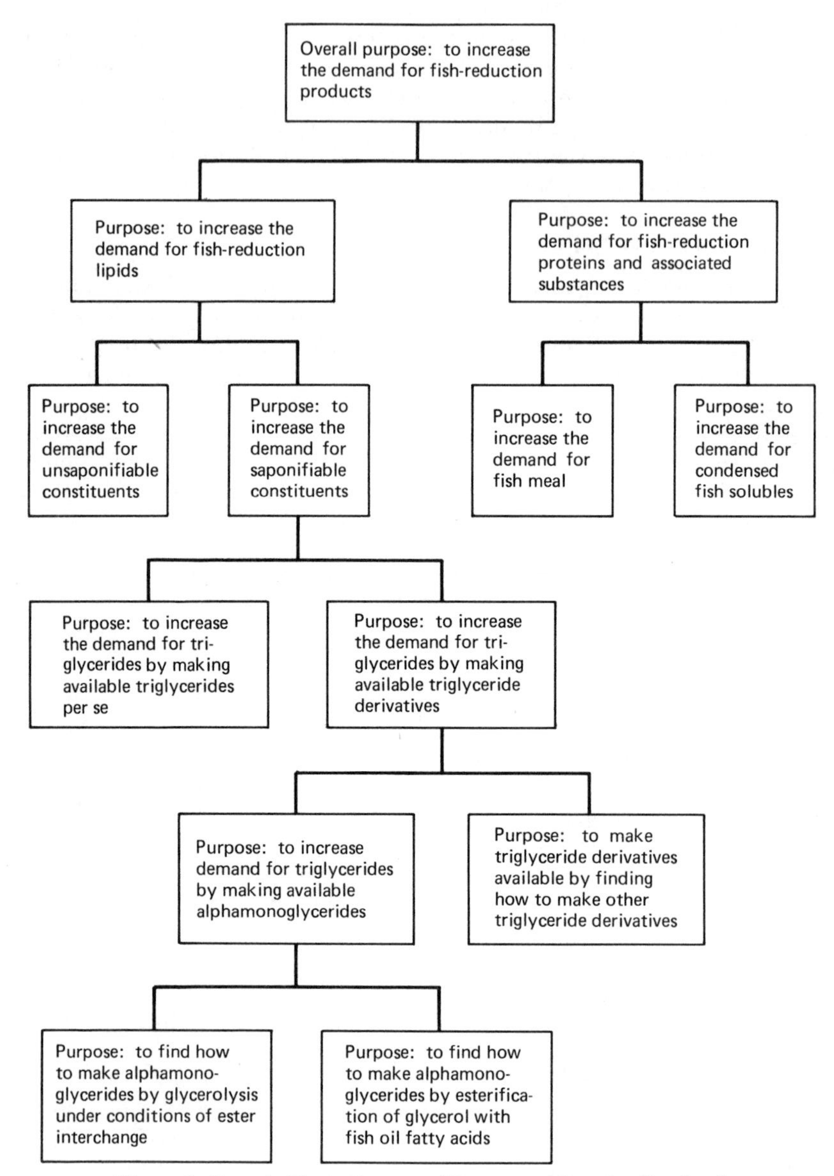


Figure 2. Branch-Chain Pattern of Purposes (Greatly Abridged)

If our research capabilities are larger, we might widen the scope of our research to include Purpose IA (Figure 5).

And if our research capabilities are larger still, we might widen the scope of our research to include Purpose I (Figure 6) and so on.

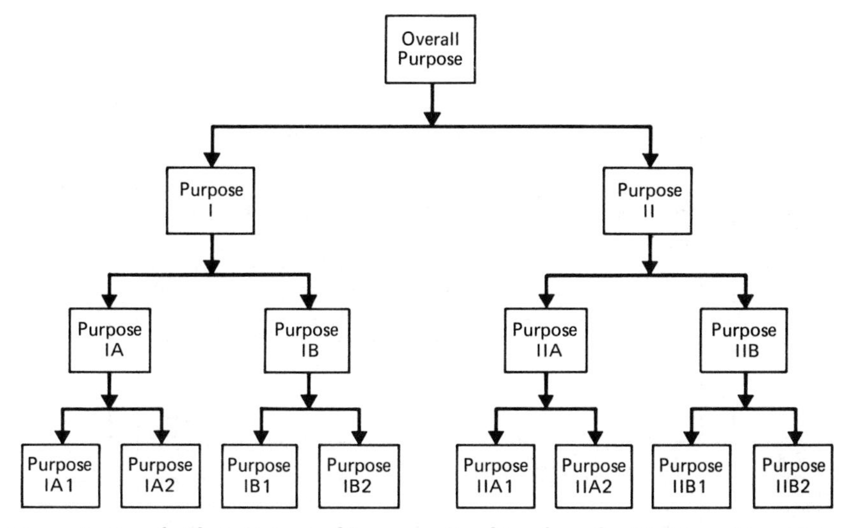


Figure 3. Branch-Chain Pattern of Purposes Resulting from the Division of Our Major Purpose

## II. Relation of Research Design to Outline of Report

From the foregoing discussion, we see the relation between the purpose of the research and the design of the research. Let us now look into the relation of the research design to the outline of the research report.

When we analyze the preceding Figure 4, 5, and 6, we see that research is made up of a number of units, such as those formed by Experiments IA1, IA2, IB1, and so on. Each of these units consists of a purpose, a procedure (that is, materials and methods) designed to meet that purpose, a set of results following from the application of the procedure, and a conclusion that derives from the results and that reflects back to the purpose of the experiment.

Evidently, since these units are basic, our report of the research should reveal them unambiguously. In fact, it should reveal unambiguously the entire structure of the research regardless of how complex it may be.

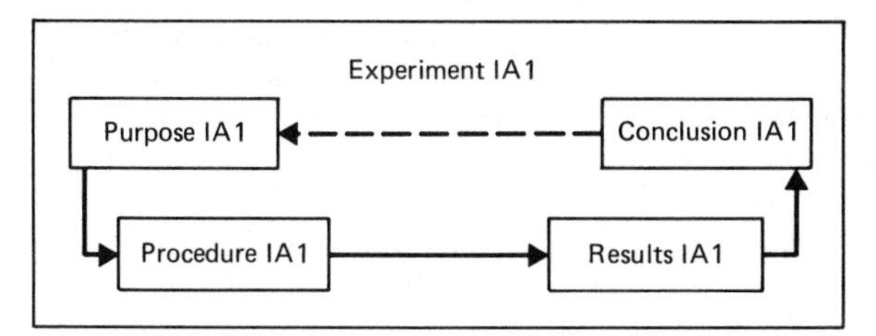


Figure 4. Research Design Required by Purpose IA1

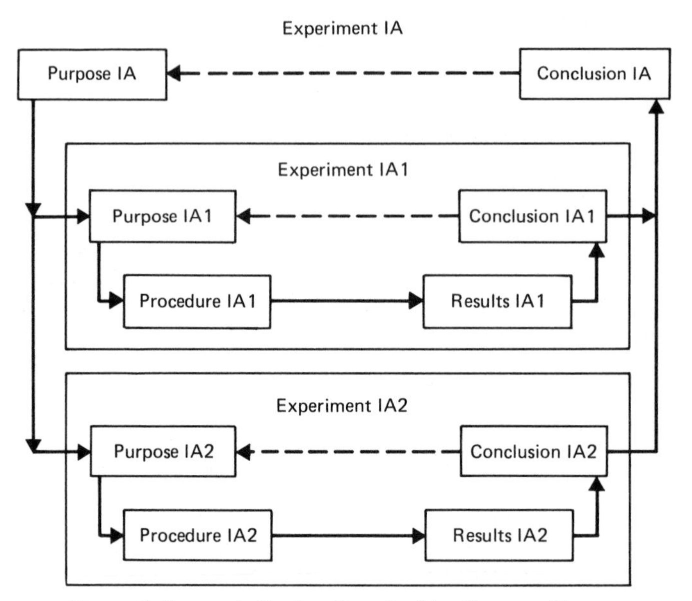


Figure 5. Research Design Required by Purpose IA

Since our purpose determines our research design and since our research design determines the outline of our report, the purpose of our research is the key to the logical construction of our outline. For that reason, let us first consider simple reports in which we have no subpurposes and then consider complex reports in which we do.

### *A. Report of Research That Has no Subpurposes*

Because the traditional outline is ideal when the research being reported has no subpurposes—that is, when only a single unit of research is being reported—it is entirely suitable for experiments such as Experiment IA1. So we use the traditional outline (Figure 7) when reporting simple research.

The outline in Figure 7 has an introduction, which gives the reader insight into the experiment by supplying background information; it then states explicitly and unambiguously the purpose of the research. The reader now knows why the research was undertaken and exactly what its purpose was. The stage is set for his consideration of the materials-and-methods section revealing how the experiment was performed. Reading this section gives him further insight into the experiment and prepares him for an analysis of the results and discussion. This analysis paves the way for his evaluation of the conclusion—his acceptance or rejection of it. The process is straightforward and efficient. This traditional outline for reporting a single unit of research, such as Experiment IA1, thus is ideal.

### *B. Report of Research That Has Subpurposes*

Most reports are based on research having subpurposes. For those reports,

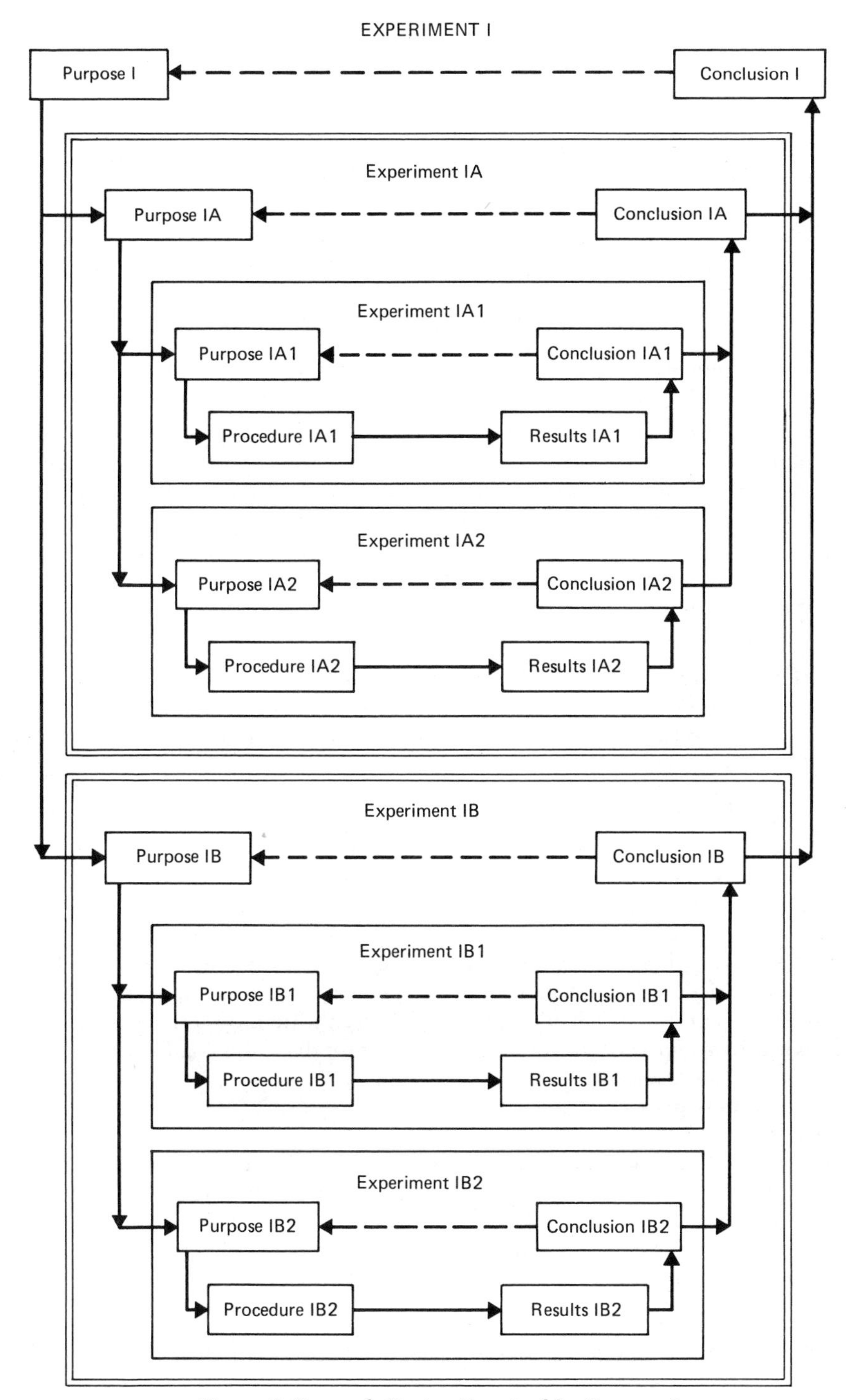


Figure 6. Research Design Required by Purpose I

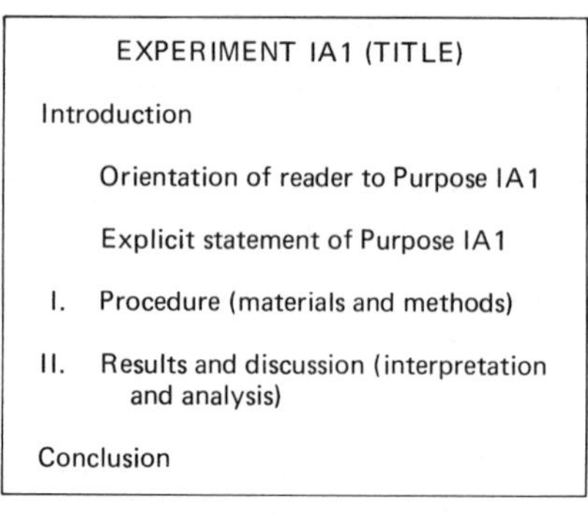
EXPERIMENT IA1 (TITLE)

Introduction

Orientation of reader to Purpose IA1

Explicit statement of Purpose IA1

I. Procedure (materials and methods)

II. Results and discussion (interpretation and analysis)

Conclusion

Figure 7. Outline for a Paper Reporting on an Experiment Based on Purpose IA1

the outline for simple research (I. Materials and Methods; II. Results and Discussion) will not suffice. Its organization is illogical for this use, since it conceals the units of research and their interrelations. How then do we outline the reports of complex research? For the answer, let us consider the reports of research having secondary purposes but no tertiary purposes and then consider the more complex reports of research having tertiary purposes as well as secondary ones.

*1. Research Having Secondary Purposes*

Figure 8 shows an outline for the report of an experiment, such as Experiment IA (Figure 5), that has secondary purposes but no purposes that are subordinate to them. In comparison with the traditional outline, this outline has the advantages that it:

a. Tells the reader not only what the overall purpose of the experiment is but also what the subpurposes are.

b. Tells him why we have those subpurposes and thus furnishes him with quick insight into them.

c. Permits the main headings of the paper to correspond to the main divisions of the research, further giving the reader quick insight.

d. Puts together all the information that belongs together. This outline thus properly unifies the reader's concept of the various experiments. His memory is not needlessly overburdened, nor is he sent on unnecessary trips backward and forward in the manuscript to find the various parts of the particular unit of research he is reading about.

e. Presents the overall conclusion in such a way that the reader does not have to infer it for himself.

f. Reveals completely and clearly what we are trying to do, why we are trying to do it, what we did, what we found out, and what we concluded from the findings.

g. Stimulates us to think deeply into the meaning of our research and thereby helps us in producing research of high quality.

*2. Research Having Tertiary Purposes*

Figure 9 shows an outline for the report of an experiment, such as Experiment I (Figure 6), that has tertiary purposes.

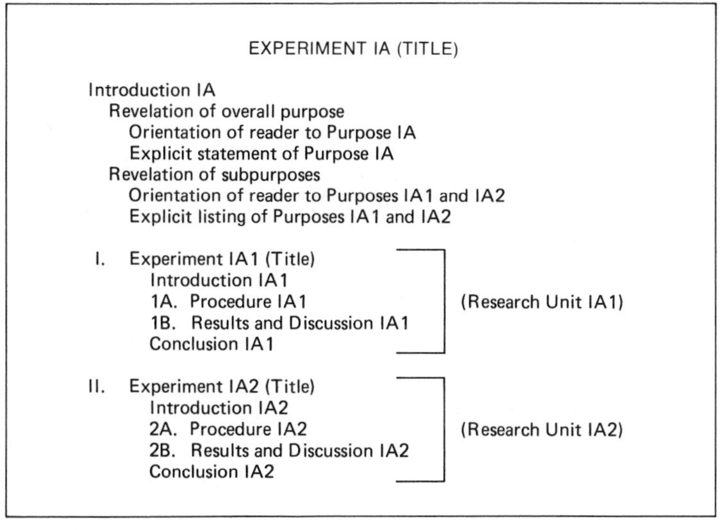


Figure 8. Outline for a Paper Reporting on an Experiment Based on Purpose IA

Often Procedures IA1, IA2, IB1, and IB2 are closely similar. On looking at the outline in Figure 9, we might therefore think that we would have to repeat the description of essentially the same procedure four times. But not so. All we need do is describe Procedure IA1 in detail and then, when describing each remaining procedure, tell only how it differed from the first.

As soon as we have completed each unit of research, we can write a paper on it, as indicated in Figure 7. Then, as we accumulate these individual papers, we can combine them as indicated in Figures 8 and 9. We thus can make as short or as long a paper as we wish, provided that we always have the appropriate unifying purposes. Accordingly, we have perfect freedom to modify our overall report to fit any changed concept we may have reached as the result of new research findings. Furthermore, we can keep the reporting of our research strictly current. We need not delay the report of one unit simply because some factor is delaying our completion of another unit.

## Summary

The outline of our research report is determined by the design of our research. This design, in turn, is determined by the purposes of our research. So, ultimately, the outline of our report is determined by the purposes of our work.

The purposes of research can be complex. Ordinarily, they form a hierarchy, ranging from an overall purpose, which covers a broad field of inquiry, to a number of subpurposes, each of which covers a relatively narrow field. Within this hierarchy, we choose a relatively general purpose or a relatively specific purpose, depending upon our time, money, and facilities.

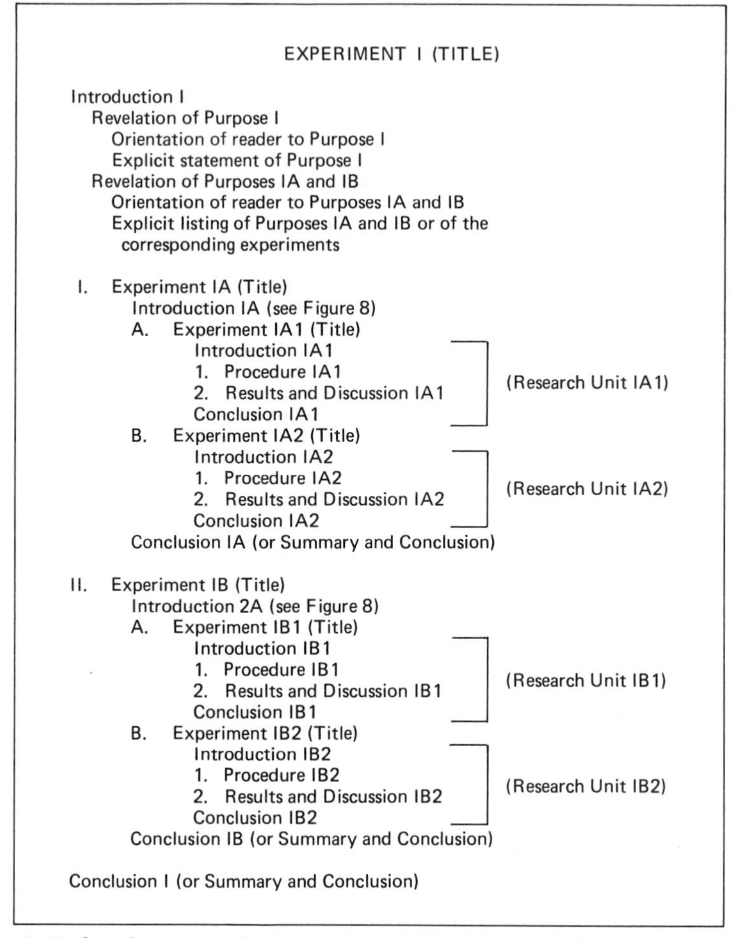


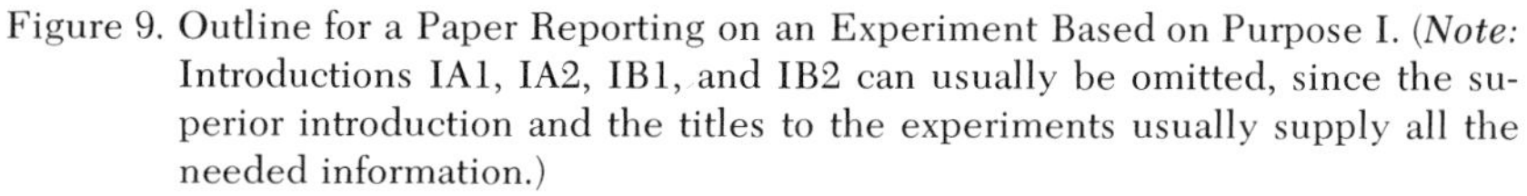
Figure 9. Outline for a Paper Reporting on an Experiment Based on Purpose I. (*Note:* Introductions IA1, IA2, IB1, and IB2 can usually be omitted, since the superior introduction and the titles to the experiments usually supply all the needed information.)

When the purpose of our work is simple—that is, when it cannot be broken into subpurposes—the design of our research is simple. In this case, our research will consist of a single unit—a purpose, a procedure to effect that purpose, a set of results obtained when the procedure is carried out, and a conclusion that derives from the results and that reflects back to the purpose. When, on the other hand, the purpose of our research is complex—that is, when it can be broken into secondary, tertiary, and quaternary subpurposes and so on, corresponding to our hierarchy of purposes—the design of our research is complex. In that case, our research will consist of a number of units that are combined and united by a purpose of the appropriate complexity.

Research reports may be divided into two main classes—those representing simple research and those representing complex research. When the research is simple, the outline of the report can be simple. That is, it can follow the traditional form—introduction, procedure (materials and methods), results and discussion, and conclusion. When, however, the research is complex, the traditional form is not adequate. In this case, the outline of the report must correspond to the outline of the research. The main divisions and subdivisions of the report then correspond to the main divisions and subdivisions of the research design. Only in those portions of the report where we present the individual units of the research do we again revert to the traditional form of introduction, procedure, results and discussion, and conclusion.

Making the outline of our report correspond to the outline of our research raises no problem except when the procedures used in our various units of research are closely similar. This problem, however, is easily solved. All we need do is to describe fully the procedure used in our first unit and then merely tell in the description of the remaining units how each of the subsequent procedures differed from the first.

By making the outline of our report correspond to the outline of our research, we give the reader immediate insight into our research and thereby enable him to read our report quickly yet understandingly.

# Appendix F

# The Decision-Making Process

## Forcing a Good Decision

*M. J. Gelpi**

Good decisions don't "just happen." They are not the result of instinct, common sense, or experience—by themselves. Rather, good decisions result from weighing the relative values of each requirement against all possible solutions.

When one requirement is all-important, the right decision may be obvious. However, most decisions are not this simple. A combination of many requirements must usually be compared with several possible solutions. It becomes more difficult under these circumstances to make decisions with a high degree of confidence. The greater the number of variables, the greater the risk of neglecting an important factor, or assigning too little importance to some key aspect of the problem.

A number of means have been devised to aid the process of making good decisions. Comprehensive detailed statistical analyses can provide a sound decision-making basis. This may not be either practical or economical. Cost analyses, weight analyses, reliability analyses, and many others can be used to guide decision making. But, it is often necessary to make decisions weighting the merits of all of these requirements and more, in a short time, using a limited amount of data. It is also often necessary to make the best decision possible with information that *is* available. In any case, the objective of the decision maker is to be certain that the information that *is* avail-

*M. J. Gelpi is on the Value Engineering Staff of the Surface Division, Westinghouse Defense and Space Center, Baltimore, Maryland. This article is reprinted by permission from *Westinghouse ENGINEER* (January, 1967), 24–25.

able is evaluated in an orderly fashion. A simple mathematical process has been devised to aid in this evaluation.

Two basic steps are involved in this mathematical procedure for decision making:

1) Determine the relative importance of each performance or specification requirement;

2) Evaluate each possible solution against each of the above requirements.

For both of these steps, evaluations are arranged so that only two alternatives are considered at a time. Every possible combination is submitted for comparison. In this way, no shades of choice are required—only a *yes* or *no* decision for each evaluation.

The decision-making process can be illustrated by applying it to a typical problem. Assume that the military services require a component of electronic equipment. In designing the equipment, the supplier attempts to develop a unit that will meet military specifications and at the same time have a minimum overall cost (initial cost plus maintenance) to the military services. In designing the equipment, a decision is required whether to use existing printed-circuit boards from a previously designed unit, new printed-circuit boards designed especially for the present unit, or a newly designed "standard universal" board. To determine the lowest overall cost to the military services, the alternatives are evaluated for these requirements:

1) Low design cost
2) Low manufacturing cost
3) Low maintenance stock cost
4) Low maintenance work cost
5) Low shock-resistance capability cost
6) Low vibration-resistance capability cost
7) Low cost for future systems.

The procedure used to establish the relative importance of each requirement is illustrated in Table I. The seven requirements are listed in the left-hand column and comparisons are made in the columns to the right. In comparing any two items, numeral one (1) indicates the more important requirement and zero (0) indicates the lesser requirement. For example, when comparing design cost to manufacturing cost in the first column, manufacturing cost is the more important consideration *for this particular project.* Similarly, for this hypothetical example, design cost is more important than maintenance stock cost, and design cost is more important than maintenance work cost. Thus, each evaluation is made by comparing only two items at a time, and every possible combination of the seven items is considered.

The total positive decisions for each requirement is tabulated in column N. The summation of these total positive decisions equals 21, which checks with the total number of possible decisions.

A *relative emphasis coefficient* for each requirement is obtained by dividing the number of positive decisions for each requirement by the total number of possible decisions, as shown in the right-hand column of the table.

TABLE I *Weighting of Requirements*

| | 1 | 2 | 3 | 4 | 5 | 6 | 7 | 8 | 9 | 10 | 11 | 12 | 13 | 14 | 15 | 16 | 17 | 18 | 19 | 20 | 21 | Positive Decisions (N) | Requirement Emphasis Coefficient (E = N/21) |
|---|---|---|---|---|---|---|---|---|---|---|---|---|---|---|---|---|---|---|---|---|---|---|---|
| 1. Design Cost | 0 | 1 | 1 | 1 | 1 | 1 | | | | | | | | | | | | | | | | 5 | 0.238 |
| 2. Manufacturing Cost | 1 | | | | | | 1 | 1 | 1 | 1 | 1 | | | | | | | | | | | 6 | 0.285 |
| 3. Maintenance Stock Cost | | 0 | | | | | 0 | | | | | 0 | 0 | 0 | 1 | | | | | | | 1 | 0.048 |
| 4. Maintenance Work Cost | | | 0 | | | | | 0 | | | | 1 | | | | 1 | 1 | 1 | | | | 4 | 0.190 |
| 5. Shock-Resistance Cost | | | | 0 | | | | | 0 | | | | 1 | | | 0 | | | 1 | 1 | | 3 | 0.143 |
| 6. Vibration-Resistance Cost | | | | | 0 | | | | | 0 | | | | 1 | | | 0 | | 0 | | 0 | 1 | 0.048 |
| 7. Future System Cost | | | | | | 0 | | | | | 0 | | | | 0 | | | 0 | | 0 | 1 | 1 | 0.048 |
| | | | | | | | | | | | | | | | | | | | | | | 21 | 1.000 |

Total number of possible decisions = $\frac{n\,(n-1)}{2}$, where $n$ = number of items under consideration; i.e., $\frac{7\,(7-1)}{2}$ = 21 total possible decisions.

The second step in the process is illustrated in Table II. For each requirement—design cost, manufacturing cost, etc.—the three possible solutions are evaluated, one against another. The number of positive decisions for each solution is converted to a *solution emphasis coefficient*, as shown.

TABLE II *Evaluation of Solutions for Each Requirement*

| | 1 | 2 | 3 | Positive Decisions | Solution Emphasis Coefficient |
|---|---|---|---|---|---|
| 1. Design Cost | | | | | |
| Existing Boards | 1 | 0 | | 1 | 0.33 |
| New Boards | 0 | | 0 | 0 | 0.0 |
| Standard Universal Boards | | 1 | 1 | 2 | 0.67 |
| 2. Manufacturing Cost | | | | | |
| Existing Boards | 1 | 1 | | 2 | 0.67 |
| New Boards | 0 | | 1 | 1 | 0.33 |
| Standard Universal Boards | | 0 | 0 | 0 | 0.0 |
| 3. Maintenance Stock Cost | | | | | |
| Existing Boards | 1 | 1 | | 2 | 0.67 |
| New Boards | 0 | | 0 | 0 | 0.0 |
| Standard Universal Boards | | 0 | 1 | 1 | 0.33 |
| 4. Maintenance Work Cost | | | | | |
| Existing Boards | 1 | 1 | | 2 | 0.67 |
| New Boards | 0 | | 1 | 1 | 0.33 |
| Standard Universal Boards | | 0 | 0 | 0 | 0.0 |
| 5. Shock-Resistance Cost | | | | | |
| Existing Boards | 1 | 1 | | 2 | 0.67 |
| New Boards | 0 | | 0 | 0 | 0.0 |
| Standard Universal Boards | | 0 | 1 | 1 | 0.33 |
| 6. Vibration-Resistance Cost | | | | | |
| Existing Boards | 0 | 0 | | 0 | 0.0 |
| New Boards | 1 | | 1 | 2 | 0.67 |
| Standard Universal Boards | | 1 | 0 | 1 | 0.33 |
| 7. Future System Cost | | | | | |
| Existing Boards | 1 | 1 | | 2 | 0.67 |
| New Boards | 0 | | 0 | 0 | 0.0 |
| Standard Universal Boards | | 0 | 1 | 1 | 0.33 |

The preferred solution is then determined by summarizing these emphasis coefficients in matrix form, as shown in Table III. The product of each requirement emphasis coefficient (Table I) and each solution emphasis coefficient (Table II) is found, and these quantities are summed for each of the three possible solutions. These sums become "figures of merit" for the solutions. In this example, the use of existing printed-circuit boards is the best decision.

TABLE III *Decision Matrix*

| | Existing Boards | New Design | Standard Universal Package |
|---|---|---|---|
| Design Cost<br>0.238 × | 0.33 = 0.078 | 0.0 = 0.0 | 0.67 = 0.160 |
| Manufacturing Cost<br>0.285 × | 0.67 = 0.190 | 0.33 = 0.094 | 0.0 = 0.0 |
| Maintenance Stock Cost<br>0.048 × | 0.67 = 0.0325 | 0.0 = 0.0 | 0.33 = 0.0159 |
| Maintenance Work Cost<br>0.190 × | 0.67 = 0.128 | 0.33 = 0.068 | 0.0 = 0.0 |
| Shock-Resistance Cost<br>0.143 × | 0.67 = 0.095 | 0.0 = 0.0 | 0.33 = 0.047 |
| Vibration-Resistance Cost<br>0.048 × | 0.0 = 0.0 | 0.67 = 0.0325 | 0.33 = 0.0159 |
| Future System Cost<br>0.048 × | 0.67 = 0.0325 | 0.0 = 0.0 | 0.33 = 0.0159 |
| | 0.556 | 0.1945 | 0.2547 |

Σ *(Weighting Emphasis Coefficient* × *Solution Emphasis Coefficient)* = *Figure of Merit*

By breaking the decision-making process down into a simple step-by-step application of logic, an orderly decision-making procedure is insured. Thus, the decision-maker knows that his solution is as good as the accuracy of his input information. In the example illustrated, the decision-maker can have a high degree of confidence in his choice because the cost of the various requirements can be either calculated or estimated rather accurately; thus, the yes-no evaluations of the various requirements can be made with high assurance. In other cases, a decision might be required where it is difficult or even impossible to assign direct costs or dollar values to all aspects of the problem. In these circumstances, the degree of confidence in the final decision must be tempered by the decision-maker's confidence in his yes-no evaluations. But by applying an ordered process in making the evaluations, the decision-maker knows that all aspects of the problem will be accounted for, and that the final decision will not be subject to inconsistencies in his analysis of the problem.

## References:

Denholm, Donald H., "Decisions for Dollars," *Proceedings of the Value Engineering Symposium,* U.S. Army Missile Command, Redstone Arsenal, Nov. 18–19 1967.

Fasal, John, "Forced Decisions for Value," *Product Engineering,* April 12, 1965.

Fallon, Carlos, "Practical Use of Decision Theory in Value Engineering," *Journal of Value Engineering,* Vol. 2, No. 3, May 15, 1964.

# Appendix G

# Approved Abbreviations of Scientific and Engineering Terms

The approved abbreviations in this appendix are those for the more commonly used technical terms only. For a complete list of approved abbreviations, you should write for *American Standard Abbreviations for Scientific and Engineering Terms** pamphlet ASA Z10.1-194. It is published by the American Society of Mechanical Engineers, 345 East 47th Street, New York, N.Y. 10022.

*Reprinted with the permission of the American Society of Mechanical Engineers.

## Abbreviations

absolute . . . . . . . . . . . . . . . abs
acre-foot . . . . . . . . . . . . . acre-ft
air horsepower . . . . . . . . air hp
alternating-current (as adjective) . . . . . . . . . . . . . . . a-c
ampere . . . . . . . . . . . . . . . amp
ampere-hour . . . . . . . . amp-hr
Angstrom unit . . . . . . . . . . . A
antilogarithm . . . . . . . . antilog
atmosphere . . . . . . . . . . . . atm
atomic weight . . . . . . . . at. wt
average . . . . . . . . . . . . . . . . avg
avoirdupois . . . . . . . . . . . avdp
azimuth . . . . . . . . . . . . . az or $\alpha$

barometer . . . . . . . . . . . . . . bar.
barrel . . . . . . . . . . . . . . . . . bbl
Baumé . . . . . . . . . . . . . . . . Bé
boiling point . . . . . . . . . . . . bp
brake horsepower . . . . . . . bhp
brake horsepower-hour . bhp-hr
Brinell hardness number . . Bhn
British thermal unit . . Btu or B

calorie . . . . . . . . . . . . . . . . . cal
candle-hour . . . . . . . . . . . . c-hr
candlepower . . . . . . . . . . . . cp
cent . . . . . . . . . . . . . . . . c or ¢
centigram . . . . . . . . . . . . . . cg
centiliter . . . . . . . . . . . . . . . cl
centimeter . . . . . . . . . . . . . cm
centimeter-gram-second (system) . . . . . . . . . . . . . cgs
coefficient . . . . . . . . . . . . . coef
cologarithm . . . . . . . . . . . colog
constant . . . . . . . . . . . . . . const
cosecant . . . . . . . . . . . . . . . csc
cosine . . . . . . . . . . . . . . . . . cos
cotangent . . . . . . . . . . . . . . . cot
counter electromotive force . . . . . . . . . . . . . . cemf

cubic . . . . . . . . . . . . . . . . . . . cu
cubic centimeter . . cu cm, cm$^3$ (liquid, meaning mililiter, ml)
cubic foot . . . . . . . . . . . . . cu ft
cubic feet per minute . . . . cfm
cubic feet per second . . . . . cfs
cubic inch . . . . . . . . . . . cu in.
cubic meter . . . . . . cu m or m$^3$
cubic micron . . cu $\mu$ or cu mu or $\mu^3$
cubic millimeter . . . . . . . cu mm or mm$^3$
cubic yard . . . . . . . . . . . cu yd

decibel . . . . . . . . . . . . . . . . . db
degree . . . . . . . . . . . . . deg or °
degree centigrade . . . . . . . . . C
degree Fahrenheit . . . . . . . . F
degree Kelvin . . . . . . . . . . . . K
degree Réaumur . . . . . . . . . . R
delta amplitude, an elliptic function . . . . . . . . . . . . . dn
direct-current (as adjective) . d-c
dozen . . . . . . . . . . . . . . . . . doz
dram . . . . . . . . . . . . . . . . . . . dr

electric . . . . . . . . . . . . . . . . elec
electromotive force . . . . . . emf
elevation . . . . . . . . . . . . . . . el
equation . . . . . . . . . . . . . . . . eq

farad . . . . . . . . . . spell out or f
feet board measure (board feet) . . . . . . . . . . . . . . . . fbm
feet per minute . . . . . . . . . fpm
feet per second . . . . . . . . . . fps
fluid . . . . . . . . . . . . . . . . . . . fl
foot . . . . . . . . . . . . . . . . . . . . ft
foot-candle . . . . . . . . . . . . . ft-c
foot-Lambert . . . . . . . . . . . ft-L

foot-pound . . . . . . . . . . . . . ft-lb
foot-pound-second (system) . . . . . . . . . . . . . . . . . fps
foot-second (see cubic feet per second)
free on board . . . . . . . . . . . fob
freezing point . . . . . . . . . . . fp

gallon . . . . . . . . . . . . . . . . . . gal
gallons per minute . . . . . gpm
gallons per second . . . . . . gps
gram . . . . . . . . . . . . . . . . . . . g
gram-calorie . . . . . . . . . . . g-cal

hectare . . . . . . . . . . . . . . . . . ha
henry . . . . . . . . . . . . . . . . . . h
hertz . . . . . . . . . . . . . . . . . Hz
high-pressure (adjective) . . h-p
horsepower . . . . . . . . . . . . . hp
horsepower-hour . . . . . . hp-hr
hour . . . . . . . . . . . . . . . . . . hr
hour (in astronomical tables) . h
hundred . . . . . . . . . . . . . . . . C
hundredweight (112 lb) . . . cwt
hyperbolic cosine . . . . . . cosh
hyperbolic sine . . . . . . . . sinh
hyperbolic tangent . . . . . tanh

inch . . . . . . . . . . . . . . . . . . . in.
inch-pound . . . . . . . . . . . in-lb
inches per second . . . . . . . . ips
indicated horsepower . . . . ihp
indicated horsepower-hour . . . . . . . . . . . . . ihp-hr
inside diameter . . . . . . . . . . ID
intermediate-pressure (adjective) . . . . . . . . . . . . . . i-p

joule . . . . . . . . . . . . . . . . . . . . j

kilocalorie . . . . . . . . . . . . . kcal
kilocycles per second . . . . . kc
kilogram . . . . . . . . . . . . . . . . kg
kilogram-calorie . . . . . . kg-cal
kilogram-meter . . . . . . . . kg-m
kilograms per cubic meter . kg per cu m or kg/m$^3$

kilograms per second . . . kgps
kiloliter . . . . . . . . . . . . . . . . kl
kilometer . . . . . . . . . . . . . . km
kilometers per second . . . kmps
kilovolt . . . . . . . . . . . . . . . . . kv
kilovolt-ampere . . . . . . . . . kva
kilowatt . . . . . . . . . . . . . . . . kw
kilowatthour . . . . . . . . . . kwhr

lambert . . . . . . . . . . . . . . . . . . L
latitude . . . . . . . . . . . . . lat or $\phi$
least common multiple . . . lcm
linear foot . . . . . . . . . . . . lin ft
liquid . . . . . . . . . . . . . . . . . . liq
liter . . . . . . . . . . . . . . . . . . . . l
logarithm (common) . . . . . . log
logarithm (natural) . . $\log_e$ or ln
longitude . . . . . . . . . long. or $\lambda$
low-pressure (as adjective) . . . . . . . . . . . . . . l-p
lumens per watt . . . . . . . . . lpw

maximum . . . . . . . . . . . . . . max
mean effective pressure . . . mep
mean horizontal candlepower . . . . . . . . . . . . . mhcp
melting point . . . . . . . . . . . mp
meter . . . . . . . . . . . . . . . . . . m
meter-kilogram . . . . . . . . m-kg
microampere . . . . . $\mu$a or mu a
microfarad . . . . . . . . . . . . . . $\mu$f
microinch . . . . . . . . . . . . . . $\mu$in.
micromicrofarad . . . . . . . . $\mu\mu$f
micromicron . . . $\mu\mu$ or mu mu
micron . . . . . . . . . . . . $\mu$ or mu
microvolt . . . . . . . . . . . . . . . $\mu$v
microwatt . . . . . . . $\mu$w or mu w
miles per hour . . . . . . . . . mph

miles per hour per second . . . . . . . . . . . mphps
milliampere . . . . . . . . . . . . . ma
milligram . . . . . . . . . . . . . . mg
millihenry . . . . . . . . . . . . . mh
millilambert . . . . . . . . . . . . mL
milliliter . . . . . . . . . . . . . . . ml
millimeter . . . . . . . . . . . . . mm
millimicron . . . . . mμ or m mu
million . . . . . . . . . . . . spell out
million gallons per day . . mgd
millivolt . . . . . . . . . . . . . . mv
minimum . . . . . . . . . . . . . . min
minute . . . . . . . . . . . . . . . . min
minute (angular measure) . . . ′
minute (time) (in astronomical tables) . . . . . . . . . . . . . . . m
molecular weight . . . . mol. wt

ohm . . . . . . . . . . spell out or Ω
ohm-centimeter . . . . . ohm-cm
ounce . . . . . . . . . . . . . . . . . oz
ounce-foot . . . . . . . . . . . . oz-ft
ounce-inch . . . . . . . . . . . oz-in.

parts per million . . . . . . . ppm
pint . . . . . . . . . . . . . . . . . . . . pt
pound . . . . . . . . . . . . . . . . . . lb
pound-foot . . . . . . . . . . . . . lb-ft
pound-inch . . . . . . . . . . . lb-in.
pounds per brake horse-power-hour . . . lb per bhp-hr
pounds per cubic foot . . . . . . . . . . . lb per cu ft
pounds per square foot . . . . psf
pounds per square inch . . . psi
pounds per square inch absolute . . . . . . . . . . . . . psia
power factor . . . spell out or pf

quart . . . . . . . . . . . . . . . . . . . qt

reactive kilovolt-ampere . kvar
reactive volt-ampere . . . . . var

revolutions per minute . . . rpm
revolutions per second . . . rps
root mean square . . . . . . . . rms

secant . . . . . . . . . . . . . . . . . sec
second . . . . . . . . . . . . . . . . . sec
second (angular measure) . . . ″
shaft horsepower . . . . . . . . shp
sine . . . . . . . . . . . . . . . . . . . . sin
sine of the amplitude, an elliptic function . . . . . . . . sn
specific gravity . . . . . . . . . sp gr
specific heat . . . . . . . . . . . sp ht
spherical candle power . . . scp
square . . . . . . . . . . . . . . . . . . sq
square centimeter sq cm or $cm^2$
square foot . . . . . . . . . . . . sq ft
square inch . . . . . . . . . . sq in.
square kilometer sq km or $km^2$
square meter . . . . . sq m or $m^2$
square micron . . . . . . . sq μ or sq mu or $\mu^2$
square millimeter . . . . sq mm or $mm^2$
square root of mean square rms

tangent . . . . . . . . . . . . . . . . tan
temperature . . . . . . . . . . . temp
tensile strength . . . . . . . . . . . ts
thousand . . . . . . . . . . . . . . . M
thousand foot-pounds . . . kip-ft
thousand pound . . . . . . . . . kip

versed sine . . . . . . . . . . . . . vers
volt . . . . . . . . . . . . . . . . . . . . . v
volt-ampere . . . . . . . . . . . . . va

watt . . . . . . . . . . . . . . . . . . . . w
watthour . . . . . . . . . . . . . . . whr
watts per candle . . . . . . . . wpc
weight . . . . . . . . . . . . . . . . . wt

yard . . . . . . . . . . . . . . . . . . . yd

# Index

*(Note: Page numbers in italics refer to illustrations and examples.)*